中国井冈山地区
陆生脊椎动物
彩色图谱

王英永　陈春泉　赵　健　吴　毅　吕植桐
杨剑焕　余文华　林剑声　刘祖尧　王　健　等著
杜　卿　张　忠　宋玉赞　汪志如　何桂强

科学出版社
北　京

内 容 简 介

　　本书基于2010年至2015年在井冈山地区开展的生物多样性综合科学考察编写而成。书中井冈山地区是由江西井冈山国家级自然保护区、南风面国家级自然保护区、七溪岭省级自然保护区、湖南桃源洞国家级自然保护区，以及风景区、林场等共同组成的一个连片区域，覆盖了罗霄山脉中段的万洋山脉主要山体，是一个相对完整的地理和生态单元。全书简要介绍了井冈山地区自然地理及生物多样性特点，以800多张照片和大量标本为凭证，对包括38种两栖动物、62种爬行动物、260种鸟类和37种哺乳动物在内的397种陆生脊椎动物的识别特征、生境与习性、地理分布及其在井冈山的种群状况和分布等进行了介绍，首次对最近7年在井冈山地区发现并发表的新种井冈攀蜥、七溪岭瘰螈、珀普短腿蟾、井冈角蟾、林氏角蟾、陈氏角蟾和井冈纤树蛙，以及该区域的新分布纪录东京湾蜓蜥进行了中文描述。全书图文并茂，具有较高的学术价值、科普价值和美学价值，对井冈山地区的自然保护、科学管理、发展规划具有指导意义。

　　本书可供动物学、生态学、林学、保护生物学等等领域的科研人员、高等学校师生，以及自然摄影爱好者，也可供政府机构和自然保护管理部门的工作者参考。

图书在版编目（CIP）数据

中国井冈山地区陆生脊椎动物彩色图谱/王英永等著. — 北京：科学出版社，2017.12
ISBN 978-7-03-054710-1

I. ① 中… II. ① 王… III. ①井冈山–陆栖–脊椎动物门–图谱 IV. ①Q959.308-64

中国版本图书馆CIP数据核字（2017）第244296号

责任编辑：王　静　王　好／责任校对：郑金红
责任印制：肖　兴／书籍设计：北京美光设计制版有限公司

科 学 出 版 社　出版
北京东黄城根北街16号
邮政编码：100717
http://www.sciencep.com

北京汇瑞嘉合文化发展有限公司　印刷
科学出版社发行　各地新华书店经销
*

2017年12月第 一 版　　开本：880×1230　A4
2017年12月第一次印刷　　印张：20 1/4
字数：686 000

定价：318.00元
（如有印装质量问题，我社负责调换）

COLORED ATLAS OF TERRESTRIAL VERTEBRATES OF THE JINGGANGSHAN REGION IN CHINA

Wang Yingyong   Chen Chunquan   Zhao Jian

Wu Yi   Lü Zhitong   Yang Jianhuan

Yu Wenhua   Lin Jiansheng   Liu Zuyao   *et al.*

Wang Jian   Du Qing   Zhang Zhong

Song Yuzan   Wang Zhiru   He Guiqiang

Science Press
Beijing

《中国井冈山地区生物多样性综合科学考察》项目组

项目组织委员会

主　　　任：龙波舟

第 一 副 主 任：陈　敏

常 务 副 主 任：陈春泉

副　 主 　任：曾以平　汪仁盛　肖邮华　曾宪文　刘森严

项目技术顾问：王伯荪　陈昌笃　汪　松　王献溥　马克平　刘信中　叶居新　杜天真

项 目 主 持 人：廖文波　王英永　彭少麟

项目组协调人：陈春泉　赵　健　宋玉赞

各专题组负责人（组长）

地质地貌组：尹国胜　　　　　　　　珍稀植物组：王　蕾

土　壤　组：曾曙才　　　　　　　　兰科植物组：杨柏云

水 资 源 组：崔大方　　　　　　　　昆 虫　组：贾凤龙

生　态　组：彭少麟　　　　　　　　鱼 类　组：欧阳珊

植　被　组：李　贞　廖文波　陈春泉　脊椎动物组：王英永

苔藓植物组：刘蔚秋　　　　　　　　孢粉与古植被组：黄康有

蕨类植物组：石祥刚　　　　　　　　子 遗 植 物 组：廖文波　金建华

种子植物组：凡　强　廖文波　　　　井 冈 山 协 调 组：陈春泉　曾宪文　宋玉赞　赵　健

陆生脊椎动物考察和研究的主要人员（按姓氏汉语拼音排序）

陈柏成　陈春泉　杜　卿　何芬奇　何桂强　金孟洁　李　锋　李　韵　李言阔　李玉龙
林剑声　刘祖尧　吕植桐　宋玉赞　汪志如　王　健　王英永　吴　毅　徐忠鲜　杨剑焕
余文华　曾宪文　曾昭驰　张　忠　张秋萍　赵　健　郑艳玲

其他参加考察或协助工作人员（按姓氏汉语拼音排序）

何海龙　黄志强　李德清　李俊杰　李润林　刘　阳　刘五井　刘信中　欧阳珊　单继红
涂晓斌　吴小平　吴英豪　赵　婷　赵小奎　郑发辉　〔日〕本川雅治　〔日〕原田正史
〔捷〕J. Hájek

项目承担单位

中山大学生命科学学院

江西省井冈山管理局

项目参加单位

江西省地质调查研究院　南昌大学　广州大学　江西省野生动植物保护管理局　江西省林业科
学研究院　江西省科学院

致谢

江西井冈山国家级自然保护区管理局　江西省遂川县林业局

江西省永新县林业局

Integrated Study on Biodiversity of the Jinggangshan Region, China Project study groups

Organizing committee

Chairman: Long Bozhou

First Vice-Chairman: Chen Min

Standing Vice Chairman: Chen Chunquan

Vice-Chairmen: Zeng Yiping, Wang Rensheng, Xiao Youhua, Zeng Xianwen and Liu Senyan

Technical consultants: Wang Bosun, Chen Changdu, Wang Song, Wang Xianpu, Ma Keping, Liu Xinzhong, Ye Juxin and Du Tianzhen

Principal investigators: Liao Wenbo, Wang Yingyong and Peng Shaolin

Project coordinators: Chen Chunquan, Zhao Jian and Song Yuzan

Sub-group principal investigators

Geology: Yin Guosheng

Soils: Zeng Shucai

Water resources: Cui Dafang

Ecology: Peng Shaolin

Vegetation: Li Zhen, Liao Wenbo and Chen Chunquan

Bryophytes: Liu Weiqiu

Pteridophytes: Shi Xianggang

Spermatophytes: Fan Qiang and Liao Wenbo

Rare plants: Wang Lei

Orchids: Yang Baiyun

Insects: Jia Fenglong

Fish: Ouyang Shan

Vertebrates: Wang Yingyong

Spores and pollen: Huang Kangyou

Relict species: Liao Wenbo and Jin Jianhua

Jinggangshan coordinators: Chen Chunquan, Zeng Xianwen, Song Yuzan and Zhao Jian

Investigators of the Terrestrial Vertebrates (in the order of Chinese pinyin)

Chen Bocheng, Chen Chunquan, Du Qing, He Fenqi, He Guiqiang, Jin Mengjie, Li Feng, Li Yun, Li Yankuo, Li Yulong, Lin Jiansheng, Liu Zuyao, Lü Zhitong, Song Yuzan, Wang Zhiru, Wang Jian, Wang Yingyong, Wu Yi, Xu Zhongxian, Yang Jianhuan, Yu Wenhua, Zeng Xianwen, Zeng Zhaochi, Zhang Zhong, Zhang Qiuping, Zhao Jian, Zheng Yanling

Other participants (in the order of Chinese pinyin)

He Hailong, Huang Zhiqiang, Li Deqing, Li Junjie, Li Runlin, Liu Yang, Liu Wujing, Liu Xinzhong, Ouyang Shan, Shan Jihong, Tu Xiaobin, Wu Xiaoping, Wu Yinghao, Zhao Ting, Zhao Xiaokui, Zheng Fahui, 〔JPN〕M. Motokawa, 〔JPN〕M. Harada, 〔CZE〕J. Hájek

Responsible institutions

School of Life Sciences, Sun Yat-sen University

Jinggangshan Administration of Jiangxi Province

Participating institutions

Geological Survey of Jiangxi Province

Nanchang University

Guangzhou University

Department of Wildlife Conservation, Jiangxi Province

Jiangxi Academy of Forestry

Jiangxi Academy of Sciences

Acknowledgments

Administration of Jinggangshan National Nature Reserve, Jiangxi Province

Suichuan Forestry Administration, Jiangxi Province

Yongxin Forestry Administration, Jiangxi Province

前 言

就某一地区而言，对其物种多样性认识的不足是生物多样性保护与管理所面临的最主要挑战。全面、系统、深入的科学考察，是比较全面了解该地区生物多样性，进而实施有针对性、富有成效保护的基础性和前提性工作。

井冈山地区跨湘赣两省，涵盖了罗霄山脉中段的万洋山脉主要山体，是由井冈山（包括井冈山国家级自然保护区、井冈山风景名胜区、林场、农田和村镇）、七溪岭省级自然保护区、南风面国家级自然保护区及湖南桃源洞国家级自然保护区共同组成的一个地理连片、相对完整生态单元，地理坐标为26°13′4″N-26°52′30″N、113°56′30″E-114°18′28″E，总面积70,874 hm²。

井冈山地区独特的自然地理区位和优越的自然条件，决定其拥有高水平的生物多样性，非凡的物种丰富度和特有性。此前，除了在上述保护区内开展过短暂科学考察外，尚未对井冈山地区展开全面系统的生物多样性考察工作，对其生物多样性的认识还相当零碎和粗浅。

宗愉和马积蕃是最早开展井冈山自然保护区爬行动物调查的学者，他们在1981-1982年调查了井冈山自然保护区大井、小井、桐木山、荆竹山等地，记录了爬行动物25种，并发表了新种井冈脊蛇（宗愉和马积蕃，1983）。1982-1983年，邹多录和钟昌富等对井冈山自然保护区的两栖动物和爬行动物做了较为系统的调查，龙宗迪等调查了鸟兽，调查结果汇编入《井冈山自然保护区考察研究》（林英，1990），首次报道了井冈山自然保护区陆生脊椎动物区系组成：两栖动物26种（含亚种）、爬行动物31种（含亚种）、鸟类94种（含亚种）、兽类42种（含亚种），共计陆生脊椎动物193种（含亚种）。其后，黄族豪等（2007a）将井冈山自然保护区的两栖动物提升至29种，但该文同时记录了鳌掌突蟾和福建掌突蟾，显然是重复记录。钟昌富（2004）、黄族豪等（2007b）先后将井冈山自然保护区爬行动物提升至36种和41种。鸟类方面，王央生（1997）、段世华等（2004）、黄族豪等（2009）先后发表了井冈山自然保护区鸟类多样性文章，这些文章均出现一些错误鸟种记录；2011年，承勇等在调查和整理文献资料的基础上，将井冈山自然保护区鸟类记录增加至196种，尽管文中也出现一些值得商榷的物种记录，却是已发表文献中数据最翔实、科学性最高的文章。哺乳动物方面，郑发辉等（2007）报道了井冈山自然保护区哺乳动物67种，但未列出物种名录。相较于井冈山自然保护区，南风面自然保护区、七溪岭自然保护区的调查工作则更显不足，所能查阅到的文献只有保护区的科考报告。总体上讲，井冈山地区陆生脊椎动物的研究还处于比较粗浅的水平，有不少鉴定错误和值得商榷的物种记录，其多样性被严重低估。

2010年8月，为全面查清井冈山地区生物多样性现实本底，受江西省井冈山管理局委托，中山大学生命科学学院承担了井冈山地区生物多样性调查工作。同年9月，陆生脊椎动物科考队成立，成员包括中山大学生命科学学院、广州大学生命科学学院和井冈山管理局等单位近30人，正式启动了对井冈山地区陆生脊椎动物的科学考察工作。

截至2015年年底，为期5年的科考工作基本结束。科考队的足迹遍及井冈山自然保护区、井冈山风景名胜区及其外围地区、遂川县南风面自然保护区和永新县七溪岭自然保护区，采集标本近

2000号，提取并保存了组织样品1000多份，录音530多段，录影320多分钟，拍摄照片（包括红外相机照片）2万余张，完成了30多个物种的分子测序工作，整合已有文献，最终确认井冈山地区陆生脊椎动物468种，包括两栖类41种、爬行类67种、鸟类290种、哺乳动物70种，发表了6个新种、2个江西省分布属新纪录，10个江西省分布种新纪录；修订了中国瘰螈、螳掌突蟾、宽头短腿蟾、淡肩角蟾、小角蟾、短肢角蟾等错误记录。上述成果汇编入2014年由科学出版社出版的《中国井冈山地区生物多样性综合科学考察》（廖文波等，2014）。

本书作为井冈山地区综合科考的成果之一，在《中国井冈山地区生物多样性综合科学考察》基础上，补充了2014年以来的最新研究成果编写而成。全书收录了调查记录到的陆生脊椎动物397种，包括38种两栖动物、62种爬行动物、260种鸟类和37种哺乳动物，使用调查期间拍摄的照片800多张。除个别物种，每一个物种有照片2-4张，以求更多地展示物种的鉴定特征、雌雄二态、成幼差别及特殊的行为习性等信息。文字描述部分包括了物种识别特征、生境与习性、种群状况、地理分布及在井冈山地区的分布等信息。

本书所使用的分类系统，两栖纲采用 amphibian species of the world 6.0（Frost, 2017）；爬行纲采用 the reptile database（Uetz and Hošek, 2017）；鸟纲采用《中国鸟类分类与分布名录》（郑光美，2011），并参考世界鸟类学家联合会世界鸟类名录7.1版（Gill and Donsker, 2017）和《中国观鸟年报》"中国鸟类名录"4.0版（董路等，2016）。最近几年，鸟类系统分类发生了较大变革，不少鸟种分类地位和学名均发生了改变。为避免混乱，本书在使用郑光美（2011）分类系统的基础上，更新了部分鸟种的学名；哺乳纲参考了《中国哺乳动物多样性及地理分布》（蒋志刚等，2015）和《中国哺乳动物彩色图鉴》（潘清华等，2007）。

本书各部分文字撰写的分工如下：

第一部分 井冈山地区自然地理及陆生脊椎动物区系概述：王英永、吕植桐

第二部分

第1章 井冈山地区两栖动物区系：王英永、赵健

第2章 井冈山地区爬行动物区系：王英永、赵健

第3章 井冈山地区鸟类区系：赵健、王英永

第4章 井冈山地区哺乳动物区系：吴毅、余文华、王英永

全书照片由林剑声、杜卿、杨剑焕、赵健、王英永、王健、刘祖尧、吕植桐、吴毅、J. Hájek、张中、汪志如、陈春泉、宋玉赞等拍摄。

在井冈山地区科学考察及本书编写过程中，我们得到了江西省野生动植物保护管理局涂晓斌先生、邵明红女士、刘信中先生，江西省井冈山管理局，井冈山国家级自然保护区，七溪岭省级自然保护区，遂川县林业局等的大力支持。李润林、李韵、李玉龙、张天度、郑艳玲等参加了部分调查工作，李德清先生提供了部分井冈山地区的景观照片，B. Bravery 先生对全书英文进行了修改。在此深表感谢。

本书如有错漏和不足之处，恳请批评指正。

王英永

2017年10月17日

Preface

The most imposing challenge confronting regional biodiversity conservation management is that species diversity of this region has not been fully revealed, and is seriously underestimated. Comprehensive, systematic and in-depth scientific investigations conducted in the region are a prerequisite to understanding the regional biodiversity and developing diversity conservation strategies.

The Jinggangshan Region in southeastern China spans Hunan and Jiangxi provinces, and covers the main massif of the Wanyang Mountains within the central section of the Luoxiao Mountains. It consists of Mount Jinggang (include Jinggangshan National Nature Reserve, Jinggangshan National Park, forestry stations, farming area, villages and towns), Qixiling Provincial Nature Reserve and Nanfengmian National Nature Reserve in Jiangxi, and Taoyuandong National Nature Reserve in Hunan, together forming a continuous geographically region and relatively complete ecological unit. The area spans 70,874 hm² from 26°13′4″N-26°52′30″N and 113°56′30″E-114°18′28″E.

The Jinggangshan Region is located a unique natural geographical location and possesses superior natural conditions, indicating high levels of biodiversity, especially species richness and endemism. However, previously rapid or partial surveys only were conducted in above nature reserves, whereas comprehensive and systematic biodiversity surveys have not been conducted in the overall region. Therefore, the biodiversity of the region is still poorly known.

Zong Yu and Ma Jifan first conducted herpetological field survey from Dajing, Xiaojing, Tongmushan and Jingzhushan in Jinggangshan Nature Reserve in 1981-1982, reported 25 reptile species and described a new species *Achalinus jinggangensis* (Zong and Ma, 1983). In 1982-1983, Zou Duolu, Zhong Changfu and others had conducted a relatively detailed survey of amphibians and reptiles, Long Zongdi had surveyed avifauna and mammal fauna from Jinggangshan National Nature Reserve. These surveys results were published in *Scientific Survey and Research on Jinggangshan Nature Reserve* (Lin Ying, 1990), where the first reported the terrestrial vertebrate fauna from Jinggangshan Nature Reserve, included 26 amphibian species (including subspecies), 31 reptile species (including subspecies), 94 aves species (including subspecies) and 42 mammal species (including subspecies). Huang Zuhao *et al.* (2007a) expanded the number of amphibian species from Jinggangshan Nature Reserve to 29, but he simaltaneously recorded *Leptolalax pelodytoides* and *L. liui* in this publication. Zhong Changfu (2004) and Huang Zuhao *et al.* (2007b) increased the number of reptile species from Jinggangshan Nature Reserve to 36 and 41 respectively. Wang Yangsheng (1997), Duan Shihua *et al.* (2004) and Huang Zuhao *et al.* (2009) studied avifauna of Jinggangshan Nature Reserve, but included some incorrect species records in their papers. Cheng Yong *et al.* (2011) increased the number of bird species from Jinggangshan Nature Reserve to 196 on the basis of their investigations and a literature review; although some species records are questionable in this paper, but the paper remains the most informative and scientific than other publications. Zheng Fahui *et al.* (2007) reported 67 mammal species inhabiting Jinggangshan Nature Reserve, but did not list each species. Compared to the Jinggangshan Nature Reserve, work within Qixiling Nature Reserve and Nanfengmian Nature Reserve is limited to few research reports. Overall, previous rudimentary and incomplete researches of terrestrial vertebrate resulted that terrestrial vertebrate diversity has been seriously underestimated.

In August 2010, the School of Life Sciences, Sun Yat-sen University was commissioned by the Jinggangshan Administration of Jiangxi Province, undertaken an investigation project of the biological diversity of the Jinggangshan Region. In September 2010, the terrestrial vertebrate expedition team was established, composed of members from the School of Life Sciences, Sun Yat-sen University and Guangzhou University, and Jinggangshan Administration.

Until the end of 2015, continual field expeditions around the overall the Jinggangshan Region had finished. A total of more than 2000 specimens and more than 1000 tissue samples for DNA was collected, more than 500 advertisement calls of batrachians and birds were recorded, and nearly 320 minutes of video and more than 20,000 photographs (including infrared Camera trapping images) were taken, DNA of more than 30 species were extracted and sequenced, and examined previous literature, finally was confirmed that 468 species of terrestrial vertebrates currently inhabit the Jinggangshan Region. These species include 41 amphibian species, 67 reptile species, 290 bird species and 70 mammal species. Among these, six new species were described; two genera and nine species were first recorded in Jiangxi. Incorrect accounts of *Paramesotriton chinensis*, *Leptolalax pelodytoides*, *Brachytarsophrys carinensis*, *Megophrys boettgeri*, *M. minor* and *M. brachykolos* were revised. All results were published in *Integrated Study on Biodiversity of Mount Jinggangshan Regions in China* by Science Press in 2014 (Liao *et al.*, 2014).

Colored Atlas of Terrestrial Vertebrates of the Jinggangshan Region in China is yet another product of the above investigations. This atlas contains more than 800 photographs across 397 terrestrial vertebrate species, including 38 amphibian species, 62 reptile species, 260 bird species and 37 mammal species. In most cases each species entry includes 2-4 photos in order to show identification characteristics, sexual dimorphism, differences between adults and larva, and special behaviors. The text includes species identification, habitat and habits, population status, geographic distribution and distribution across the Jinggangshan Region.

The amphibian classification system used in this atlas is based on the *Amphibian Species of the World 6.0*, an Online Reference (Frost, 2017). The reptile system is based on the *Reptile Database* (Uetz and Hošek, 2017). Avian classification is based on *A Checklist on the Classification and Distribution of the Birds of China* (Zheng, 2011), also refer to *IOC World Bird List* (V 7.1) (Gill and Donsker, 2017) and the *CBR Checklist of Birds of China* (V 4.0) (Dong, 2016). Mammals were classified according to *A Guide to the Mammals of China* (Smith and Xie, 2009) and *A Field Guide to the Mammals of China* (Pan *et al.*, 2007).

The tabula authors of this atlas.

Part 1. Introduction to Terrestrial Vertebrate Fauna of the Jinggangshan Region: Wang Yingyong and Lü Zhitong

Part 2.

Chapter 1 Amphibian Fauna of the Jinggangshan Region: Wang Yingyong and Zhao Jian

Chapter 2 Reptile Fauna of the Jinggangshan Region: Wang Yingyong and Zhao Jian

Chapter 3 Avifauna of the Jinggangshan Region: Zhao Jian and Wang Yingyong

Chapter 4 Mammal Fauna of the Jinggangshan Region: Wu Yi, Yu Wenhua and Wang Yingyong

All species pictures were photographed by Lin Jiansheng, Du Qing, Yang Jianhuan, Zhao Jian, Wang Yingyong, Wang Jian, Liu Zuyao, Lü Zhitong, Wu Yi, J. Hájek, Zhang Zhong, Wang Zhiru, Chen Chunquan and Song Yuzan.

During field expeditions and preparation of this atlas, we were supported by Tu Xiaobing, Shao Minghong and Liu Xinzhong from the Department of Wildlife Conservation of Jingxi Province. Support was also provided by Jinggangshan Administration of Jiangxi Province, Jinggangshan National Nature Reserve, and Qixiling Provincial Nature Reserve and the Forestry Bureau of Suichuan County. We are also grateful to Li Runlin, Li Yun, Li Yulong, Zhang Tiandu and Zheng Yanling for assistance with field expeditions and to Li Deqing who provided some photos of the Jinggangshan Region. We thank Mr B. Bravery for editing English text with tremendous skill.

Wang Yingyong

Oct 17, 2017

目 录

第2章　井冈山地区爬行动物区系　59

第3章　井冈山地区鸟类区系　103

第4章 井冈山地区哺乳动物区系 263

Contents

Preface

Part 1
Introduction to Terrestrial Vertebrate Fauna of the Jinggangshan Region

Part 2
Terrestrial Vertebrate Fauna of the Jinggangshan Region

Chapter 2   Reptile Fauna of the Jinggangshan Region   59

Chapter 3   Avifauna of the Jinggangshan Region   103

Chapter 4   Mammal Fauna of the Jinggangshan Region   263

Part **1**

井冈山地区自然地理
及陆生脊椎动物
区系概述

Introduction to
Terrestrial Vertebrate
Fauna of the
Jinggangshan Region

第 1 章 自然地理概况

1.1 自然地理区位和范围

井冈山地区位于罗霄山脉中段的万洋山脉。罗霄山脉是中国大陆地形第三阶梯内东南丘陵盆地区的一条南北走向的大型山脉，其地理位置独特，北镶长江，南接南岭，鄱阳湖和洞庭湖分列东西，是鄱阳湖水系和洞庭湖水系的分界性山脉。

本书界定的井冈山地区是以江西井冈山为核心，向外延展，跨湘赣两省，涵盖了罗霄山脉中段的万洋山脉主要山体，包括 4 个自然保护区，即东坡的江西井冈山国家级自然保护区、七溪岭省级自然保护区和南风面国家级自然保护区，西坡的湖南桃源洞国家级自然保护区，地理坐标为 26°13′4″N-26°52′30″N、113°56′30″E-114°18′28″E，总面积 70,874 hm² （图 1-1 和图 1-2）。

1.2 地质地貌

井冈山地区地貌类型属褶皱断块的中山地貌，微地貌形态包括山体（山脊）、山坡、坡脚（麓）、峡谷、构造盆地和岩溶地貌等类型。主要山峰有南风面、鄣峰、笔架山、五指峰、江西坳、平水山等，海拔均在 1500 m 以上。主峰南风面，海拔 2120 m；最低点为七溪岭的龙源口，海拔 200 m。主要基岩有

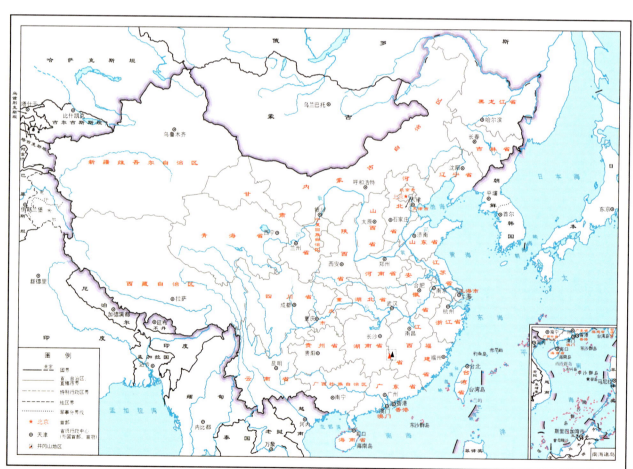

图 1-1　井冈山地区地理位置
Figure 1-1　Location of the Jinggangshan Region.

花岗岩、硅质板岩、板岩及沙砾岩等。地层出露主要有古生代的寒武纪、奥陶纪、泥盆纪和新生代的第四纪地层。该区域地处欧亚大陆东南部，中国扬子古板块与华夏古板块结合带的南东侧，属于华南加里东造山带、华夏构造域（图1-3和1-4）。

1.3 水资源和土壤

井冈山地区构成罗霄山脉中段的主峰地段及两侧集水区，是赣江水系、湘江水系的分界岭。在江西境内为赣江水系，包括禾水、蜀水和遂川江3个一级支流，有7条主要河流的集水面积都超过30 km²，其中，龙源口集水面积81 km²，石溪79 km²，行洲河98 km²，小溪洞47 km²，湘洲河43 km²等；在湖南境内为湘江水系，有一级支流洣水，其他如斜濑水、河漠水、沔水、东风河等长度超过5 km或集雨面积超过10 km²的河流49条，是湘东南重要的水源林区。主要土壤类型有：山地红壤、山地红黄壤、山地黄壤、山地黄棕壤、山地草甸，在部分山顶还发育山地沼泽土共6个土类（图1-5和图1-6）。

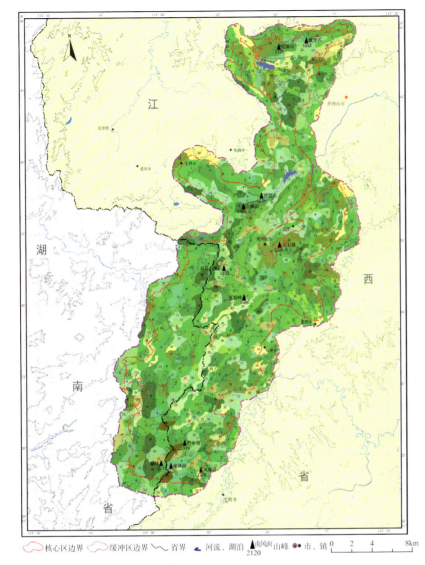

图1-2 井冈山地区的范围和植被图
Figure 1-2　The boundary and vegetation maps of the Jinggangshan Region

图1-3 井冈山地区的地形地貌（摄影／李德清）
Figure 1-3　The main terrain and landscapes of the Jinggangshan Region（Photo by Li Deqing）

图 1-4 井冈山地区的地形地貌（摄影 / 李德清）
Figure 1-4   The main terrain and landscapes of the Jinggangshan Region（Photo by Li Deqing）

图 1-5 急流（摄影 / 李德清）
Figure 1-5   Showing montane rushing streams in the Jinggangshan Region（Photo by Li Deqing）

图 1-6 缓溪（摄影 / 李德清）
Figure 1-6   Showing slow flowing streams in the Jinggangshan Region（Photo by Li Deqing）

1.4 气候

　　井冈山地区属中亚热带季风气候区。年平均气温为14.2℃，最热7月平均气温为23.9℃，7月极端最高气温36.7℃；最冷1月平均气温为3.4℃，1月极端最低气温为−11.0℃。年降水量1400-2100 mm，最大降水量为2878.8 mm（2002年），平均年蒸发量978.8 mm。随环境地形梯度的变化，气候垂直差异明显，导致热能和水分在时空上有明显差异，形成了中亚热带、北亚热带、暖温带3个垂直气候亚带，山地气温垂直递减率为0.44-0.57℃/100 m（图1-7和图1-8）。

1.5 植被类型与植物多样性

1.5.1 植被类型复杂多样

　　井冈山地区植被类型非常丰富，包括14个植被型，58个群系组，90个群系，180个群丛（李贞，2014）。其中，自然植被为12个植被型，83个群系，167个群丛。在自然群丛中有52个为珍稀孑遗植物群丛，总面积达785.75 hm²。自沟谷、低地到低山、中山，发育形成典型的垂直植被带谱。依次是沟谷季雨林、季风常绿阔叶林—山地常绿阔叶林—山地常绿落叶阔叶混交林—山地常绿针阔叶混交林—山地常绿矮林或落叶灌草丛—中山草甸等(图1-9-图1-11)。

图 1-7　井冈山地区独特的气候条件，极易形成壮观的云海景观（摄影 / 李德清）
Figure 1-7　The unique climate in the Jinggangshan Region often forms a spectacular sea of clouds（Photo by Li Deqing）

图 1-8　井冈山地区雪景（摄影 / 李德清）
Figure 1-8　Snow in the Jinggangshan Region（Photo by Li Deqing）

图 1-9　井冈山常绿阔叶林（摄影 / 李德清）
Figure 1-9　Typical evergreen broad leaved forest in the Jinggangshan Region（Photo by Li Deqing）

图 1-10　山间苔藓（摄影 / 廖文波）
Figure 1-10　Montane Bryophytes（Photo by Liao Wenbo）

图 1-11　南风面山顶草甸（摄影／李德清）
Figure 1-11　Mountain meadow on top of Nanfengmian（Photo by Li Deqing）

1.5.2 植物物种多样性水平高，是第三纪、第四纪冰期避难所

井冈山地区记录的高等植物多达323科1290属3745种，保存了大量原始古老的孑遗物种和区域特有种，是亚洲大陆东部冰期物种"自北向南"迁移的避难所，也是间冰期"自南向北"扩张等历史演化过程的策源地。

井冈山地区特有现象明显，保存有东亚特有科7科、中国特有科5科、中国特有属44属、中国特有种1253种。各类珍稀濒危保护植物多达203种，包括苔藓植物2科2属2种、蕨类植物3科3属4种、裸子植物8科18属23种、被子植物35科90属174种。保存的原始、孑遗物种有77科122属181种（廖文波等，2014）。

第2章 陆生脊椎动物区系特征

2.1 物种多样性的高丰度地区

井冈山地区地理区位独特，水热条件良好，生态环境和植被类型复杂多样，是欧亚大陆东南部陆生脊椎动物重要栖居地，是动物南北迁徙、东西扩散的关键地区，是欧亚大陆东南部陆生脊椎动物物种多样性的高丰度区。截至2015年年底，共记录了陆生脊椎动物468种，包括41种两栖类、67种爬行类、290种鸟类和70种哺乳动物。

井冈山地区是8个物种的模式产地。这8个物种分别是模式产地在七溪岭自然保护区的七溪岭瘰螈、模式产地在井冈山自然保护区、桃源洞自然保护区的珀普短腿蟾、模式产地在井冈山的井冈角蟾、林氏角蟾、陈氏角蟾、井冈纤树蛙、井冈攀蜥和井冈脊蛇。

2010-2015年共发现10个江西省新纪录种、2个新纪录属。新纪录种分别是两栖纲的宜章臭蛙和崇安湍蛙、爬行纲的北部湾蜓蜥、鸟纲云南柳莺和绿背姬鹟、哺乳纲的无尾蹄蝠、褐扁颅蝠、毛翼管鼻蝠、水甫管鼻蝠和泰坦尼亚彩蝠。新纪录属分别是纤树蛙属和攀蜥属。

2.2 井冈山地区是陆生脊椎动物东西扩散、南北迁徙的重要通道，汇聚了多种区系成分

纤树蛙属已知分布区包括越南北部、泰国东北部、中国西藏东南部、云南南部、广西中南部及湖南南部的宜章；井冈山是其最东和最北的分布纪录。短腿蟾属已知的分布区由中南半岛至中国云南和广西西部，延伸到四川南部、广西北部至贵州、广东南岭；井冈山地区是其最北和最东的已知分布记录。攀蜥属已知分布区包括2个间断地理区域，其中之一是从秦岭向南，越过中国西南山区、喜马拉雅南坡和东坡、中南半岛至马来半岛的广大大陆地

区；另一个分布区是中国台湾岛和琉球群岛（Ota *et al.*, 1998）；井冈山是上述2个分布区外的第三个独立分布点，在遗传地理学研究方面具有非常重要的价值。黄岗臭蛙和天目臭蛙被描述为独立物种后，花臭蛙的分布区只限于湖北宜昌至重庆巫山、湖北恩施等地，分子系统学研究表明，花臭蛙分布于包括井冈山的整个罗霄山脉，是该种分布的最东界。东京湾蜓蜥已知的分布区包括越南北部、中国海南、广西东南部六万山、广东西部云开山脉、广东中北部英德、江西九连山和井冈山；井冈山是其最北的分布记录。

井冈山地区记录的棘腹蛙、粉红胸鹨、淡绿鹎鹛、蓝鹎、黄眉林雀等物种，主要分布区在西南山地，同时间断分布于东南山地的武夷山脉，井冈山（罗霄山脉）位于这两个分布区的中间地带，成为这些物种在这两片分布区进行东西扩散的重要"跳板"和"廊桥"。此外，罗霄山脉（井冈山地区）是中国中东部地区候鸟迁徙的地标性山脉，大量鹭类、秧鸡类、鸻类、莺类等沿罗霄山脉进行南北迁徙。

2.3 特有性突出

井冈山地区拥有中国特有的陆生脊椎动物57种。其中，两栖纲有24种，占井冈山两栖类58.5%；

爬行纲有18种，占井冈山爬行类的26.9%；鸟纲9种，占井冈山鸟类的3.1%；哺乳纲6种，占井冈山哺乳动物的8.6%（表1-1）。

上述58种中国特有动物中，井冈脊蛇、七溪岭瘰螈、林氏角蟾、陈氏角蟾、井冈纤树蛙和井冈攀蜥共6种是井冈山地区特有种；井冈角蟾、弓斑肥螈是罗霄山脉特有种；珀普短腿蟾、寒露林蛙是罗霄山脉和南岭特有种，宜章臭蛙除了湖北五峰的分布点外，其分布区集中在南岭和罗霄山脉。

2.4 保存了大量国家级保护动物和珍稀濒危物种

井冈山地区珍稀濒危陆生脊椎动物共计97种。其中，国家重点保护野生动物名录57种，包括国家I级保护野生动物4种，国家II级保护野生动物53种；12种列入《濒危野生动植物种国际贸易公约》附录I，即CITES附录I，38种列入附录II，6种列入附录III。《IUCN濒危物种红色名录》受胁物种23种，包括极危（CR）等级3种，濒危（EN）等级7种，易危（VU）等级13种。《中国物种红色名录》受胁物种54种，其中，3种为极危（CR）等级，10种为濒危（EN）等级，41种为易危（VU）等级（表1-2）。

表 1-1　井冈山地区中国特有陆生脊椎动物和区域特有动物
Table 1-1　Chinese and regional endemic terrestrial vertebrates in the Jinggangshan Region

序号 No.	中国特有种 Endemic species to China	井冈山地区特有种 To the Jinggangshan Region	罗霄山脉特有种 To Luoxiao Mountains	罗霄山脉和南岭特有种 To Luoxiao and Nanling Mountains
1	大鲵 *Andrias davidianus*			
2	七溪岭瘰螈 *Paramesotriton qixilingensis*	√		
3	弓斑肥螈 *Pachytriton archospotus*		√	
4	东方蝾螈 *Cynops orientalis*			
5	崇安髭蟾 *Leptobrachium liui*			
6	福建掌突蟾 *Leptolalax liui*			
7	珀普短腿蟾 *Brachytarsophrys popei*			√
8	井冈角蟾 *Megophrys jinggangensis*		√	
9	陈氏角蟾 *Megophrys cheni*	√		
10	林氏角蟾 *Megophrys lini*	√		
11	三港雨蛙 *Hyla sanchiangensis*			
12	寒露林蛙 *Rana hanluica*			√
13	长肢林蛙 *Rana longicrus*			
14	弹琴蛙 *Nidirana adenopleura*			
15	阔褶水蛙 *Hylarana latouchii*			
16	竹叶臭蛙 *Odorrana versabilis*			
17	花臭蛙 *Odorrana schmackeri*			
18	宜章臭蛙 *Odorrana yizhangensis*			
19	武夷湍蛙 *Amolops wuyiensis*			
20	福建大头蛙 *Limnonectes fuluoxiaoensis*			

序号 No.	中国特有种 Endemic species to China	井冈山地区特有种 To the Jinggangshan Region	罗霄山脉特有种 To Luoxiao Mountains	罗霄山脉和南岭特有种 To Luoxiao and Nanling Mountains
21	九龙棘蛙 *Quasipaa jiulongensis*			
22	小棘蛙 *Quasipaa exilispinosa*			
23	经甫树蛙 *Rhacophorus chenfui*			
24	井冈纤树蛙 *Gracixalus jinggangensis*	√		
25	崇安草蜥 *Takydromus sylvaticus*			
26	北草蜥 *Takydromus septentrionalis*			
27	蹼趾壁虎 *Gekko subpalmatus*			
28	井冈攀蜥 *Japalura jinggangensis*	√		
29	宁波滑蜥 *Scincella modesta*			
30	股鳞蜓蜥 *Sphenomorphus incognitus*			
31	井冈脊蛇 *Achalinus jinggangensis*	√		
32	锈链腹链蛇 *Amphiesma craspedogaster*			
33	颈棱蛇 *Macropisthodon rudis*			
34	饰纹小头蛇 *Oligodon ornatus*			
35	挂墩后棱蛇 *Opisthotropis kuatunensis*			
36	山溪后棱蛇 *Opisthotropis latouchii*			
37	台湾钝头蛇 *Pareas formosensis*			
38	福建钝头蛇 *Pareas stanleyi*			
39	纹尾斜鳞蛇 *Pseudoxenodon stejnegeri*			
40	乌梢蛇 *Ptyas dhumnades*			
41	赤链华游蛇 *Sinonatrix annularis*			
42	环纹华游蛇 *Sinonatrix aequifasciata*			
43	白眉山鹧鸪 *Arborophila gingica*			
44	灰胸竹鸡 *Bambusicola thoracica*			
45	黄腹角雉 *Tragopan caboti*			
46	白颈长尾雉 *Syrmaticus ellioti*			
47	华南冠纹柳莺 *Phylloscopus goodsoni*			
48	棕噪鹛 *Garrulax poecilorhynchus*			
49	褐顶雀鹛 *Alcippe brunnea*			
50	黄腹山雀 *Parus venustulus*			
51	蓝鹀 *Latoucheornis siemsseni*			
52	长吻鼹 *Euroscaptor longirostris*			
53	西南鼠耳蝠 *Myotis altarium*			
54	藏酋猴 *Macaca thibetana*			
55	中华鬣羚 *Capricornis milneedwardsii*			
56	红腿长吻松鼠 *Dremomys pyrrhomerus*			
57	巢鼠 *Micromys minutus*			

表 1-2 井冈山地区珍稀濒危物种名录
Table 1-2 Rare and endangered species inhabiting the Jinggangshan Region

序号 No.	物种 Species	IUCN 红色名录 IUCN Red List	中国红色名录 China Red List	国家重点保护野生动物名录 China Key List of wildlife	CITES 附录 CITES App.
1	大鲵 *Andrias davidianus*	CR	CR	II	I
2	长肢林蛙 *Rana longicrus*	VU			
3	虎纹蛙 *Hoplobatrachus chinensis*		VU	II	II
4	棘腹蛙 *Quasipaa boulengeri*	EN	VU		
5	九龙棘蛙 *Quasipaa jiulongensis*	VU	VU		
6	小棘蛙 *Quasipaa exilispinosa*	VU	VU		
7	棘胸蛙 *Quasipaa spinosa*	VU	VU		
8	乌龟 *Mauremys reevesii*	EN	EN		III

续表

序号 No.	物种 Species	IUCN 红色名录 IUCN Red List	中国红色名录 China Red List	国家重点保护野生动物名录 China Key List of wildlife	CITES 附录 CITES App.
9	平胸龟 *Platysternon megacephalum*	EN	EN		I
10	中华鳖 *Pelodiscus sinensis*	VU	VU		
11	崇安草蜥 *Takydromus sylvaticus*		VU		
12	井冈脊蛇 *Achalinus jinggangensis*	CR	VU		
13	王锦蛇 *Elaphe carinata*		VU		
14	玉斑锦蛇 *Euprepiophis mandarinus*		VU		
15	黑眉锦蛇 *Orthriophis taeniurus*		VU		
16	乌梢蛇 *Ptyas dhumnades*		VU		
17	灰鼠蛇 *Ptyas korros*		VU		
18	银环蛇 *Bungarus multicinctus*		VU		
19	舟山眼镜蛇 *Naja atra*		VU		II
20	白头蝰 *Azemiops feae*		VU		
21	尖吻蝮 *Deinagkistrodon acutus*		VU		
22	短尾蝮 *Gloydius brevicaudus*		VU		
23	海南虎斑鳽 *Gorsachius magnificus*	EN	EN	II	
24	鸳鸯 *Aix galericulata*			II	
25	黑冠鹃隼 *Aviceda leuphotes*			II	II
26	凤头蜂鹰 *Pernis ptilorhynchus*			II	II
27	黑翅鸢 *Elanus caeruleus*			II	II
28	蛇雕 *Spilornis cheela*			II	II
29	凤头鹰 *Accipiter trivirgatus*			II	II
30	赤腹鹰 *Accipiter soloensis*			II	II
31	松雀鹰 *Accipiter virgatus*			II	II
32	日本松雀鹰 *Accipiter gularis*			II	II
33	雀鹰 *Accipiter nisus*			II	II
34	苍鹰 *Accipiter gentilis*			II	II
35	普通鵟 *Buteo japonicus*			II	II
36	林雕 *Ictinaetus malaiensis*			II	II
37	白腹隼雕 *Aquila fasciata*			II	II
38	鹰雕 *Spizaetus nipalensis*			II	II
39	红隼 *Falco tinnunculus*			II	II
40	灰背隼 *Falco columbarius*			II	II
41	燕隼 *Falco subbuteo*			II	II
42	游隼 *Falco peregrinus*			II	I
43	白眉山鹧鸪 *Arborophila gingica*		VU		
44	黄腹角雉 *Tragopan caboti*	VU	VU	I	I
45	白鹇 *Lophura nycthemera*			II	
46	勺鸡 *Pucrasia macrolopha*			II	
47	白颈长尾雉 *Syrmaticus ellioti*			I	I
48	花田鸡 *Coturnicops exquisitus*	VU	VU	II	
49	斑尾鹃鸠 *Macropygia unchall*			II	
50	小鸦鹃 *Centropus bengalensis*			II	
51	褐翅鸦鹃 *Centropus sinensis*			II	
52	草鸮 *Tyto capensis*			II	II
53	领角鸮 *Otus bakkamoena*			II	II
54	红角鸮 *Otus sunia*			II	II
55	黄嘴角鸮 *Otus spilocephalus*			II	II
56	鹰鸮 *Ninox scutulata*			II	II

序号 No.	物种 Species	IUCN 红色名录 IUCN Red List	中国红色名录 China Red List	国家重点保护野生动物名录 China Key List of wildlife	CITES 附录 CITES App.
57	褐林鸮 *Strix leptogrammica*			II	II
58	雕鸮 *Bubo bubo*			II	II
59	长耳鸮 *Asio otus*			II	II
60	短耳鸮 *Asio flammeus*			II	II
61	领鸺鹠 *Glaucidium brodiei*			II	II
62	斑头鸺鹠 *Glaucidium cuculoides*			II	II
63	仙八色鸫 *Pitta nympha*	VU	VU	II	II
64	小燕尾 *Enicurus scouleri*		VU		
65	白喉林鹟 *Rhinomyias brunneatus*	VU	VU		
66	画眉 *Garrulax canorus*				II
67	红嘴相思鸟 *Leiothrix lutea*				II
68	麻雀 *Passer rutilans*		EN		
69	黄胸鹀 *Emberiza aureola*	EN			
70	长吻鼹 *Euroscaptor longirostris*		VU		
71	中蹄蝠 *Hipposideros larvatus*		VU		
72	无尾蹄蝠 *Coelops frithi*		VU		
73	中华鼠耳蝠 *Myotis chinensis*		VU		
74	藏酋猴 *Macaca thibetana*		VU	II	II
75	穿山甲 *Manis pentadactyla*	CR	EN	II	II
76	斑林狸 *Prionodon pardicolor*		VU	II	I
77	果子狸 *Paguma larvata*				III
78	大灵猫 *Viverra zibetha*		EN	II	III
79	小灵猫 *Viverricula indica*		VU	II	III
80	黄喉貂 *Martes flavigula*			II	III
81	食蟹獴 *Herpestes urva*				III
82	猪獾 *Arctonyx collaris*		VU		
83	水獭 *Lutra lutra*		EN	II	I
84	狼 *Canis lupus*		VU	II	II
85	豺 *Cuon alpinus*	EN	EN	II	II
86	貉 *Nyctereutes procyonoides*		VU		
87	豹猫 *Felis bengalensis*		VU		II
88	金猫 *Felis temmincki*		CR	II	I
89	云豹 *Neofelis nebulosa*	VU	EN	I	I
90	豹 *Panthera pardus*	EN	CR	I	I
91	毛冠鹿 *Elaphodus cephalophus*		VU		
92	赤麂 *Muntiacus muntjak*		VU		
93	獐 *Hydropotes inermis*	VU	VU		
94	水鹿 *Rusa unicolor*	VU	VU	II	
95	中华鬣羚 *Capricornis milneedwardsii*		VU	II	I
96	斑羚 *Naemorhedus griseus*	VU	EN	II	I
97	豪猪 *Hystrix hodgsoni*		VU		

Chapter 1 Physical Geography

1.1 Geological location and area

The Jinggangshan Region is located in the middle section of the Luoxiao Mountains. The Luoxiao Mountains is situated in the middle section of the hilly basin area of southeastern China within the third scale terrace of China mainland terrain, and runs in a north-south direction, north to the Yangtze River, the south to the Nanling Mountains, the east adjacent to Poyang Lake, west adjacent to the Dongting Lake, becoming a watershed between water systems associated with Poyang Lake and Dongting Lake.

This atlas defines the Jinggangshan Region as follows: a continuous region is composed of Jinggangshan National Nature Reserve, Qixiling Provincial Nature Reserve, Nanfengmian National Nature Reserve, Taoyuandong National Nature Reserve and their surrounding areas. It covers main area of the Wanyang Mountains along the border between the Hunan and Jiangxi, and occupies an area of 70,874hm². The geographical coordinates of the Jinggangshan Region are 26°13′4″N-26°52′30″N, and 113°56′30″E-114°18′28″E (Figure 1-1 and Figure 1-2).

1.2 Geology and geomorphology

The geomorphology of the Jinggangshan Region is a middle mountain landform controlled by fold blocks. The topography includes mountain massifs, slopes, valleys, structural basins and karst landforms. The main peaks are Nanfengmian, Lingfeng, Bijiashan, Wuzhifeng, Jiangxi'ao and Pingshuishan, all of which exceed 1500 m above sea level. The highest peak, Nanfengmian, at an elevation of 2120.4 m above sea level; the lowest point, Longyuankou in Mount Qixiling, at an elevation of 200 m above sea level. The main bedrocks are granite, siliceous slate, slate and sandstone. The geologic ages of the main strata are Cambrian, Ordovician and Devonian in the Paleozoic, and Quaternary in the Cenozoic. This region lies in southeastern Eurasia, and southeast of junction of the Yangtze Plate and Cathaysian Plate, which are part of the South China Caledonides and Cathaysian tectonic domain (Figure 1-3 and Figure 1-4).

1.3 Hydrology and Soil

The Jinggangshan Region includes the main peaks in the middle of Luoxiao Mountains and the catchment areas on the both sides. This makes it a watershed between the Ganjiang River and Xiangjiang River systems. The Ganjiang River system in Jiangxi comprises three primary tributaries (i.e., Heshui, Shushui and Suichuangjiang Rivers), other seven main rivers with a catchment area of more than 30 km², including Longyuankou River (drainage area of 81 km²), Shixi River (79 km²), Xingzhou River (98 km²), Xiaoxidong (47 km²), Xiangzhou River (43 km²) etc. The Xiangjiang River system in Hunan comprises a primary tributary (Mishui River), and 49 other rivers with length longer than 5 km, or with catchment areas greater than 10 km² (e.g. Xielaishui, Hemoshui, Mianshui and Dongfenghe Rivers), making it one of the most important forested areas for water source conservation in southern Hunan. Main soil types include six types: mountain red soils, mountain red-yellow soils, mountain yellow soils, mountain yellow-brown soils, mountain meadow soils, and mountain boggy soils (only on some of the mountain tops) (Figure 1-5 and Figure 1-6).

1.4 Climate

The Jinggangshan Region lies in the mid-subtropical monsoon climate zone. The average annual temperature is 14.2 °C. The hottest month is July, with an average temperature of 23.9 °C and a highest temperature of 36.7 °C; the coldest month is January with an average temperature of 3.4 °C and a lowest temperature of −11.0 °C. The annual precipitation here is 1400-2100 mm, the highest was 2878.8 mm in 2002. The average annual evaporation discharge is 978.8 mm. The climate changes with altitude, resulting in obvious differences in heat energy and moisture over time and space and forming three vertical climate zones: mid-subtropical zone, north subtropical zone and warm temperate zone. The lapse rate of temperature in these mountains is 0.44-0.57 °C/100 m (Figure 1-7 and Figure 1-8).

1.5 Vegetation types and plant diversity

1.5.1 Diversity of vegetation

There are 14 vegetation types, 58 formation-groups, 90 formations and 180 associations in the Jinggangshan Region (Li, 2004); the natural vegetation is divided into 12 vegetation types, 83 formations and 167 associations. There are 52 rare and endangered or relict plant associations within the natural associations, covering 785.75 hm^2. A typical vertical vegetation spectrum exists from valleys and lowlands to low mountains and middle mountains, the sequence of which is valley monsoon forest, valley monsoon evergreen broadleaved forest, mountain evergreen broadleaved forest, mountain evergreen and deciduous broadleaved mixed forest, mountain coniferous and broadleaved mixed forest, mountain evergreen elfin forest or deciduous scrubland and middle mountain meadow (Figure 1-9-Figure 1-11).

1.5.2 High levels of plant diversity, relict species and refuge in the Tertiary and Quaternary Ice ages

A total of 3745 species of higher plant from the Jinggangshan Region belonged to 1290 genera, 323 families. There are a large number of original, relict and regional endemic species. The area was a refuge for species migrating from north to south on the East Asian mainland during the Tertiary glacial period, and a source of historical evolution processes such as species dispersion northward during interglacial periods.

Endemism in the Jinggangshan Region is obvious. Seven families are endemic to East Asian, five families endemic to China, 44 genera endemic to China, and 1253 species endemic to China. Rare and endangered plant species number as many as 203, including two species of Bryophytes (2 genera, 2 families), four species of Ferns (3 genera, 3 families), 23 species of Gymnosperms (18 genera, 8 families), and 174 species of Angiosperms (90 genera, 35 families). There are 77 families, 122 genera and 181 species of original or relict plant species in the region (Liao *et al.*, 2014).

Chapter 2 Terrestrial Vertebrate Fauna

2.1 A zone of high species diversity

The Jinggangshan Region is located in an uncommon geographical region with advantageous hydrothermal condition, complicated ecological environments, diverse vegetation types, and high levels of plant diversity. It is a crucial area for Eurasia terrestrial vertebrates carried out north-south migration and east-west dispersion. As of end of 2015, 468 terrestrial vertebrate species had been recorded in the area, including 41 amphibian species, 67 reptile species, 290 bird species and 70 mammal species. Therefore, the Jinggangshan Region represents a biodiversity hotspot within southeastern Eurasia with high levels of terrestrial vertebrate diversity.

The type localities of eight species are within the Jinggangshan Region. *Paramesotriton qixilingensis* Yuan *et al.*, 2014 was originally described from Qixiling Nature Reserve, other seven species, i.e., *Brachytarsophrys popei* Wang and Zhao, 2014, *Megophrys jinggangensis* (Wang, 2012), *Megophrys lini* (Wang and Yang, 2014), *Megophrys cheni* (Wang and Liu, 2014), *Gracixalus jinggangensis* Zeng, Zhao, Chen, Chen, Zhang and Wang, 2017, *Japalura jinggangensis* Wang, sp. nov. and *Achalinus jinggangensis* (Zong and Ma, 1983) were originally described from the Jinggangshan Region.

From 2010 to 2015, 10 species and two genera were found from the Jinggangshan Region as first records for Jiangxi Province, including *Odorrana yizhangensis* Fei, Ye, and Jiang, 2007, *Amolops chunganensis* (Pope, 1929), *Sphenomorphus tonkinensis* Nguyen *et al.*, 2011, *Phylloscopus yunnanensis* (La Touche, 1922), *Ficedula elisae* (Weigold, 1922), *Coelops frithii* Blyth, 1848, *Tylonycteris robustula* (Thomas, 1915), *Murina shuipuensis* Eger and Lim, 2011, *Harpiocephalus harpia* (Temminck, 1840) and *Kerivoula titania* Gray, 1842, and two genera, i.e., *Gracixalus* Delorme, Dubois, Grosjean and Ohler, 2005 and *Japalura* Gray, 1853.

2.2 An important passage for east-west dispersal and north-south migration of terrestrial vertebrates and the convergence of multiple fauna components

The description of *Gracixalus jinggangensis* from in the Jinggangshan Nature Reserve represents the easternmost and northernmost distribution records of the genus *Gracixalus*, which is previously known to distribute in south and central Guangxi, Yizhang County in Hunan, southern Yunnan, southeastern Xizang, China, and northern Vietnam and northeastern Thailand. The description of *Brachytarsophrys popei* brings distribution area of the genus *Brachytarsophrys* from the Indo-China Peninsula, Yunnan and western Guangxi, Guizhou, southern Sichuan, northern Guangxi, and the Nanling Mountains in Guangdong, extending to the Jinggangshan Region. Similarly, the Jinggangshan Nature Reserve, as type locality of *Japalura jinggangensis*, is situated in the mid-point of two substantially isolated geographical ranges of the genus *Japalura*, one of which covering a broad mainland area from the Qinling Mountains, through the southwestern China, extending to southern and eastern Himalayas, Indo-China Peninsula and Malay Peninsula; the other one composed of several islands of Taiwan and Ryukyu Islands (Ota *et al.*, 1998). *Odorrana huanggangensis* and *Odorrana tianmuii* were divided from *Odorrana schmackeri* complex and were described as two distinct new species resulted that the distribution area of *Odorrana schmackeri* is restricted to Yichang and Enshi in Hubei to Wushan in Chongqing. Howover, the populations from the Luoxiao Mountains is classified as *Odorrana schmackeri* on basis of our molecular data. The record of *Sphenomorphus tonkinensis* from Jinggangshan Nature Reserve brings its distribution area from northern Vietnam, China Hainan, through the Liuwan Mountains of Guangxi, Yunkai Mountains in the border between Guangdong and Guangxi, northern Guangdong, extending to the Jinggangshan Region, which should be its north most distribution.

Quasipaa boulengeri, *Anthus roseatus*, *Pteruthius xanthochlorus*, *Latoucheornis siemsseni*, and *Sylviparus modestus* recorded in the Jinggangshan Region are mainly distributed in southwest mountain areas

and discontinuously in the Wuyi Mountains. The Jinggangshan Region(Luoxiao Mountains) is located in the intermediate zone between these two mountain areas, and is an important springboard and bridge for east-west dispersal of aforementioned species between these two distribution areas. In addition, the Luoxiao Mountains (the Jinggangshan Region) is a landmark mountain range for migratory birds in eastern Central China, including numerous egrets, rails, thrushes and warblers, who migrate in north and south direction along the Luoxiao Mountains.

2.3 Significant endemism

There are 57 Chinese endemic terrestrial vertebrate species in the Jinggangshan Region, including 24 amphibian speceis (58.5% of amphibians from the Jinggangshan Region); 18 reptile species (26.9% of reptiles from the Jinggangshan Region); nine bird species (3.1% of birds from the Jinggangshan Region), and six mammal species (8.6% of mammals from the Jinggangshan Region) (Table 1-1). Of which, six species (*Achalinus jinggangensis*, *Paramesotriton qixilingensis*, *Megophrys lini*, *Megophrys cheni*, *Gracixalus jinggangensis* and *Japalura jinggangensis*) are defined as micro-endemic to the Jinggangshan Region; *Megophrys jinggangensis* and *Pachytriton archospotus* are defined as endemic to the Luoxiao Mountains; *Brachytarsophrys popei* and *Rana hanluica* are endemic to the Luoxiao Mountains and Nanling Mountains; and the distribution of *Odorrana yizhangensis* is concentrated in the Luoxiao Mountains and Nanling Mountains (plus a group in Wufeng, Hubei).

2.4 National protected animals and rare endangered species

There are 97 rare and endangered terrestrial vertebrate species in the Jinggangshan Region. Of which, 57 are listed in the "China Key List of Wildlife", including four listed in China Key List: I, and 53 listed in China Key List: II ; 12 species are listed in CITES (the Convention on International Trade in Endangered Species of Wild Fauna and Flora) App. I, 38 species are listed in CITES App. II, and six species are listed in CITES App. III; 23 species are listed in "IUCN

Red List of Threatened Species", including two as critically endangered (CR), seven as endangered (EN) and 13 as vulnerable (VU); 54 species are listed in the "China Species Red List", including three as critically endangered (CR), 10 as endangered (EN) and 41 as vulnerable (VU) (Table 1-2).

Part 2

井冈山地区陆生
脊椎动物区系

Terrestrial Vertebrate
Fauna of the
Jinggangshan
Region

第 1 章

井冈山地区
两栖动物区系

Chapter 1

Amphibian Fauna of the Jinggangshan Region

　　基于 2010−2015 年在井冈山地区密集开展的两栖爬行动物调查和广泛采样，共记录两栖动物 2 目 9 科 25 属 41 种，包括最近几年所发表的 6 个新种，即七溪岭瘰螈、珀普短腿蟾、井冈角蟾、林氏角蟾、陈氏角蟾和井冈纤树蛙；2 个江西省分布新纪录种，宜章臭蛙和崇安湍蛙，纤树蛙属是江西省新纪录属。中国特有两栖动物 24 种，国家 Ⅱ 级重点保护野生动物 2 种，CITES 附录 Ⅰ 和附录 Ⅱ 各 1 种，IUCN 红色名录极危（CR）1 种，濒危（EN）1 种。

　　本书共收录井冈山地区两栖动物 38 种。大鲵和经甫树蛙的记录依据分别来自井冈山自然保护区馆藏标本和历史文献资料，我们的调查并未发现这 2 个物种；东方蝾螈只有照片凭证，未采集到标本。上述 3 个物种均未列入本书。

　　本书所使用的缩写词说明如下：SVL= 头休长，即吻端至泄殖腔长；TAL= 尾长；HDL= 头长，即吻端至口角长；HDW= 头宽，即左右上颌与颧弓关节处宽度；SNT= 吻长，即吻端至眼前角长度；ED= 眼径，即眼前角至后角距离；TMD= 鼓膜径；HND= 手长，即桡尺骨远端至第三指末端的长度；RAD= 桡尺骨（小臂）长；FTL= 足长，即从胫骨远端到第三趾末端；TIB= 胫骨长；IND= 鼻间距；IOD= 眶间距；TMD= 鼓膜径；a.s.l.= 海拔。

Based on continual herpetological field surveys coupled with extensive sampling in the Jinggangshan Region conducted from 2010 to 2015, have resulted in discovery of 41 species of amphibians, belonged to 25 genera, nine families and two orders. Six new species, *Paramesotriton qixilingensis* Yuan, Zhao, Jiang, Hou, He, Murphy, and Che, 2014, *Brachytasophrys popei* Wang and Zhao, 2014, *Megophrys jinggangensis* (Wang, 2012), *Megophrys lini* (Wang and Yang, 2014), *Megophrys cheni* (Wang and Liu, 2014), and *Gracixalus jinggangensis* Zeng, Zhao, Chen, Chen, Zhang and Wang, 2017 have been described in the last several years. Two species, *Odorrana yizhangensis* Fei, Ye, and Jiang, 2007 and *Amolops chunganensis* (Pope, 1929) and the genus *Gracixalus* were found at Mount Jinggang as first records in Jiangxi province. Among surveyed amphibian species, 23 species are endemic to China, two are listed in the China Key List II, one is listed in CITES App. I and one in CITES App. II, and one is listed on the IUCN Red List as critically endangered (CR) and one as endangered (EN).

In this atlas we selected 38 known amphibian species from the Jinggangshan Region. *Andrias davidianus* was recorded from collection in Jinggangshan National Nature Reserve, and *Rhacophorus chenfui* was recorded from the literature and not found in our surveys. We only obtained voucher photos of *Cynops orientalis* and did not collect voucher specimens. Therefore, these three species are not included in this book.

Abbreviations in this book are SVL = snout-vent length; TAL= tail length from tip of the tail to the anterior end of the vent; HDL = head length from tip of snout to the articulation of the jaw; HDW = head width, between left and right articulations of the quadratojugal and maxilla; SNT = snout length, from tip of snout to the anterior corner of the eye; ED = eye diameter, from the anterior corner of the eye to posterior corner of the eye; IND = internasal distance; IOD = interorbital distance; TMD = tympanum diameter; HND = hand length, from distal end of radioulna to tip of distal phalanx of III; RAD = radioulna length; FTL = foot length, from distal end of tibia to tip of distal phalanx of III; TIB = tibial length; a.s.l.= above sea level.

有尾目 CAUDATA
蝾螈科 Salamandridae

1.1 七溪岭瘰螈 *Paramesotriton qixilingensis* Yuan, Zhao, Jiang, Hou, He, Murphy, and Che, 2014

识别特征　体型较大。头体长短于尾长。成年雌性头体长 66.8-74.4 mm，尾长 73.1-80.7 mm；成年雄性头体长 65.9-67.7 mm，尾长 71.7-73.1 mm。头扁平，长稍大于宽，头两侧各有一个腺脊。吻钝。鼻孔接近吻端。唇褶发达，位于上颌两侧。有喉褶。犁骨齿"Λ"形。舌卧于口底，两侧游离。头部平滑，有疣粒；体背和体侧皮肤粗糙，散布瘰粒。体背中央脊崎发达，无背侧崎。四肢稍长，贴体相向时彼此重叠。前肢四指，后肢五趾，无蹼。尾侧扁，末端尖。泄殖腔在雄性显著肿胀，有乳状突；雌性稍肿胀。全身底色暗棕色至黄褐色。头体腹面有橘黄色或橘色云斑。

生境与习性　栖息于有岩沙的山涧溪流。

分布　中国特有种。仅分布在模式产地江西七溪岭自然保护区。

种群状况　七溪岭自然保护区常见。

Identification　Body size large; snout-vent length shorter than tail length, adult females measuring 66.8-74.4 mm SVL, 73.1-80.7 mm TAL, adult males measuring 65.9-67.7 mm SVL, 71.7-73.1 mm TAL; head depressed and flattened, slightly longer than width; a glandular ridge present on both sides of head; snout truncated; nostrils close to the tip of the snout; labial fold evident on lateral side of upper jaw; gular fold present; vomerine tooth patch Λ-shaped; tongue adhering to mouth floor with free lateral margins; dorsal surface of head smooth with tubercles; skin of body rough, scattered with warts on dorsal and lateral surface; vertebral ridge conspicuous; lateral ridge absent; limbs long, hind limb overlapping forelimb when adpressed; four fingers, five toes, without webbing; tail laterally compressed and extremity pointed; cloaca significantly swollen with papillae in males, slightly swollen in females. Ground color dark brown to olive-brown; ventral surface of head and body with orange-yellow or orange blotches.

Habitat and Behavior　Inhabits montane streams with rocks and sand.

Distribution　Endemic to China. Found only from the type locality, Qixiling Nature Reserve, Jiangxi.

Status of Population　Common species in Qixiling Nature Reserve.

1.1a

1.1b

1.1c

1.1d

1.1e

1.2 弓斑肥螈 *Pachytriton archospotus* Shen, Shen, and Mo, 2008

识别特征 体型中等。头扁平，长大于宽。吻短于眶间距。在繁殖季节雄性肛区稍肿胀，生殖乳突长而密。皮肤光滑，肋沟没有或不显。体侧及尾侧有沟状横斑纹。背面棕黑色或灰棕色，有清晰或不清晰的黑色小点斑。腹面橘色，有不规则黑棕色斑点。

生境与习性 栖息在海拔 800-1600 m 的山间多岩石溪流。

分布 中国特有种。分布在中国湘赣交界的罗霄山脉中段和南段，至少包括井冈山、南风面自然保护区、齐云山自然保护区、八面山自然保护区和桃源洞自然保护区。

种群状况 地区性常见种。

Identification Body moderate; head depressed and flattened, longer than wide; snout shorter than interorbital distance; cloacal region slightly swollen in male during breeding season; papillae on cloacal wall long and dense. Skin smooth; costal grooves indistinct or absent; sides of trunk and tail with transverse sulcus-shaped streaks. Brown-black or grey-brown above, with distinct or indistinct small black spots; orange below, with irregular black-brown spots.

Habitat and Behavior Inhabits montane rocky streams between 800-1600 m a.s.l.

Distribution Endemic to China. Distributed in middle and south areas of Luoxiao Mountains, situated in border between Jiangxi and Hunan provinces, China, at least including Mt. Jinggang, Nanfengmian Nature Reserve, Qiyunshan Nature Reserve, Bamianshan Nature Reserve and Taoyuandong Nature Reserve.

Status of Population Regional common species.

无尾目 ANURA
蟾蜍科 Bufonidae

1.3 中华蟾蜍 *Bufo gargarizans* Cantor, 1842

识别特征 体型大，成年雄性头体长79-100 mm，雌性98-121 mm。头宽大于头长。鼻间距小于眶间距。无上颌齿和犁骨齿。后肢短，跟部不相遇，胫跗关节前伸达肩部。上眼睑无明显疣粒。除头顶光滑无骨质棱脊和疣粒外，皮肤粗糙，背表面密布瘰粒，腹表面密布疣粒。耳后腺大，椭圆形。无股腺，无跗褶。趾缘膜显著，第四趾半蹼。雄性无声囊，内侧三指有黑色强婚刺。体色随季节变异较大。雄性背面一般为墨绿色、灰绿色至褐绿色；雌性多为棕黄色。体侧有黑褐色纵行宽纹，通常纹上方大疣白色。体侧一般无棕红色斑纹。腹面乳黄色与棕色或黑色形成花斑。

生境与习性 栖息于陆地多种生境。

分布 东北、华北、华中、华东、向南至广东和广西、向西分布至贵州、四川。俄罗斯阿穆尔河流域和萨哈林岛（库页岛）、朝鲜半岛、日本宫古岛、琉球群岛。在井冈山地区广布于各种生境。

种群状况 常见种。

Identification Body large, measuring 79-100 mm SVL in adult males, 98-121 mm SVL in adult females; head width larger than head length; internasal distance less than interorbital distance; maxillary teeth and vomerine teeth absent; hind limbs short, the heels not meeting and tibio-tarsal articulation reaching the shoulder; upper eyelid without distinct tubercles; except top of head smooth without bony ridge and tubercles, skin rough with densely covered warts on dorsal surface and tubercles on the ventral surface; parotoid gland large, oval; no tarsal fold; no femoral gland on rear of thigh; lateral fringes on toes distinct, half web at IV toe; no vocal sac in males. Usually dark green, grey green to brownish green above in males, brownish yellow above in females; usually a wide dark brown longitudinal strip, no brown red markings on the side of body; wart above the strip large, white; cream and brown or black below, forming piebald pattern.

Habitat and Behavior Inhabits diverse biotopes.

Distribution NE, N, C and E China, south to Guangxi and Guangdong, west to Guizhou and Sichuan. Amur River Basin and Sakhalin Island, Russia, Korean Peninsula, Miyakojima Japan, Ryukyu Islands. Widely occurs in the Jinggangshan Region.

Status of Population Common species.

1.3a

1.3c

1.3b

1.3d

1.4 黑眶蟾蜍 *Duttaphrynus melanostictus* (Schneider, 1799)

识别特征　雄蛙约 76 mm，雌蛙约 106 mm。头部具有黑色骨质棱，鼓膜大而明显。背部和体侧密布瘰粒和疣粒，腹部和四肢密布疣粒，疣粒上有黑色角质刺。雄蛙第 1、第 2 指基部有黑色婚垫。体色多变。

生境与习性　栖息于不同海拔的多种生境中。除繁殖期在水中生活外，一般多在陆地活动。

分布　江西、湖南、贵州、四川、云南、广西、海南、广东、福建、浙江、台湾、香港和澳门。国外分布于巴基斯坦、尼泊尔、孟加拉国、印度、斯里兰卡、安达曼群岛、马尔代夫、马来西亚、苏门答腊、爪哇、加里曼丹岛和巴厘岛，其中一些地方海拔超过 2000 m。引种到苏拉威西、东帝汶和巴布亚新几内亚。在井冈山记录于新城区、黄坳等地。

种群状况　常见种。

Identification　A moderate-sized bufonid with approximately 76 mm SVL in males, approximately 106 mm SVL in females; presence of two black bony ridges near orbits on top of head; tympanum large and distinct; skin rough with densely covered warts on surface of dorsum and flank of body; tubercles with black spines on the ventral surface of body and limbs; black nuptial pads on bases of first and second fingers. Color varied.

Habitat and Behavior　Inhabits various habitats at different altitudes. Terrestrial, breeds in pools.

Distribution　Jiangxi, Hunan, Guizhou, Sichuan, Yunnan, Guangxi, Hainan, Guangdong, Fujian, Zhenjiang, Taiwan, Hong Kong, and Macau. Throughout Pakistan, Nepal, Bangladesh, India, Sri Lanka, Andaman Islands, Maldives, Malaysia, Sumatra, Java, Kalimantan Island, and Bali, up to 2000 m a.s.l. in some places. Introduced onto Sulawesi, East Timor, and Papua New Guinea. Recorded from the Xinchengqu, Huang'ao in Mt. Jinggang.

Status of Population　Common species.

1.4a

1.4b

无尾目 ANURA
角蟾科 Megophryidae

1.5 崇安髭蟾 *Leptobrachium liui* (Pope, 1947)

识别特征　体型中等，成年雄性头体长 68-91 mm，雌性 57-77 mm。头平扁，宽大于长。吻棱发达。颊部倾斜。鼓膜隐蔽。无犁骨齿。后肢短，跟部不相遇，胫跗关节前伸达肩部。皮肤颗粒状，身体和四肢背面有网状皮肤棱。雄性上唇边缘有一个强圆锥状的黑色角质刺，雌性代之的是一个橘红色小疣粒。体背浅褐色、染紫色，并散布有黑色斑，四肢有黑色横斑纹。腹面灰白色，密布白色小颗粒。虹膜上半部分黄绿色，下半部分深棕色，瞳孔直立黑色。雄性有单个声囊。

生境与习性　栖息于海拔 200-1600 m 的山地亚热带常绿阔叶林，常见于森林地面。繁殖季节为 12 月至翌年 1 月，只见于山涧溪流。卵灰白色。蝌蚪大，有"Y"形浅色斑纹在尾和身体交界处，三年后变态成为幼体。

分布　中国特有种。分布于江西、浙江、福建、广西、广东和湖南。在井冈山记录于五指峰、八面山、荆竹山、西坪、茅坪等地。

种群状况　常见种。

Identification　A moderate-sized moustache toad with 68-91 mm SVL in adult males, 57-77 mm SVL in adult females; head depressed, the head broader than head length; canthus rostralis developed; loreal region oblique; tympanum hidden; no vomerine teeth; hind limbs short, the heels not meeting, tibio-tarsal articulation reaching the shoulder; skin granular, with reticular ridges on dorsum of body and limbs; the presence of, single strongly conical black keratose spine on edge of upper lip in adult males; replaced with single red-orange small tubercle in females. Above light brown with a purple, and with scattered black spots and black transverse bands on limbs; pale grey below, with densely arranged white granules; the upper half part of iris yellowish-green, lower half part dark brown; pupil vertical and black. Single vocal sac in males.

Habitat and Behavior　Inhabits montane subtropical evergreen broadleaved forest at 200-1600 m a.s.l. It is more frequently found on forest floor, but during breeding season from December to next January only found in mountainous streams. Ova grey-white; tadpoles great, with light-colored Y-shaped marking in the border region of the tail and body, metamorphosis into juveniles after three years.

Distribution　Endemic to China. Jiangxi, Zhejiang, Fujian, Guangxi, Guangdong and Hunan. Recorded from Wuzhifeng, Bamianshan, Jingzhushan, Xiping and Maoping in Mt. Jinggang.

Status of Population　Common species.

1.5a　1.5b　1.5c　1.5d

1.6 福建掌突蟾 *Leptolalax liui* Fei and Ye, 1990

识别特征　小型掌突蟾，最大头体长不到 30 mm。头长几等于头宽。吻高，圆形。吻棱稍圆。颊部倾斜。鼓膜清晰。没有犁骨齿。有大而圆形的内掌突和外掌突。内跖突大，卵圆形，没有外跖突。脚趾缘膜较宽。背部皮肤粗糙，有痣粒和疣粒，在四肢上形成纵行皮肤棱。颞褶清晰，在前肢基部上方有一个圆形突起的橘红色腺体。腹面皮肤光滑。突起的胸腺较大，长椭圆形，靠近腋窝。大腿后面突起的股腺大于趾端。腹侧腺排成一纵行，彼此平行。体背棕色或深棕色，两眼间有深色三角形斑，背上有 "W" 形斑，均镶浅色边缘。腹面白色。

生境与习性　栖息于海拔 300-1300 m 的亚热带常绿阔叶林。更常见于溪流中岩石上、溪流附近的落叶层、植物茎秆上。

分布　中国特有种。井冈山的掌突蟾种群与福建掌突蟾武夷山（模式产地）种群有较大的形态差异和遗传分化，因此，有必要重新评估其分类地位。在此，暂时将其定为福建掌突蟾，记录于五指峰、双溪口、西坪、茅坪、大院和神农谷等地。

种群状况　常见种。

Identification　A small-sized metacarpal-tubercled toad, with maximum SVL less than 30 mm; head length approximately equal to head width; snout deep, rounded; canthus rostralis slightly rounded; loreal region oblique; tympanum distinct; no vomerine teeth; the presence of a large, round inner palmar tubercle and outer palmar tubercle; inner metatarsal tubercle large, oval; outer metatarsal tubercle absent; toes with broad lateral fringes; dorsal skin rough, with granules, tubercles, and forming longitudinal ridges on limbs; supratympanic fold distinct; single orbicular, raised orange supra-axillary gland above the insertion of arm; ventral skin smooth; raised pectoral gland large, long elliptic, closer to axilla; raised femoral gland on rear of thigh larger than tips of toes; ventrolateral glands in a longitudinal line parallel to each other. Above brown or dark brown, with a dark interorbital triangular marking and W-shaped dorsal marking bordered by a light color; below white.

Habitat and Behavior　Inhabits moist subtropical evergreen broadleaved forests at elevations between 300-1300 m a.s.l. It is more frequently found ensconced among rocks in streams and on leaf litter, stem or branch near streamlets.

Distribution　Endemic to China. The *Leptolalax* population of the Jianggangshan Region is remarkably distinguished from *Leptolalax liui* from Wuyi Mountain, Fujian (type locality) in having discrete morphological differences and genetic divergence; thus, re-evaluation of the taxonomic status of *Leptolalax* population from the Jianggangshan Region is necessary. Herein, this *Leptolalax* population provisionally is designated as *Leptolalax liui*, and recorded from Wuzhifeng, Shuangxikou, Xiping, Maoping, Dayuan, Shennonggu and other places.

Status of Population　Common species.

1.7 珀普短腿蟾 *Brachytasophrys popei* Wang and Zhao, 2014

识别特征　一种相对小型的短腿蟾，雌性头体长 86.2 mm，雄性 70.7-83.5 mm。头大，极度平扁，宽是长的 1.2 倍，近于头体长的 1/2。吻棱不发达。鼓膜隐蔽。犁骨齿在两列发达凸起的犁骨棱上；犁骨棱长，远超内鼻孔后缘水平线，彼此间距约为犁骨棱长的 1.5 倍。舌深缺刻。后肢跟部不相遇；胫附关节前伸达口角处。胫骨短，仅为头体长的 36%-43%。脚趾蹼发达，雄性可至各趾 1/3 至 3/4，雌性 1/4 至 1/2；各趾均有缘膜，雄性缘膜非常宽，雌性稍逊。各指基部有不甚清晰的关节下瘤。上眼睑具疣粒，其中之一扩大成为浅黄色发达的钝圆锥状"角"。雄性第 1 指和第 2 指基部背面有细密小黑色婚刺。雄性单声囊。雌性在输卵管内成熟卵浅黄色。蝌蚪腹部有一条白色横纹，体侧各有 1 条纵走白色纹。

生境与习性　栖息于山溪溪流，周围是亚热带常绿阔叶林。海拔 900-1300 m。它们多被发现于山溪石下。7-9 月能听到短促响亮的重复叫声，每个鸣段包含 12-17 个音符，音符间隔约为 0.41 s。繁殖季节至少在 7-9 月。

分布　中国特有种。分布于罗霄山脉中段和南岭。模式产地：湖南桃源洞自然保护区（26° 30′ 8.79″ N，114° 03′ 38.27″ E；1045 m a.s.l.），正模标本：SYS a001867。

种群状况　区域性常见种。

Identification　A relatively small-sized short-legged toad, measuring 86.2 mm SVL in adult female, 70.7-83.5 mm SVL in adult males; head enormous, extremely depressed, nearly 1.2 times as broad as long, nearly one-half as broad as SVL; canthus rostralis not developed; tympanum hidden; vomerine teeth present on two prominently elevated ridges, extending behind far beyond the posterior level of the choanae, widely separated by a distance nearly 1.5 times the length of one; margin of tongue deeply notched behind; heels not meeting; tibio-tarsal articulation reaching the commissure of the jaw; tibia relatively short, 0.36-0.43 times as long as SVL; toes nearly 1/3 to 3/4 webbed in males, 1/4 to 1/2 webbed in females; the web extending as a fringe along either side, the fringe significantly wide in males, slightly wide in females; subarticular tubercle indistinct at base of each finger; upper eyelid with tubercles, one of which is enlarged and remarkably prominent, bluntly conical light-yellow horn; tiny, black nuptial spines densely covering the dorsal surface of the first finger and second finger base in adult males; single vocal sac in males; gravid females bear pure yellowish oocytes; tadpoles with a transverse white stripe on ventral surface and two longitudinal white stripes along the sides of the body.

Habitat and Behavior　Inhabits montane streams surrounded by moist subtropical evergreen broadleaved forests at 900-1300 m a.s.l. It is more frequently found ensconced among rocks in a stream. During July to September males emit a series of short loud notes repeated about twenty times with slight pauses between from hidden positions. The breeding season is at least from July to September.

Distribution　Endemic to China. Distributed in the middle section of the Luoxiao Mountains and Nanling Mountains. Type localicy: Taoyuandong Nature Reserve (26° 30′ 8.79″ N, 114° 03′ 38.27″ E; 1045 m a.s.l.), Hunan. Holotype: SYS a001867.

Status of Population　Regional common species.

1.7a

1.7b

1.7c

1.7d

1.8 井冈角蟾 *Megophrys jinggangensis* (Wang, 2012)

识别特征　小型角蟾，成年雌性头体长 38.4-41.6 mm，雄性 35.1-36.7 mm。头长几等于头宽。鼓膜大，约为眼径的 80%。有犁骨齿。舌端有缺刻。指序 II < I < IV < III。趾指侧缘膜小。趾基部有厚的肉质蹼。背有疣粒和两个肿胀的背侧皮肤褶。体侧有分散的脓包状的大疣粒。上眼睑上有几个大疣粒，其中之一成为非常发达的角状突，位于上睑边缘。背浅棕色，4 条纵行的深棕色宽纹彼此平行。两眼间有一个暗棕色三角形斑。四肢和趾指背面浅棕色有暗棕色横纹。腹表面发灰，有黑色和棕色斑点。雄性有单个咽下声囊，繁殖季节成年雄性第 1 指背和第 2 指基背部有细密小黑色婚刺。

生境与习性　栖息于山间水势较缓的溪流，四周有亚热带常绿阔叶林，海拔 400-1250 m。繁殖季节应该在 9 月之前。

分布　中国特有种。中国湘赣交界的罗霄山脉中段和北段，模式产地：江西井冈山（26°33′06.30″ N，114°09′17.60″ E；845 m a.s.l.），模式标本：SYS a001430。

种群状况　区域性常见种。

Identification　A small-sized species with adult females measuring 38.4-41.6 mm SVL and males measuring 35.1-36.7 mm SVL; head length approximately equal to head width; tympanum large and distinct, approximately 0.8 times eye diameter; vomerine teeth on two weak ridges; tongue not notched behind; relative finger length II < I < IV < III; weak lateral fringes present on digits; toe bases with thick, fleshy web; dorsum with tubercles and swollen dorsolateral folds; large pustules scattered on flanks; several large tubercles on the upper eyelid, one of them horn-like, remarkably prominent at the edge of the eyelid; dorsum light brown, four wide longitudinal dark brown stripes parallel to each other, a dark brown triangle between eyes; dorsal limbs and digits light brown with dark brown transverse bands; ventral surface grayish with black and brown spots; males with a single subgular vocal sac; tiny, black nuptial spines densely covering the dorsal surface of the first finger and second finger base in adult males during breeding season.

Habitat and Behavior　Inhabits small, slow-moving montane streams surrounded by moist subtropical evergreen broadleaved forests at elevations between 400-1250 m a.s.l. The breeding season of the new species is likely before September.

Distribution　Endemic to China. Distributed in central and northern areas of the Luoxiao Mountains, situated on the border between Jiangxi and Hunan, Type locality: Mount Jinggang (26°33′06.30″N, 114°09′17.60″E; 845 m a.s.l.). Holotype: SYS a001430.

Status of Population　Regional common species.

1.9 林氏角蟾 *Megophrys lini* (Wang and Yang, 2014)

识别特征　小型角蟾，成年雄性头体长 34.1-39.7 mm，雌性 37.0-39.9 mm。头长几等于头宽。没有犁骨齿，舌端不具缺刻。鼓膜清晰，中等大小。后肢较长，跟部重叠，胫跗关节前伸达眼前角。指序 II ≤ I < IV < III。趾指侧缘膜宽，仅在趾基部扩大成为蹼。趾指基部关节下瘤清晰。背皮肤光滑，有小瘰粒，通常在背上有几条弯曲的弱皮肤棱，体侧有几个疣粒。腹表面平滑。上眼睑边缘有一个小角状突。颞褶窄细，浅色。背面浅棕色或橄榄色，有镶有浅色边缘的深色三角形眶间斑和"X"形背斑。雄性有单个咽下声囊，繁殖季节成年雄性第 1 指背和第 2 指背部有细密小黑色婚刺。雌蛙输卵管内卵为纯黄色。

生境与习性　栖息于山间急流，周围为潮湿的亚热带常绿阔叶林，海拔 1100-1620 m。繁殖季节可能在 9 月之后。

分布　中国特有种。分布于中国湘赣交界的罗霄山脉中段，包括江西井冈山自然保护区、遂川南风面自然保护区和湖南炎陵桃源洞自然保护区。模式产地：江西井冈山的八面山（26° 34′ 37.97″ N，114° 06′ 6.43″ E；1369 m a.s.l.），正模标本：SYS a001420。

种群状况　区域性常见种。

Identification　A small-sized species with 34.1-39.7 mm SVL in adult males, 37.0-39.9 mm SVL in adult females; head length approximately equal to head width; vomerine teeth absent; margin of the tongue smooth, not notched behind; hind limbs elongated, the heels overlapping and tibio-tarsal articulation reaching the anterior corner of the eye; relative finger length II ≤ I < IV < III; lateral fringes on the digits wide, toes with more extended webbing at their bases; subarticular tubercle on each digit distinct; dorsal skin smooth with scattered granules, usually a few curved weak ridges on back, several tubercles on flanks; ventral surface smooth; a small horn-like tubercle at the edge of the eyelid; supratympanic fold narrow, light color; light brown or olive above, a dark interorbital triangular marking and X-shaped dorsal marking bordered with a light edge; males with a single subgular vocal sac; tiny, black nuptial spines densely covering the dorsal surface of the first finger and second finger base in adult males during breeding season; gravid females bear pure yellowish oocytes.

Habitat and Behavior　Inhabits rushing mountain streams surrounded by moist subtropical evergreen broadleaved forests at elevations between 1100-1620 m a.s.l. The breeding season likely begins after September.

Distribution　Endemic to China. Distributed in central areas of the Luoxiao Mountains, situated on the border between Jiangxi and Hunan, including Jinggangshan Nature Reserve, Nanfengmian Nature Reserve and Taoyuandong Nature Reserve. Type locality: the Bamianshan (26° 34′ 37.97″ N, 114° 06′ 6.43″ E; 1369 m a.s.l.), Mt. Jinggang, Jiangxi. Holotype: SYS a001420.

Status of Population　Regional common species.

1.9a 　 1.9b 　 1.9c 　 1.9d

1.10 陈氏角蟾 *Megophrys cheni* (Wang and Liu, 2014)

识别特征 小型角蟾，成年雌蛙头体长 31.8-34.1 mm，雄蛙 26.2-29.5 mm。头长几等于头宽。无犁骨齿。舌端具缺刻。鼓膜清晰或不清晰，通常鼓膜上部隐藏在颞褶之下。后肢长，跟部重叠较多，胫跗关节前伸达鼻孔至吻端区域。指序 I < II < IV < III。趾指缘膜宽，趾基部有不发达的蹼。每趾基部关节下瘤清晰。背面和体侧皮肤光滑有疣粒，通常形成两个背侧疣粒行，彼此平行，在背部两个背侧疣粒行之间形成一个 "X" 形弱脊。在胫部大疣粒横向排列成 4-5 行。上眼睑边缘有一个小角状突。颞褶肿胀，浅色。腹面光滑。背面红棕色或橄榄棕色，背部有深色网状斑纹，四肢背面有深色横斑纹。雄性有单个咽下声囊，繁殖季节成年雄性第 1 指背和第 2 指基背部有细密小黑色婚刺。

生境与习性 栖息在海拔 1200-1530 m 的山间沼泽，周围是亚热带常绿阔叶林。

分布 中国特有种。分布于中国湘赣边界的罗霄山脉中段，包括井冈山自然保护区和桃源洞自然保护区。模式产地：江西井冈山的荆竹山（26° 29′ 45.95″ N，114° 04′ 45.66″ E；1210 m a.s.l.）。正模标本：SYS a001873。

种群状况 区域性常见种。

Identification A small-sized species with 31.8-34.1 mm SVL in adult females, 26.2-29.5 mm SVL in adult males; head length approximately equal to head width; vomerine teeth absent; margin of tongue notched behind; tympanum distinct or indistinct, usually its upper part hidden under the supratympanic fold; hind limbs elongated, the heels more overlapping and tibio-tarsal articulation reaching the region between the nostril and tip of snout; relative finger length I < II < IV < III; lateral fringes on digits wide, toes with rudimentary webbing at their bases; subarticular tubercle on each toe distinct; skin of all upper surfaces and flanks smooth with tubercles, usually forming a dorsolateral row of tubercles of parallel to contralateral rows, an X-shaped weak ridge between the dorsolateral tubercle rows on dorsum; large tubercles arranged in four or five transverse rows on shanks; a small horn-like tubercle at the edge of the eyelid; supratympanic fold swollen, light color; ventral surface smooth, red-brown or olive-brown above with dark reticular marking on dorsum, dorsal surface of the limbs with dark transverse bands; males with a single subgular vocal sac; tiny, black nuptial spines densely covering the dorsal surface of the first finger and second finger base in adult males during breeding season.

Habitat and Behavior Inhabits mountainous swamps surrounded by moist subtropical evergreen broadleaved forests at elevations between 1200-1530 m a.s.l.

Distribution Endemic to China. Distributed in central areas of the Luoxiao Mountains, situated on the border between Jiangxi and Hunan, including Jinggangshan Nature Reserve and Taoyuandong Nature Reserve. Type locality: the Jingzhushan, Mt. Jinggang Jiangxi (26°29′45.95″N, 114°04′45.66″E; 1210 m a.s.l.). Holotype: SYS a001873.

Status of Population Regional common species.

1.10a

1.10b

1.10c

无尾目 ANURA
雨蛙科 Hylidae

1.11 中国雨蛙 *Hyla chinensis* Günther, 1858

识别特征 体较小，雄蛙头体长 30-33 mm，雌蛙头体长 29-38 mm。吻棱明显。背面观吻端圆，侧面观高厚而平切。颊部略倾斜。皮肤光滑，颞褶细；喉部光滑，体腹面密布疣粒。指趾端有吸盘和边缘沟。指基部具微蹼，外侧 3 趾具 2/3 蹼。跟部重叠，胫跗关节前伸达鼓膜或眼。背绿色，头部、前臂和胫部镶黑色细线纹；吻端肉色，向后延伸经颊部、过眼、颞区而止于前肢基部上方；上唇、眼下区域浅黄绿色；体侧及腹面浅黄色；体侧及股部前后有数量不等的黑斑；跗足部和腕掌部肉色。雄性具单声囊。

生境与习性 栖息于海拔 100-1200 m 山区。夜晚常见于植物叶片及茎秆上。

分布 华中、华东及华南。越南北部。在井冈山记录于龙市等地。

种群状况 常见种。

Identification A small-sized tree toad, with 30-33 mm SVL in males, 29-38 mm SVL in females; canthus rostralis distinct; snout rounded in dorsal view, deep and truncated in profile view; loreal region slightly oblique; dorsal skin smooth; supratympanic fold narrowed; throat smooth, posteriorly ventral skin densely arranged small tubercles; tips of digits with well-developed disks and marginal grooves; fingers with rudimentary webbings at their bases; toes II, III and IV two-thirds webbed; heels overlapping and tibiotarsal articulation reaching the tympanum and eye. Upper surface green, bordered with black lines on the head, arms and shanks; flesh-colored on the tip of snout, posteriorly extended across loreal region, eye and tempord region, terminating at the above insertion of forelimbs; upper lip, regions below eyes with yellowish-green patches; flanks of body and ventral surface yellowish; flanks, frontal and rear of thighs with a certain number of black spots; tarsal-foot and wrist-palm flesh-colored. Single vocal sac in males.

Habitat and Behavior Inhabits mountainous areas at elevations between 100-1200 m a.s.l. More frequently found on a leaf, stem or branch of plants near rice paddies, streams and pools.

Distribution C, E and S China. N Vietnam. Recorded from Longshi in Mt. Jinggang.

Status of Population Common species.

1.11a　1.11b　1.11c　1.11d

1.12 三港雨蛙 *Hyla sanchiangensis* Pope, 1929

识别特征 雄蛙头体长约 30 mm，雌蛙头体长 36-41 mm。吻棱明显。吻端圆，高厚而平切。颊部略倾斜。皮肤光滑，颞褶细；喉部光滑，有少量疣粒；体腹面密布疣粒。指趾端有吸盘和边缘沟。指间有蹼，趾间至少 2/3 蹼。跟部重叠，胫跗关节前伸达眼。背绿色，边缘锯齿状镶黑色细线纹；吻端棕色，向后延伸经颊部、过眼、颞区，与体侧棕色区域贯通；上唇、颊部及眼下区域有浅黄绿色斑块；体侧棕色，腋窝、上臂基部、体侧后段及腹股沟、大腿前后黄色，胫跗关节处、跗足背面有黄色斑块，所有黄色部分有黑色斑点；腹面后胸部棕黄色，向后浅黄白色；四肢腹面和手足部棕色。雄性具单声囊。

生境与习性 栖息于海拔 400-1600 m 丘陵和山区。夜晚常见稻田里和小水沟或水塘附近，攀附于植物叶片及茎秆上。

分布 江西、湖南、福建、安徽、浙江、贵州、广东北部和广西北部。在井冈山记录于大井、白银湖、湘洲等地。

种群状况 常见种。

Identification A small-sized tree toad, with approximately 30 mm SVL in males, 36-41 mm SVL in females; canthus rostralis distinct; snout rounded in dorsal view, deep and truncated in profile view; loreal region slightly oblique; dorsal skin smooth; supratympanic fold narrowed; throat smooth with a few tubercles, posteriorly ventral skin densely with tubercles; tips of digits with well-developed disks and marginal grooves; fingers webbed; toes at least two-thirds webbed; heels overlapping and tibio-tarsal articulation reaching the eye. Upper surface green with serrated edge, bordered with black lines; brown on the tip of snout, posteriorly extended across loreal regions, eyes and temporal regions, communicated with brown regions on sides of body; upper lip, loreal regions and the regions below eyes with yellowish-green patches; sides of body brown; axillae, bases of arms, posterior regions of sides of body, groins, frontal and rear of thighs, dorsal surface of tibio-tarsal articulations and acrotarsiums yellow with black spots; throat, chest brown, posteriorly becoming cream-white; ventral surface of limbs, feet and hands brown. Single vocal sac in males.

Habitat and Behavior Inhabits mountainous and hilly areas at elevations between 400-1600 m a.s.l. It is more frequently found on a leaf, stem or branch of plants near rice paddies and pools.

Distribution Jiangxi, Hunan, Fujian, Anhui, Zhejiang, Guizhou, northern Guangdong and northern Guangxi. Recorded from Dajing, Baiyinhu and Xiangzhou in Mt. Jinggang.

Status of Population Common species.

无尾目 ANURA
蛙科 Ranidae

1.13 长肢林蛙 *Rana longicrus* Stejneger, 1898

识别特征 体纤细，雄蛙头体长约 40 mm，雌蛙约 50 mm。腿细长，跟部重叠，胫跗关节前伸超过吻端。指趾端钝圆；关节下瘤发达而显著。趾间具 1/3-1/2 蹼。皮肤光滑，通常肩上方背部中央有"八"字形黑色短肤棱，体侧有少许疣粒；背侧褶细直，由眼后直达胯部。背橘红色、黄褐色或绿褐色，两眼间有深色横纹，通常横纹内有横向肤棱；颞部三角区深色；四肢背面有深褐色横纹；腹面白色。雄性第一指背面有婚垫；无声囊。

生境与习性 栖息于海拔 200 m 以上的平原、丘陵和山区。

分布 中国特有种。分布于福建、江西、广东和台湾中部和北部。在井冈山地区广泛分布。

种群状况 区域性常见种。IUCN 红色名录：易危。

Identification Body slender with approximately 40 mm SVL in males, approximately 50 mm SVL in females; legs slender and long, heels overlapping, tibio-tarsal articulation forward beyond the tip of snout; tips of digits blunt and rounded; subarticular tubercle well-developed and distinct; toes 1/3 to 1/2 webbed; skin smooth, usually a reverse V-shaped black dermal ridge on dorsal surface above shoulder; several tubercles on flanks of body; two dorsolateral folds narrow and straight, from posterior edges of eyelids to above insertions of thighs. Red-orange, yellow-brown or green-brown above; a transverse dark stripe between eyes, usually a dermal ridge in this dark stripe; temporal regions dark-colored; dorsal surface of limbs with darker brown transverse bands; ventral surface white. The presence of nuptial pads on dorsal surface of first finger in males. No vocal sac.

Habitat and Behavior Inhabits plains, hills and mountains at elevations above 200 m a.s.l.

Distribution Endemic to China. Fujian, Jiangxi, Guangdong and northern and central Taiwan. Widely occurs in the Jinggangshan Region.

Status of Population Regional common species. IUCN Red List: VU.

1.13a　1.13b　1.13c　1.13d

1.14 寒露林蛙 *Rana hanluica* Shen, Jiang, and Yang, 2007

识别特征　体中等, 略粗壮于长肢林蛙。雄蛙头体长约 55 mm, 雌蛙约 65 mm。腿细长, 跟部重叠较多, 后肢贴体前伸, 胫跗关节达到或超过吻端。指趾端钝圆, 关节下瘤发达而显著。趾间近满蹼。皮肤光滑, 体背及体侧有疣粒, 部分疣粒黑色; 通常肩上方背部中央有 "Λ" 形黑色短肤棱, 背侧褶细直, 在眼后直达胯部。体背面橘红色、黄褐色或绿褐色, 两眼间有深色横纹; 四肢背面有深褐色横纹; 腹面浅黄色或乳白色。雄性第一指有 4 团婚垫, 上有白色婚刺; 无声囊。

生境与习性　栖息于海拔 300-1300 m 的丘陵和山区。

分布　中国特有种。分布于南岭到罗霄山脉, 在井冈山地区见于五指峰和南风面自然保护区等地。

种群状况　区域性常见种。

Identification　Body moderate, stouter than *Rana longicrus*, with approximately 55 mm SVL in males, approximately 65 mm SVL in females; legs slender and long, heels most overlapping, tibio-tarsal articulation forward reaching the tip of snout or beyond; tips of digits blunt and rounded; subarticular tubercle well-developed and distinct; toes with nearly full webbings; skin smooth, with several tubercles on dorsum and flanks of body, some of which are black; usually a Λ-shaped black dermal ridge on dorsal surface above shoulder; two dorsolateral folds narrow and straight, from posterior edges of eyelids to above insertions of thighs. Red-orange, yellow-brown or green-brown above; a transverse dark stripe between eyes; dorsal surface of limbs with darker brown transverse bands; ventral surface yellowish or cream. The presence of four patches of nuptial pads on dorsal surface of first finger, bearing white nuptial spines in males. No vocal sac.

Habitat and Behavior　Inhabits plains, hills and mountains at elevations between 300-1300 m a.s.l.

Distribution　Endemic to China. Distributed in the Nanling Mountains to Luoxiao Mountains, Recorded from Wuzhifeng and Nanfengmian Nature Reserve in the Jinggangshan Region.

Status of Population　Regional common species.

1.14a

1.14b

1.14c

1.14d

1.15 黑斑侧褶蛙 *Pelophylax nigromaculatus* (Hallowell, 1861)

识别特征　体较粗壮，成年雄蛙头体长约 70 mm，成年雌蛙约 75 mm。后肢短，跟部不相遇，胫跗关节前伸达鼓膜和眼之间。指趾末端尖；指间无蹼；趾间蹼延伸至趾端，但凹陷较深，第四趾蹼达远端关节下瘤。背部皮肤粗糙，背侧褶发达，甚宽，其间有长短不一的纵行肤棱；胫部有纵肤棱。肩上方无扁平腺体。体和四肢腹面光滑。体背褐色、蓝绿色、深绿色或黄绿色，有黑色斑点；有些个体有绿色或浅黄色脊线纹。四肢有深色横纹。体腹面白色，四肢腹面肉色。有一对颈侧外声囊。

生境与习性　栖息于海拔 2200 m 以下平原、山区和丘陵。常见于水田、池塘、沼泽。

分布　除新疆、西藏、青海、台湾和海南外，见于中国其余各省。国外分布于俄罗斯、朝鲜半岛和日本。记录于井冈山湘洲、黄坳、新城区等地。

种群状况　常见种。

Identification　Body robust, with approximately 70 mm SVL in adult males, approximately 75 mm SVL in adult females; legs short, heels not meeting, tibio-tarsal articulation forward reaching the region between tympanum and eye; tips of digits pointed; fingers not webbed; toes with nearly full webbings depressed deeply; skin rough, with developed and thick dorsolateral folds; longitudinal dermal ridges between the dorsolateral folds and on shanks; no flattened suprabrachial gland; ventral surface of body and limbs smooth. Dorsum blue-green or darker green or yellowish-green, with black spots; some with a green or yellowish vertebral line; dorsal limbs with black transverse bands; ventral surface of body white; ventral surface of limbs flesh-colored. Males with a pair of external lateral vocal sacs.

Habitat and Behavior　Inhabits plains, hills and mountains at elevations below 2200 m a.s.l. More frequently found in rice paddies, marshes, ponds.

Distribution　Except Xinjiang, Xizang, Qinghai, Taiwan and Hainan, distributed in other provinces in China. Russia, Korean Peninsula and Japan. Recorded from Xiangzhou, Huang'ao and Xinchengqu in Mt. Jinggang.

Status of Population　Common species.

1.15a

1.15b

1.16 弹琴蛙 *Nidirana adenopleura* (Boulenger, 1909)

识别特征　体较粗壮，成蛙头体长 50-60 mm。后肢较长，胫跗关节前伸达鼻孔或吻端。指末端略膨大；指间无蹼；趾末端有吸盘和腹侧沟，趾间具 1/3-1/2 蹼。背部皮肤光滑，密布小粒，体侧及体后段、四肢背面有疣粒。背侧褶宽窄适度；胫部有纵肤棱。内跗褶显著，有扁平肩上腺体。体和四肢腹面光滑。体背棕色、灰棕色、棕红色或棕黄色，两眼间有一浅色点斑；脊线纹浅色或橘红色，由肛上方向前延展。四肢有深色横纹。体腹面灰白色。有一对咽下外声囊。

生境与习性　栖息于海拔 30-1800 m 的山区和丘陵。常见于水田、池塘、沼泽。

分布　中国特有种。分布于湖南、江西、安徽、江苏、浙江、福建、台湾。在井冈山地区广泛分布。

种群状况　常见种。

Identification　Body moderately robust, with approximately 50-60 mm SVL in adult individuals; legs longer, tibio-tarsal articulation forward reaching the region between nostril and tip of snout; tips of fingers slightly enlarged; fingers not webbed; toes 1/3-1/2 webbed; disks and circummarginal grooves of tips of toes present; skin smooth above with densely arranged granules; posterior dorsum and flanks with spinous tubercles; dorsolateral folds moderately thick; shanks with longitudinal dermal ridges; internal tarsal fold distinct; the presence of flattened suprabrachial gland; ventral surface of body and limbs smooth. Brown, grey-brown, reddish-brown or yellowish-brown above, with a light-colored spot between eyes; vertebral line light-colored or orange, extending forward from above anal region; limbs with dark transverse bands; ventral surface of body grey-white. Males with a pair of external subgular vocal sacs.

Habitat and Behavior　Inhabits plains, hills and mountains at elevations between 30-1800 m a.s.l. More frequently found in paddies, marshes, ponds.

Distribution　Endemic to China. Distributed in Hunan, Jiangxi, Anhui, Jiangsu, Zhejiang, Fujian, Taiwan. Widely occurs in the Jinggangshan Region.

Status of Population　Common species.

1.17 阔褶水蛙 *Hylarana latouchii* (Boulenger, 1899)

识别特征   体型稍小的水蛙，雄蛙头体长 36-40 mm，雌蛙 42-53 mm。胫跗关节前伸达眼部。指末端钝圆；指间无蹼；趾末端有吸盘和腹侧沟，趾间半蹼。背部皮肤多刺粒，体后段背部、体侧及四肢背面有疣粒。通常背侧褶甚宽阔；胫部有纵肤棱。内跗褶显著；口角有 2 个颌腺。体和四肢腹面光滑，乳黄或灰白色。体背棕色、棕红色或棕黄色。四肢有深棕色横纹。大腿后方有黑色云斑。有一对咽侧内声囊。

生境与习性   栖息于海拔 30-1800 m 的平原、丘陵和山区。常见于山区的水田、池塘和水沟。

分布   中国特有种。主要分布于江西、湖南、安徽、江苏、浙江、福建、台湾、广东、香港、广西、贵州（荔波）等地。在井冈山地区广泛分布。

种群状况   常见种。

Identification   A slightly small-sized frog, with 36-40 mm SVL in males, 42-53 mm in females; tibio-tarsal articulation forward reaching the eyes; tips of fingers rounded; fingers not webbed; toes one-half webbed; disks and circummarginal grooves of tips of toes present; dorsal skin granular; posterior dorsum and flanks of body, dorsal surface of limbs with tubercles; dorsolateral folds significantly thick; shanks with longitudinal dermal ridges; internal tarsal fold distinct; two rictal glands on the corner of mouth; ventral surface of body and limbs smooth, yellowish or grey-white. Dorsal surface brown, reddish-brown or yellowish-brown; limbs with dark transverse bands; rear of thigh with black blotches. Males with a pair of internal lateral vocal sacs.

Habitat and Behavior   Inhabits plains, hills and mountains at elevations between 30-1800 m a.s.l. More frequently found in mountainous paddies, marshes, ponds.

Distribution   Endemic to China. Jiangxi, Hunan, Anhui, Jiangsu, Zhejiang, Fujian, Taiwan, Guangdong, Hong Kong, Guangxi and Guizhou (Libo). Widely occurs in the Jinggangshan Region.

Status of Population   Common species.

1.17a

1.17b

1.18 沼水蛙 *Hylarana guentheri* (Boulenger, 1882)

识别特征　体中等，最大头体长可达 100 mm。皮肤光滑，背侧褶发达，自眼睑后直达胯部并与对侧的背侧褶平行；无颞褶；胫部有纵行的肤棱。胫跗关节前达鼻眼之间。第四趾蹼达远端关节下瘤，其余各趾全蹼。体棕色或灰棕色。颌腺浅黄色。背侧褶下缘有黑色纵纹，体侧有不规则黑斑。雄性肱腺肾形。有 1 对咽侧下声囊。

生境与习性　栖息于池塘、水田、溪流及水洼地。白天隐伏，夜间活动。繁殖季节雄蛙往往停在水草上鸣叫求偶。

分布　广泛分布于台湾和长江以南各省区。国外见于越南、老挝。在井冈山见于茨坪、下庄、长古岭、湘洲等地。

种群状况　常见种。

Identification　Body moderate, with maximum SVL 100 mm; skin smooth; dorsolateral dermal fold developed, extended from posterior edge of eyelid to above insertion of thigh of parallel to contralateral fold; no supratympanic fold; several longitudinal dermal ridges on dorsal surface of shanks; tibio-tarsal articulation reaching the region between eye and nostril; toes with full webbings, but web of toe IV only reaching distal subarticular tubercle. Brown or grey-drown above; maxillary gland yellowish; lower edge of dorsolateral fold bordered with black longitudinal stripe; several large, irregular black spots scattered on flank of body. Presence of a pair of humeral glands, a pair of lateral subgular vocal sacs in males.

Habitat and Behavior　Inhabits ponds, rice paddies, streams and pools. Nocturnal. Usually males ensconce on a floating leaved plant and call to attract females during breeding season.

Distribution　Widely distributed in the Taiwan and southern region of the Yangtze River of China.Vietnam and laos. Recorded from Ciping, Xiazhuang, Changguling and Xiangzhou in Mt. Jinggang.

Status of Population　Common species.

1.19 竹叶臭蛙 *Odorrana versabilis* (Liu and Hu, 1962)

识别特征 中等体型臭蛙，雄蛙头体长 68-80 mm，雌蛙 71-87 mm。头扁平。吻长而扁，背视圆形。腿长，跟部明显重叠，胫跗关节前伸达到或超过吻端。指趾均具吸盘和腹侧沟。指间无蹼。趾间近满蹼。背部皮肤有细密颗粒，繁殖期非常光滑，上唇缘有锯齿状突；颌腺 2 个；体背后段、体侧上部及股后有分散疣粒，近腹缘疣粒密集；有背侧褶，由眼后直达肛上区域。腹面皮肤光滑，胸部以后有小疣粒。背面棕色，有时为绿色，或有绿色斑点；两眼间有一浅黄色点斑；有一条黑色纵纹起于吻端，沿吻棱下缘、过眼、沿背侧褶下缘而延伸至胯部；除上臂外，四肢背面有深棕色横纹。腹面乳白色。雄性第一指背有白色婚垫，1 对咽下声囊。

生境与习性 栖息于海拔 500-1350 m 的山区森林。常见于溪流及其附近。

分布 中国特有种。分布于江西、湖南、广东、广西和贵州。在井冈山记录于八面山、荆竹山、河西垄等地。

种群状况 常见种。

Identification Moderate-sized cascade frog, with 68-80 mm SVL in males, 71-87 mm SVL in females; head depressed; snout longer and depressed, rounded in dorsal view; leg longer; the heels most overlapping; tibio-tarsal articulation reaching tip of snout or beyond it; fingers not webbed; feet nearly fully webbed; disks of digits moderately expanded and ventrally with circummarginal grooves; dorsal skin with weak granulations, smooth during the breeding season; tubercles arranged at edge of upper lip and forming serrated lip edge; two maxillary glands; posterior dorsum, upper region of flanks of body and rear of thighs with scattered small tubercles; lower region of flank densely arranged large yellowish tubercles; dorsolateral fold present, extending from posterior edge of eyelid to the above anal region; ventral skin smooth, with small tubercles on the chest and belly. Dorsum brown, sometimes green or with green mottling; single small rounded yellowish spot between eyes; a longitudinal black stripe beginning from tip of snout, posteriorly extending along lower edge of the canthus rostralis, through eye, along lower edge of dorsolateral fold to the crotch; except arm, limbs with dark brown transverse bands; venter creamy white. Males with white nuptial pads on dorsal surface of first finger, paired subgular vocal sacs.

Habitat and Behavior Inhabits montane forests at elevations between 500-1350 m a.s.l. More frequently found in mountainous streams and nearby areas.

Distribution Endemic to China. Jiangxi, Hunan, Guangdong, Guangxi and Guizhou. Recorded from Bamianshan, Jingzhushan and Hexilong in Mt. Jinggang.

Status of Population Common species.

1.19a

1.19b

1.19c

1.20 大绿蛙 *Odorrana graminea* (Boulenger, 1900)

识别特征　成年雌蛙头体长约为雄性的 2 倍，通常雄蛙头体长 42-53 mm，雌蛙 78-100 mm。腿细长，跟部重叠较多，胫跗关节前伸超过吻端。指趾均具吸盘和腹侧沟。指间无蹼。趾间满蹼。背面皮肤光滑，无疣粒；颌腺 2 个；体侧颗粒状有疣粒；略具背侧褶。腹面皮肤光滑。背面绿色，多有深褐色斑点；两眼间有一个不清晰的浅黄色圆形点斑；头侧及体侧上部深棕色，体侧下部色浅，唇及颌腺乳白色；四肢棕色，具深棕色横纹。股后浅黄色有棕色大理石斑纹。腹面乳白色，四肢腹面乳黄色，有时有深色斑纹。雄性第一指背面具白色婚垫，1 对咽下声囊。

生境与习性　栖息于海拔 300 m 以上的山区森林。常见于溪流及其附近。

分布　广泛分布于热带、亚热带中国、越南、缅甸、老挝等地。广布于井冈山地区。

种群状况　常见种。

Identification　Adult females have SVL approximately twice that of males; usually male SVL 42-53 mm, female SVL 78-100 mm; leg slender and longer; heels mostly overlapping; tibiotarsal articulation forward beyond tip of snout; fingers not webbed; feet fully webbed to disks; disks of digits moderately expanded and with ventral circummarginal grooves; dorsal skin smooth without tubercles; two maxillary glands; flanks weakly granular with several tubercles; weak dorsolateral folds present; venter smooth. Upper dorsum green, usually with dark brown spots; single small, faint, rounded yellowish spot between eyes; sides of head, upper region of flanks dark brown, lower region of flanks light-colored; lip-stripe and maxillary glands creamy white; dorsal limbs and digits brown with dark brown transverse bands; rear of thighs yellowish with brown marbling; venter creamy white; ventral limbs creamy yellow, sometimes with dark mottling. Males with velvety nuptial pads on dorsal surface of first finger, and paired subgular vocal sacs.

Habitat and Behavior　Inhabits montane forests at elevations above 300 m a.s.l. More frequently found in mountainous streams and nearby areas.

Distribution　Widely distributed in tropical and subtropical China and Vietnam, Myanmar and Laos. Widely recorded from the Jinggangshan Region.

Status of Population　Common species.

1.20a

1.20b

1.21 花臭蛙 *Odorrana schmackeri* (Boettger, 1892)

识别特征 成年雌蛙头体长约为雄性的 2 倍，雌蛙最大头体长 85 mm。腿细长，跟部重叠较多，胫跗关节前伸达眼和鼻孔之间。指趾均具吸盘和腹侧沟。指间无蹼。趾间满蹼，达趾端。背面皮肤鲨鱼皮状，满布粗糙颗粒，背和体侧有脓包状突起；无背侧褶；颞褶略显。腹面皮肤光滑。刚完成变态的幼体背部浅绿色，有不显的暗斑，皮肤散布很多大疣粒；四肢和指趾有深色横纹。成体背面绿色，有深棕色大斑，形成绿色网纹；两眼间有一浅色点斑；上下唇浅色，有黑色竖斑；有时头侧及体侧为深棕色，有少量绿色块斑；四肢背面绿色，具深棕色横纹，有时绿色不显；股后有棕色大理石斑纹。腹面乳白色。雄性有 1 对咽下声囊，繁殖期第一指背面具白色婚垫，未见胸腹部的白色刺。成熟卵动物极灰黑色，植物极白色。

生境与习性 栖息于海拔 150 m 以上的山区森林。常见于溪流及其附近。

分布 中国特有种。分布于河南、陕西、甘肃、四川、贵州、湖北、湖南和江西。广布于井冈山地区。

种群状况 井冈山地区常见种。

Identification Adult females have SVL approximately twice that of males, with maximum SVL of 85 mm; leg slender and longer; heels mostly overlapping; tibio-tarsal articulation forward reaching between eye and nostril; digits with disks and ventral circummarginal grooves; fingers not webbed; toes fully webbed to tips of toes; dorsal skin shagreened with heavy granulations; dorsum and flanks with large pustules; dorsolateral fold absent; supratympanic fold slightly distinct; ventral skin smooth. In froglings after metamorphosis, light-green above, with large tubercles, limbs and digits with faint transverse bands; in adult frogs, dorsum green with large red-brown spots and forming green reticular marking; a rounded light-colored spot between the eyes; lips light-colored with dark cross-bars; sometimes sides of head and flanks of body red-brown, with a few green marks; dorsal limbs green with red-brown transverse bands, sometimes green indistinct; rear of thigh marbled dark brown and yellowish-green; white below. Males with paired subgular vocal sacs, and white nuptial pad on dorsal surface of first finger; no white spines on chest and belly during the breeding season; gravid females bear white oocytes with melanic poles.

Habitat and Behavior Inhabits montane forests at elevations above 150 mm a.s.l. More frequently found in mountainous streams and nearby areas.

Distribution Endemic to China. Henan, Shaanxi, Gansu, Sichuan, Guizhou, Hubei, Hunan and Jiangxi. Widely recorded from the Jinggangshan Region.

Status of Population Common species.

1.21a

1.21b

1.21c

1.22 宜章臭蛙 *Odorrana yizhangensis* Fei, Ye, and Jiang, 2007

识别特征 成年雌蛙头体长约为雄性的 1.3 倍，雄蛙头体长 52.6-57.4 mm。腿细长，跟部重叠较多，胫跗关节前伸达吻端。指趾均具吸盘和腹侧沟，吸盘大，三角形。指间无蹼，趾间满蹼，达趾端。背面皮肤鲨鱼皮状，满布粗糙颗粒。体侧有疣粒。无背侧褶；颞褶细弱。腹面皮肤光滑，下颌缘及胸腹部具白色刺粒。成体体侧及背面绿色，有棕红色大斑，形成绿色网纹；两眼间有一浅色点斑；四肢和指趾背面绿色，具深棕色横纹。腹面乳白色，咽喉部色深，胸部有黑褐色斑点。雄性第一指背面具发达白色婚垫，1 对咽下声囊。

生境与习性 栖息于海拔 1100 m 以上的山区溪流。

分布 中国特有种。目前已知分布于湖南莽山自然保护区、广东南岭自然保护区、湖北五峰和罗霄山脉。在井冈山地区记录于荆竹山和梨树洲。

种群状况 常见种。

Identification Adult females from Mt. Jinggang have SVL approximately 1.3 times that of males; males 52.6-57.4 mm SVL; leg slender and longer; heels mostly overlapping; tibio-tarsal articulation forward reaching the tip of the snout; digits with greatly enlarged, triangular disks and ventral circummarginal grooves; usually disks of fingers larger than that of toes; hands not webbed; feet fully webbed to tips of toes; dorsal skin shagreened with heavy granulations; flanks with tubercles; dorsolateral fold absent; supratympanic fold thin; ventral skin smooth; edge of lower lip, chest and belly with white spinous granules; dorsum green with large red-brown spots and forming green reticular marking; rounded light-colored spot between the eyes; dorsal surface of limbs and digits green with red-brown transverse bands; creamy white below, throat dark-colored, chest and belly with black-brown spots. Males with developed white nuptial pad on dorsal surface of first finger, and paired subgular vocal sacs.

Habitat and Behavior Inhabits mountainous streams at elevations above 1100 m a.s.l.

Distribution Endemic to China. Distributed in Mangshan Nature Reserve of Hunan, Nanling Nature Reserve of Guangdong, Wufeng county of Hubei, and Luoxiao Mountains. Recorded from the Jingzhushan and Lishuzhou in the Jinggangshan Region.

Status of Population Common species.

1.23 崇安湍蛙 *Amolops chunganensis* (Pope, 1929)

识别特征 体型较小，雄性头体长 38-39.5 mm。吻较长；吻棱发达。颊部略倾斜；腿细长，跟部重叠较多，胫跗关节前伸达眼和吻端之间。指趾均具吸盘和腹侧沟，趾吸盘小于指吸盘。指间无蹼。趾间蹼达趾端，仅第 4 趾蹼达第 3 关节下瘤。背面皮肤光滑，无疣粒。背侧褶细直；颞褶不显。腹面皮肤光滑。通常背面棕黄色，有黑色斑点；一条过眼黑纹自吻端沿吻棱下方、背侧褶下方延至肩区；一条乳白线纹沿上唇延至上臂基部上方；乳白线纹和黑纹之间在颞区至肩上区域深棕色；四肢具灰黑色细横纹。腹面乳黄色。雄性第一指背面具有细密婚刺的白色婚垫，1 对咽下声囊。

生境与习性 在井冈山地区，栖息于海拔 1000 m 以上潮湿的亚热带常绿阔叶林间山区溪流。

分布 中国特有种。分布于陕西南部、甘肃南部、四川东部、贵州、广西北部、广东南岭、湖南西部和东部、江西（井冈山）、福建（武夷山）。在井冈山地区记录于荆竹山、水口、梨树洲等地。

种群状况 区域性常见种。

Identification Body slightly small, with 38-39.5 mm SVL in males; snout longer; canthus rostralis developed; loreal region slightly oblique; leg slender and longer; heels mostly overlapping; tibio-tarsal articulation forward reaching between eye and tip of snout; digits with disks and circummarginal grooves; disks of toes smaller than those of fingers; hands not webbed; feet fully webbed to tips of toes, only fourth toe webbed to third subarticular tubercle; dorsal skin smooth without tubercles; dorsolateral fold narrow and straight; supratympanic fold indistinct; venter smooth. Usually dorsum yellowish-brown with black spots; a black stripe across the eye extended from tip of snout to the shoulder area along canthus rostralis and dorsolateral fold; a creamy white stripe extended along upper lip to above the insertion of arm; temporal and shoulder area dark brown between creamy white and black stripes; limbs with thin grey black transverse bands; venter creamy yellow. Males with developed white nuptial pad bearing heavy fine nuptial spines on dorsal surface of first finger, and paired subgular vocal sacs.

Habitat and Behavior Individuals from the Jinggangshan Region inhabit mountainous streams surrounded by moist subtropical evergreen broadleaved forests at elevations above 1000 m a.s.l.

Distribution Endemic to China. Widespread in southern Shaanxi, southern Gansu, eastern Sichuan, Guizhou, northern Guangxi, Guangdong (Nanling Mountains), eastern and western Hunan, Jiangxi (Mt. Jinggang), and Fujian (Wuyi Mountain). Recorded from Jingzhushan, Shuikou, and Lishuzhou in the Jinggangshan Region.

Status of Population Regional common species.

1.23a 1.23b 1.23c 1.23d

1.24 华南湍蛙 *Amolops ricketti* (Boulenger, 1899)

识别特征　体型中等，雌蛙略大于雄蛙，头体长平均约 55 mm。体较平扁；头亦平扁，头宽大于头长。吻棱清晰。颊部略倾斜。鼓膜小，清晰或不显。犁骨棱上有发达犁骨齿。跟部重叠，胫跗关节前伸达眼。吸盘发达具腹侧沟，指吸盘大于趾吸盘；指间无蹼。趾间满蹼；第 1 和第 5 趾游离侧有缘膜。背面皮肤鲨鱼皮状，满布粗糙颗粒，体侧有脓包状突起；无背侧褶；颞褶略显，颌腺 1-2 个。腹面皮肤颗粒状或有细皱纹。体背面灰绿色或黄绿色，有浅色蠹状纹，四肢有深色横纹。腹面黄白色，咽胸部有深色大理石斑纹。雄性第 1 指基部具发达婚垫，上有乳白色粗密强婚刺。无声囊。

生境与习性　栖息于海拔 200-1500 m 山区的湍急溪流中。

分布　江西、湖南、安徽、浙江、江苏、福建、广东、广西、云南、四川和贵州等地。越南中部和北部。见于整个井冈山地区。

种群状况　常见种。

Identification　Body moderate and dorsoventrally compressed, with mean SVL of approximately 55 mm; females slightly larger than males; head depressed, HL larger than HW; canthus rostralis distinct; loreal region slightly oblique; tympanic small, distinct or indistinct; vomerine ridges bearing developed teeth; heels overlapping; tibio-tarsal articulations forward reaching the eyes; disks enlarged with ventral circummarginal grooves; disks on fingers larger than those on toes; hands not webbed; feet fully webbed; free sides of first and fifth toes with fringes; dorsal skin shagreened with heavy granulations; flanks with pustules; dorsolateral fold absent; supratympanic fold slightly distinct; maxilla glands one or two; ventral skin granular or with thin wrinkles. Dorsum grey-green or yellowish-green, with light-colored wormlike marks; limbs with dark-colored transverse bands; venter yellowish-white; throat and chest with dark marbling. Males with nuptial pad bearing strong, white nuptial spines on dorsal surface of first finger; vocal sac absent.

Habitat and Behavior　Inhabits mountainous streams at elevations between 200-1500 m a.s.l.

Distribution　Jangxi, Hunan, Anhui, Zhejiang, Jiangsu, Fujian, Guangdong, Guangxi, Yunnan, Sichuan, Guizhou and other provinces. N and C Vietnam. Recorded from overall the Jinggangshan Region.

Status of Population　Common species.

1.25 武夷湍蛙 *Amolops wuyiensis* (Liu and Hu, 1975)

识别特征 雄性头体长平均约42 mm，雌性约49 mm。体较平扁；头亦平扁，头宽大于头长。吻棱清晰。颊部略倾斜。鼓膜不清晰。犁骨棱细长，无犁骨齿。跟部重叠，胫跗关节前伸达眼前。吸盘发达具腹侧沟，指吸盘大于趾吸盘；指间无蹼。趾间满蹼；第5趾游离侧有缘膜。背面皮肤鲨鱼皮状，满布粗糙颗粒，并有分散疣粒，体侧脓包状突起较多；无背侧褶；颞褶略显，颌腺2个。腹面皮肤有扁平疣粒。体背面灰绿色或黄绿色，有黑棕色大斑，四肢有深色横纹。腹面黄白色，喉胸部有灰黑色云斑。雄性第1指基部具发达婚垫，有黑色稍显稀疏圆锥状强婚刺。有1对咽下声囊。

生境与习性 栖息于山区中海拔100-1300 m较宽的湍急溪流内及其附近，溪流边树木杂草繁茂。

分布 中国特有种。分布于江西（井冈山、三清山、大鄣山和武夷山脉），浙江、福建北部和安徽东部。在井冈山仅记录于小溪洞。

种群状况 常见种。

Identification Body moderate and dorsoventrally compressed, with mean SVL of 42 mm in males and 49 mm in females; head depressed, HL larger than HW; canthus rostralis distinct; loreal region slightly oblique; tympanic indistinct; vomerine ridges narrower and longer, lacking vomerine teeth; heels overlapping; tibio-tarsal articulations forward reaching the anterior region of the eyes; disks enlarged with ventral circummarginal grooves; disks on fingers larger than those on toes; hands not webbed; feet fully webbed; free side of fifth toes with fringes; dorsal skin shagreened with heavy granulations and scattered tubercles; flanks with heavy pustules; dorsolateral fold absent; supratympanic fold slightly distinct; maxilla glands two; ventral skin with flattened tubercles. Dorsum grey-green or yellowish-green, with large black-brown spots; limbs with dark-colored transverse bands; venter yellowish-white; throat and chest with grey-black botches. Males with nuptial pad bearing strong, conical, black nuptial spines on dorsal surface of first finger, and paired subgular vocal sacs.

Habitat and Behavior Inhabit mountainous streams surrounded by dense forests and shrubs at elevations between 100-1300 m a.s.l.

Distribution Endemic to China. Distributed in Jiangxi (Mt. Jinggang, Sangqingshan, Dazhangshan and Wuyi Mountains), Zhejiang, northern Fujian and eastern Anhui. Only recorded from Xiaoxidong in Mt. Jinggang.

Status of Population Common species.

1.25a 1.25b 1.25c

无尾目 ANURA
叉舌蛙科 Dicroglossidae

1.26 泽陆蛙 *Fejervarya multistriata* (Hallowell, 1861)

识别特征　小型蛙类，成体一般不超过 50 mm。吻尖，吻棱不显，颊部倾斜。鼓膜大而清晰。前后肢相对较短，跟部相遇或不相遇，胫跗关节前达肩部到眼后方。指趾端尖，不膨大。趾间半蹼。皮肤粗糙，体背满布长短不一的纵肤褶和疣粒。无背侧褶，颞褶清晰。体色多变，两眼间常有深色横纹，肩部常有 "W" 形深色斑纹；背常有浅色脊线。雄性具单个咽下声囊。

生境与习性　生活于平原、丘陵和 2000 m 以下山区的稻田、沼泽、水塘、水沟等静水域或其附近的旱地草丛，偶见于山溪边。

分布　分布于包括井冈山地区全境的热带、亚热带中国。国外分布于日本和东南亚地区。

种群状况　常见种。

Identification　Body small, usually SVL less than 50 mm in adults; snout pointed, canthus rostralis indistinct; loreal region oblique; limbs relatively short; heels not meeting; tibio-tarsal articulation reaching between shoulder and eye; tips of digits pointed, not enlarged; feet one-half webbed; dorsal skin rough, with longitudinal, long or short dermal ridges and tubercles; dorsolateral fold absent; supratympanic fold distinct. Dorsum colored variably, with a dark-colored transverse stripe between the eyes and a W-shaped stripe between shoulders; usually present light-colored vertebral line. Males with single subgular vocal sac.

Habitat and Behavior　Inhabits plain, hill and mountain areas at elevations below 2000 m a.s.l. More frequently found in rice paddies, marshes, ponds, ditches and near grass-shrub, dry lands; occasionally found in mountain streams.

Distribution　Distributed in tropical and subtropical China, including all of the Jinggangshan Region. Japan and SE Asia.

Status of Population　Common species.

1.27 虎纹蛙 *Hoplobatrachus chinensis* (Osbeck, 1765)

识别特征 体型中等大小，最大头体长接近 110 mm。吻钝尖，吻棱圆钝，颊区倾斜。前后肢相对较短，跟部相遇或略重叠，胫跗关节前达鼓膜。指趾端尖，不膨大。趾间全蹼，第 1 和第 5 趾外侧缘膜发达。皮肤粗糙，体背满布长短不一的纵肤褶和疣粒。颞褶发达，清晰；无背侧褶。胫部疣粒排列成行。背黄绿色、灰棕色或墨绿色，有不规则深色斑纹；四肢背面有深色横纹；腹面白色，喉胸部有深色斑纹。雄性有一对咽下声囊。

生境与习性 栖息于山区和丘陵地区的农田、池塘、水坑内，常隐藏在土洞、灌丛间和杂草中。

分布 主要分布在长江以南地区，以及台湾和海南。在井冈山记录于湘洲、新城区等地。

种群状况 IUCN 红色名录：易危。国家 II 级保护野生动物。

Identification Body moderate, SVL maximum 110 mm; snout blunted and pointed, canthus rostralis bluntly rounded; loreal region oblique; limbs relatively short; heels meeting or just overlapping; tibio-tarsal articulation reaching the tympanum; tips of digits pointed, not enlarged; feet fully webbed; free sides of first and fifth toes with fringes; dorsal skin rough, with longitudinal, long or short dermal ridges and tubercles; tubercles arranged in rows on shanks; supratympanic fold distinct and developed; dorsolateral fold absent. Dorsum yellowish-green, grey-brown or dark green, with irregular dark markings; dorsal limbs with dark-colored transverse bands; venter white. Males with paired subgular vocal sacs.

Habitat and Behavior Inhabits rice paddies, ponds and ditches in hilly and mountainous areas, often hidden in soil burrows, shrubs or weeds.

Distribution Mainly in the south of the Yangtze River, Taiwan and Hainan. Recorded from Xiangzhou and Xinchengqu in Mt. Jinggang.

Status of Population IUCN Red List: VU; China Key List: II.

1.27

1.28 福建大头蛙 *Limnonectes fujianensis* Ye and Fei, 1994

识别特征 雄蛙显著大于雌蛙，雌蛙最大头体长 55 mm，雄性最大头体长 65 mm；成年雄蛙头甚大，枕部隆起，头长大于头宽而略小于头体长的一半；雌蛙头部相对小，枕部低平，吻钝尖，吻棱不显，颊区甚倾斜。鼓膜隐蔽于皮下。前后肢短粗，跟部不相遇，雄蛙胫跗关节前伸达眼后角，雌蛙达肩部。指趾端略膨大成球状。指间无蹼；趾间半蹼；第 1 和第 5 趾外侧有缘膜。体背皮肤粗糙，有圆疣和皮肤棱，尤以眼后颞区上方 1 对彼此平行的长肤棱，背部肩上方有一黑色"Λ"形皮肤棱为其显著特征。无背侧褶；颞褶较发达；腹面皮肤光滑。背面灰棕色或黑褐色，两眼间有深色横纹；唇缘有深色纵纹、四肢背面有黑横纹。无声囊。

生境与习性 多栖息于山区路边和田间的小水沟或浸水塘内。

分布 中国特有种。分布于湖南、江西、安徽、浙江、江苏、福建、广东、广西、香港、澳门和台湾。广泛分布于井冈山地区。

种群状况 常见种。

Identification Body size in males larger than those in females, SVL maximum 55 mm in females and 65 mm in males; head significantly great, and occipital region uplifted in males; snout blunted and pointed, canthus rostralis indistinct; loreal region extremely oblique; tympanum hidden; limbs relatively short; heels not meeting; tibio-tarsal articulation reaching the posterior corner of eye in males, reaching shoulder in females; tips of digits enlarged, rounded; hands not webbed; feet one-half webbed; free sides of first and fifth toes with fringes; dorsal skin rough, with tubercles and dermal ridges; having two significant diagnosis characters: two longitudinal, elongated dermal ridges parallel to each other above temporal region behind the eyes, and Λ-shaped dark-colored ridge above shoulder; supratympanic fold developed; dorsolateral fold absent; ventral skin smooth. Dorsum grey-brown or dark brown, sometimes with irregular dark markings; dorsal limbs with dark-colored transverse bands; lip with dark-colored vertical bars; venter white. No vocal sacs.

Habitat and Behavior Inhabits hill and mountain areas. More frequently found in pools and small ditches near roads and farmland.

Distribution Endemic to China. Hunan, Jiangxi, Anhui, Zhejiang, Jiangsu, Fujian, Guangdong, Guangxi, Hong Kong, Macao and Taiwan. Widely recorded from the Jinggangshan Region.

Status of Population Common species.

1.29 棘腹蛙 *Quasipaa boulengeri* (Günther, 1889)

识别特征　体肥硕，成年雄蛙头体长 68-92 mm，雌蛙 82-105 mm。头宽大于头长。吻端圆，吻棱钝圆，颊区甚倾斜。鼓膜不甚清晰。后肢相对长而粗壮，跟部相遇或重叠，雄蛙胫跗关节前伸达吻端。指趾端略膨大成球状。关节下瘤发达。指间无蹼；趾间近满蹼；第 1 和第 5 趾外侧有薄缘膜。背皮肤粗糙，密布疣粒，疣粒上有黑刺，有些疣粒排成纵行而形成皮肤棱。无背侧褶；颞褶发达，有刺疣；头后部两眼后角之间有一条横行的皮肤褶并形成一条横沟；繁殖期雄蛙胸腹部满布疣粒，其上有黑色锥状角质刺。背面棕灰色，两眼间有深色横纹，有时体背有红色云斑，四肢背面有灰黑横纹。雄性内侧 3 指有黑色圆锥状强婚刺，第 1 指婚垫发达，密布婚刺，第 2 指婚垫不发达，有稀疏婚刺，第 3 指婚垫不显，仅有几粒婚刺。有单个咽下声囊。

生境与习性　栖息在海拔 300-1900 m 山区溪流。

分布　江西、湖南、广西、云南、贵州、重庆、四川、甘肃、陕西、山西、湖北。越南。广泛分布于井冈山地区。

种群状况　常见种。IUCN 红色名录：易危。中国物种红色名录：易危。

Identification　Body stout, with 68-92 mm SVL in adult males, 82-105 mm in females; HW broader than HL; snout rounded; canthus rostralis bluntly rounded; loreal region extremely oblique; tympanum indistinct; limbs relatively short and robust; heels meeting or just overlapping; tibio-tarsal articulation reaching the eye; tips of digits enlarged, rounded; subarticular tubercles developed; hands not webbed; feet nearly fully webbed; free sides of first and fifth toes with fringes; dorsal skin rough, densely covered with tubercles with black spines; some of which arranged in longitudinal rows and forming dermal ridges supratympanic fold developed, covered with spiny tubercles; dorsolateral fold absent; ventral skin smooth in females; heavy tubercles with black spines on chest and belly in males during breeding season. Dorsum brown or dark brown; a dark-colored stripe between the eyes; sometimes with irregular grey-black markings on dorsum; dorsal limbs with dark-colored transverse bands; lip with dark-colored vertical bars; venter white. Males with strongly conical black nuptial spines on dorsal surface of first to third fingers: nuptial pad developed with dense nuptial spines on first finger, nuptial pad undeveloped with scattered nuptial spines on second finger; nuptial pad invisible only with several nuptial spines on third finger; presence of a single subgular vocal sac.

Habitat and Behavior　Inhabits mountainous streams at elevations between 300-1900 m a.s.l.

Distribution　Jiangxi, Hunan, Guangxi, Yunnan, Guizhou, Chongqing, Sichuan, Gansu, Shaanxi, Shanxi and Hubei. Vietnam. Recorded from overall the Jinggangshan Region.

Status of Population　Common species. IUCN Red List: VU. China Species Red List: VU.

1.29a　1.29b　1.29c　1.29d

1.30 九龙棘蛙 *Quasipaa jiulongensis* (Huang and Liu, 1985)

识别特征　体肥硕，成年个体头体长约74 mm。头宽大于头长。吻端钝尖，吻棱清晰，颊区倾斜。鼓膜隐蔽。前后肢短粗，跟部相遇或重叠，雄蛙胫跗关节前伸达吻端。指趾端略膨大成球状。关节下瘤发达。指间无蹼；趾间近满蹼；第1和第5趾外侧有发达缘膜。背皮肤粗糙，散布有黑刺的疣粒。眼后方有横肤褶和肤沟。无背侧褶；颞褶发达，其上有刺疣；四肢背面有纵走肤褶，其上有刺疣；雌蛙腹面皮肤光滑，繁殖期雄蛙胸部满布有黑刺的疣粒。体背面黑褐色或棕色，背两侧有对称排列的4-5个黄色斑，眼后方有深色横纹。腹白色，喉和前胸部有黑色云斑。雄性内侧2指有黑色小婚刺：第1指婚垫较发达，散布婚刺，第2指有10余枚婚刺。具单个咽下声囊。

生境与习性　栖息于海拔800 m以上山区的小溪中，溪旁树木繁茂。

分布　中国特有种。分布于江西、浙江和福建。在井冈山仅记录于八面山。

种群状况　IUCN 红色名录：易危。中国物种红色名录：易危。

Identification　Body stout, SVL approximately 74 mm in adults; HW greater than HL; snout bluntly pointed; canthus rostralis distinct; loreal region oblique; tympanum hidden; limb short and robust; heels meeting or just overlapping; tibio-tarsal articulation reaching the tip of snout; tips of digits enlarged, rounded; subarticular tubercles developed; hands not webbed; feet nearly fully webbed; free sides of first and fifth toes with fringes; dorsal skin rough with scattered tubercles, which bearing black spines; a transvers dermal ridge and groove extended from posterior edge of upper eyelid to contralateral one; dorsolateral fold absent; supratympanic fold developed with spiny tubercles; dorsal surface of limbs with longitudinal dermal ridges with spiny tubercles; ventral skin smooth in females; chest with heavy tubercles with black spines in males during breeding season. Dorsum brown or dark brown with symmetrically arranged four to five large light-colored spots in two dorsolateral rows; a dark-colored stripe between the eyes; sometimes with irregular grey-black markings on dorsum; dorsal limbs with dark-colored transverse bands; lip with dark-colored vertical bars; venter white; throat and anterior chest with black blotches. Males with conical, small black nuptial spines on dorsal surface of first and second fingers: nuptial pad developed with scattered small nuptial spines on first finger, nuptial pad undeveloped with more than ten small nuptial spines on second finger; presence of a single subgular vocal sac.

Habitat and Behavior　Inhabits mountainous streamlets surrounded by moist subtropical evergreen broadleaved forests at elevations above 800 m a.s.l.

Distribution　Endemic to China. Jiangxi, Zhejiang and Fujian. Recorded from Bamianshan in Mt. Jinggang.

Status of Population　IUCN Red List: VU. China Red Species List: VU.

1.30b

1.30a

1.30c

1.31 小棘蛙 *Quasipaa exilispinosa* (Liu and Hu, 1975)

识别特征 体型中等，成年个体头体长约 60 mm。头宽略大于头长。吻端钝圆，吻棱不显，颊区倾斜。鼓膜隐约可见。前后肢短粗，跟部刚刚重叠，胫跗关节前伸达眼部。指趾端略膨大成球状。关节下瘤发达。指间无蹼；趾间近满蹼，第 4 趾蹼深缺刻。背皮肤粗糙，散布疣粒。颞褶发达；雌蛙腹面皮肤光滑，繁殖期雄蛙胸腹部满布有黑刺的疣粒。体背面黑褐色或棕色，散有不规则黄色斑，眼后方有深色横纹。下腹及后肢腹面蜡黄色。雄性内侧 3 指有黑色强婚刺。具单个咽下声囊。

生境与习性 栖息于海拔 300-1300 m 的山区的小溪中，溪旁树木繁茂。

分布 中国特有种。分布于江西、浙江、福建和广东。井冈山记录于荆竹山。

种群状况 井冈山地区种群数量很少，属稀见物种。IUCN 红色名录：易危；中国物种红色名录：易危。

Identification Body moderate, SVL approximately 60 mm in adults; HW slightly greater than HL; snout blunt; canthus rostralis indistinct; loreal region oblique; tympanum barely visible; limb short and robust; heels just overlapping; tibio-tarsal articulation reaching the eye; tips of digits enlarged, rounded; subarticular tubercles developed; hands not webbed; feet nearly fully webbed, but web of toe IV depressed deeply; dorsal skin rough with scattered tubercles; supratympanic fold developed; ventral skin smooth in females; heavy tubercles with black spines on chest and belly in males during breeding season. Dorsum brown or dark brown with scattered irregular yellow or red-brown spots; a dark colored stripe between the eyes; posterior venter and ventral surface of hind limbs yellow. Males with strong conical black nuptial spines on dorsal surface of first to third fingers and with a single subgular vocal sac.

Habitat and Behavior Inhabits mountainous streamlets surrounded by moist subtropical evergreen broadleaved forests at elevations between 300-1300 m a.s.l.

Distribution Endemic to China. Jiangxi, Zhejiang, Fujian and Guangdong. Only recorded from Jingzhushan in Mt. Jinggang.

Status of Population Rare species in the Jinggangshan Region. IUCN Red List: VU; China Species Red List: VU.

1.32 棘胸蛙 *Quasipaa spinosa* (David, 1875)

识别特征　体肥硕，雄蛙体长约 80 mm，雌蛙头体长可超过 120 mm。头宽大于头长。吻端圆，吻棱钝圆，颊区倾斜。鼓膜隐蔽。前后肢短粗，跟部相遇或略重叠，雄蛙胫跗关节前伸达眼部。指趾端略膨大成球状。关节下瘤发达。指间无蹼；趾间近满蹼；第 1 和第 5 趾外侧有发达缘膜。雌蛙背部皮肤较光滑，有零星疣粒，其上常有 1 枚黑刺；雄蛙背部皮肤稍粗糙，有疣粒和皮肤褶；无背侧褶；颞褶稍发达，有零星刺疣；腹面皮肤光滑，有疣粒；雄蛙繁殖期胸部满布有黑刺的疣粒。背面棕色或黑棕色，两眼间有深色横纹，体背常有灰黑色云斑，唇缘有深色纵纹，四肢背面有灰黑横纹。内侧 2 指有黑色相对稍强的婚刺；第 1 指婚垫不甚发达，散布婚刺，第 2 指有少量稀疏排列的婚刺。具单个咽下声囊。

生境与习性　栖息于海拔 1500 m 以下山溪中。

分布　分布于贵州、云南、安徽、江苏、浙江、江西、福建、广东、香港和广西。国外分布于越南北部。广布于井冈山地区。

种群状况　常见种。IUCN 红色名录：易危。中国物种红色名录：易危。

Identification　Body stout, SVL about 80 mm in adult males, sometimes exceeding 120 mm in adult females; HW greater than HL; snout rounded; canthus rostralis bluntly rounded; loreal region oblique; tympanum hidden; limb short and robust; heels meeting or overlapping; tibio-tarsal articulation reaching the eye; tips of digits enlarged, rounded; subarticular tubercles developed; hands not webbed; feet nearly fully webbed; free sides of first and fifth toes with fringes; skin of dorsum and flanks smooth with scattered tubercles generally bearing black spines in females, rough with tubercles and longitudinal dermal ridges in males; dorsolateral fold absent; supratympanic fold slightly developed with scattered spiny tubercles; ventral skin smooth with scattered tubercles; of which with black spines on chest in males during breeding season. Dorsum brown or dark brown, with a dark-colored stripe between the eyes; sometimes with irregular grey-black markings on dorsum; dorsal limbs with dark-colored transverse bands; lip with dark-colored vertical bars; venter creamy white. Males possess developed nuptial pad with moderately conical black nuptial spines on dorsum of first finger, indistinct nuptial pad with several nuptial spines on dorsum of second finger; and a single subgular vocal sac.

Habitat and Behavior　Inhabits mountainous streams at elevations below 1500 m a.s.l.

Distribution　Guizhou, Yunnan, Anhui, Jiangsu, Zhejiang, Jiangxi, Fujian, Guangdong, Hong Kong and Guangxi. N Vietnam. Widely recorded from the Jinggangshan Region.

Status of Population　Common species. IUCN Red List: VU. China Species Red List: VU.

1.32a

1.32b

1.32c

无尾目 ANURA
树蛙科 Rhacophoridae

1.33 井冈纤树蛙 *Gracixalus jinggangensis* Zeng, Zhao, Chen, Chen, Zhang and Wang, 2017

识别特征　体型较小，头体长 27.9-33.8 mm。头宽略大于头长。无犁骨齿。头、体、四肢背面皮肤粗糙，散布小疣粒；上眼睑和背部无棘刺。腹面皮肤颗粒状。胫部无突起物；指基部微蹼；趾间蹼较发达，第 5 趾内侧蹼至远端关节下瘤上缘，第 4 趾两侧蹼至中间关节下瘤上缘，蹼式：I (2)- (2½) II (1⅔)-(3) III (2)-(3) IV (3)-(2) V；关节下瘤清晰，第 1 趾至第 5 趾依次为 1 个、1 个、2 个、3 个、2 个。体背棕色至米黄色，有一个倒 "Y" 形深色斑从两眼间延伸至背部。雄性具 1 个咽下声囊，第 1、第 2 指基部背面有隆起的婚垫，其上有勉强可见的颗粒状婚刺。

生境与习性　栖息于海拔 1100-1340 m 的竹林。7 个模式标本发现在竹竿上，距地面 1.0-2.0 m 的；6 个模式标本发现在虎杖、山茶和荨麻等植物叶片上，距地面 0.4-1.2 m。卵、蝌蚪和繁殖习性未知。

分布　中国特有种。正模标本：SYS a004811，成年雄性。2016 年 5 月采集于江西井冈山荆竹山（26° 29′ 28.53″ N，114° 04′ 32.94″ E；1208 m a.s.l.）。

种群状况　稀见物种。

Identification　Body size small, SVL 27.9-33.8 mm; head width slightly greater than head length; no vomerine teeth; upper eyelid and dorsum lacking spines; skin of dorsal and lateral surface of head, body and limbs rough, sparsely scattered with tubercles; ventral skin granular; tibiotarsal projection absent; fingers with weak webbings at finger bases; toes with moderately developed webbings, webbing formula I (2), (2½) II (1⅔), (3) III (2), (3) IV (3), (2) V; subarticular tubercles distinct, formula 1, 1, 2, 3, 2; brown to beige above, with an inverse Y-shaped dark brown marking extended from interorbital region to dorsum of body; males possess a single subgular vocal sac, bulged nuptial pads with barely visible minute granules on dorsal surface of the bases of first and second fingers.

Habitat and Behavior　It is restricted to a bamboo forest at elevations between 1100-1340 m a.s.l. Seven type specimens were found on the alive and withered bamboo poles at heights between 1.0-2.0 m, six on the leaves of plants of genera *Reynoutria*, *Camellia* and *Urtica* etc. at heights between 0.4-1.2 m. Eggs and tadpoles are not found, breeding habits are unknown.

Distribution　Endemic to China. Holotype: SYS a004811, adult male. Collected on May 2016 from the Jingzhushan in Mt. Jinggang.

Status of Population　Rare species.

1.33a

1.33b

1.33c

1.33d

1.34 布氏泛树蛙 *Polypedates braueri* (Vogt, 1911)

识别特征 雄蛙体长约 45 mm，雌蛙体长约 60 mm。体较扁，头扁平。吻背面观钝尖，侧面观钝圆；吻棱发达。鼓膜大而显著。瞳孔水平椭圆形。后肢细长，跟部重叠，胫跗关节前伸达眼和鼻孔之间。仅趾间有蹼，指趾端具发达吸盘。体背皮肤光滑，有小痣粒；腹面有扁平疣粒。无背侧褶。颞褶发达。体色多变，通常浅棕色，一般具深色"X"形斑或纵条纹。大腿后面具网状斑纹。雄性第 1、第 2 指具乳白色婚垫。具 1 个咽下声囊。

生境与习性 栖息于水坑、沟渠、山溪、田间、灌木草丛中。雌蛙产卵于泡沫状卵泡中，卵泡乳黄色，常挂于水边植物上、也见于草丛或泥窝内。

分布 中国热带、亚热带地区（包括云南、四川、广西、广东、福建、浙江、江西、安徽和台湾）。在井冈山广泛分布。

种群状况 常见种。

Identification SVL approximately 45 mm in males, 60 mm in females; body compressed dorsoventrally; head depressed; snout bluntly pointed in dorsal view, bluntly rounded in profile view; canthus rostralis developed; tympanum large and distinct; pupil horizontal oval; leg slender and long; heels overlapping; tibio-tarsal articulation forward reaching between nostril and eye; only feet webbed; digits with developed disks; dorsal skin smooth with granulations; venter with flattered tubercles; dorsolateral fold absent; supratympanic fold developed. Color variable, generally brown or dark brown, usually a dark-colored X-shaped mark or longitudinal stripes on the dorsum of body; rear of thigh with reticular markings. Males with creamy white nuptial pads on first and second fingers, and a single subgular vocal sacs.

Habitat and Behavior Inhabits puddles, pools, ditches, streams, and rice paddies surrounded by broad leaf plants and shrubs, where usually females lay eggs in foam nests hanging on leafs of plants above the water surface. Foam nest creamy yellowish.

Distribution Ropical and subtropical China (including Yunnan, Sichuan, Guangxi, Guangdong, Fujian, Zhejiang, Jiangxi , Anhui and Taiwan).Widely occurs in the Jinggangshan Region.

Status of Population Common species.

1.34a

1.34b

1.34c

1.35 大树蛙 *Rhacophorus dennysi* Blanford, 1881

识别特征　雄性体长约 80 mm，雌性约 90 mm。体略扁平，头扁平。吻端低平而钝尖，吻棱发达。鼓膜大而显著。瞳孔平置椭圆形。后肢长，跟部不相遇或刚刚相遇，胫跗关节前伸达眼或鼻眼之间。指趾端具吸盘，趾吸盘小于指吸盘。第3、第4指间半蹼。趾间满蹼。体背皮肤光滑，有小痣粒；腹面有扁平疣粒。无背侧褶。颞褶发达。体背绿色，有镶黄色细纹的棕黑色斑点；一般沿体侧有成行的镶有黑色边缘线的白色大斑点，有时为白色纵纹；前臂后侧、跗部后侧均有一条白色或粉红色镶有黑色线纹的纵走宽纹，分别延伸到第4指和第5趾远端。肛上方也有一条白色宽纹。腹面灰白色；蹼具网状斑纹。雄性第1、第2指有浅色婚垫。具单个咽下声囊。

生境与习性　栖息于山区溪流、稻田、水坑附近的灌木或草丛中。产卵于泡沫状胶质卵泡中，卵泡乳黄色，常挂于水边植物上。

分布　华南，西至四川和贵州。缅甸、越南北部和西北部、老挝。在井冈山地区广泛分布。

种群状况　常见种。

Identification　SVL approximately 80 mm in males, 90 mm in females; body slightly compressed dorsoventrally; head depressed; snout gradually depressed to its tip, and bluntly pointed in profile view; canthus rostralis developed; tympanum large and distinct; pupil horizontal oval; leg slender and long; heels not meeting or just meeting; tibio-tarsal articulation forward reaching the eye or between nostril and eye; hand with one-half webbing between third and fourth fingers; feet webbed; digits with developed disks, which on fingers are larger than those on toes; dorsal skin smooth with small granulations; venter with flattered tubercles; dorsolateral fold absent; supratympanic fold developed. Green above, with dark-colored spots edged by yellow curve on dorsum, usually with several white spots edged by a black curve arranged in a row, sometimes a wide white stripe on flank; rear of forearm and shank with wide longitudinal white stripes extending to the distal tip of the fourth finger and fifth toe, respectively; supraanal region with a transverse white stripe; venter grey-white; webs with reticular markings. Males with light-colored nuptial pads on first and second fingers, and a single subgnlar vocal sac.

Habitat and Behavior　Inhabits mountainous puddles, pools, ditches, streams, rice paddies and adjacent areas, where usually the females lay eggs in foam nest hanging on leafs of plants above the water surface. Foam nest creamy yellowish.

Distribution　S China west to Sichuan and Guizhou. Myanmar, N and NE Vietnam, Laos. Widely recorded from the Jinggangshan Region.

Status of Population　Common species.

1.35a　1.35b　1.35c

无尾目 ANURA
姬蛙科 Microhylidae

1.36 粗皮姬蛙 *Microhyla butleri* Boulenger, 1900

识别特征　体小，头体长约 22 mm。吻端钝尖。鼓膜不显。无犁骨齿。指趾端有小吸盘。指间无蹼；趾间微蹼。跟部重叠，胫跗关节前达眼部。体背粗糙，多疣粒，疣粒排成纵行。背灰棕色或红棕色，有独特的镶有黄边的黑棕色大斑。四肢有黑色斑纹。腹面白色，喉部有小黑斑点。雄蛙具单个咽下声囊。

生境与习性　生活于丘陵山区的水田、水沟、草地等环境。

分布　从云南和四川东部至浙江。印度东北部、缅甸、越南、泰国、马来半岛和新加坡。广泛分布于井冈山地区。

种群状况　常见种。

Identification　Body small, SVL approximately 22 mm; snout bluntly pointed; tympanum indistinct; vomerine teeth absent; small disks on tips of digits; hands not webbed; feet weakly webbed; heels overlapping; tibio-tarsal articulation forward reaching the eye; dorsal skin rough with many tubercles in rows. Dorsum grey-brown or red-brown, with unique large dark brown marks with yellow edge; limbs with transverse black bands; venter white, throat with black spots. Single subgular vocal sac in males.

Habitat and Behavior　Inhabits rice paddies, ditches, grasses in mountains and hills.

Distribution　Yunnan and eastern Sichuan to Zhejiang. NE India, Myanmar, Vietnam, Thailand, Malay Peninsula and Singapore. Recorded from overall the Jinggangshan Region.

Status of Population　Common species.

1.36a

1.36b

1.37 饰纹姬蛙 *Microhyla ornata* (Duméril and Bibron, 1841)

识别特征　体小，头体长约 22 mm。头小、吻尖，整个身体背面观略呈三角形。无犁骨齿。吻棱不显。鼓膜不显。跟部重叠，胫跗关节前达肩部或肩部前方。趾端无吸盘。趾间有蹼迹。背面皮肤略显粗糙，有小疣排列成行。腹面光滑。背棕色或深棕色，前后有 2 个深棕色 "Λ" 形斑。四肢有深色横纹。腹面白色，雌蛙咽喉部有深灰色小斑点，雄蛙咽喉部黑色。具单个咽下声囊。

生境与习性　生活于平原、丘陵和山区水田，水沟等环境。

分布　华中至华南地区，北至山西和陕西。泰国和印度尼西亚，东南穿过马来半岛至新加坡。广泛分布于井冈山地区。

种群状况　常见种。

Identification　Body small, SVL approximately 22 mm; snout pointed; body triangular in dorsal view; vomerine teeth absent; tympanum indistinct; heels overlapping; tibio-tarsal articulation forward reaching the shoulder or before shoulder; no disks on tips of digits; hands not webbed; toes with a mere rudiment of web; dorsal skin slightly rough with small tubercles in rows; venter smooth. Dorsum brown or red-brown, with two Λ-shaped dark-colored marks arranged one after another; limbs with transverse black bands; venter white, throat with dark grey spots in females, black in males. Single subgular vocal sac in males.

Habitat and Behavior　Inhabits rice paddies, ditches and grasses in plains, mountains and hills.

Distribution　S and C China, north to Shanxi and Shaanxi. Thailand and Indonesia, and southeast through the Malay Peninsula to Singapore. Recorded from overall the Jinggangshan Region.

Status of Population　Common species.

1.37a　1.37b　1.37c　1.37d

1.38 小弧斑姬蛙 *Microhyla heymonsi* Vogt, 1911

识别特征 体小、头体长稍长于 20 mm。头小，吻尖。无犁骨齿。鼓膜不显。颊部几垂直。跟部重叠，胫跗关节前达眼部。趾端微有吸盘。趾间有蹼迹。皮肤光滑。背面浅灰色或浅褐色，自吻端至肛部有一浅色细脊线，在脊线两侧有 2 对前后排列黑色弧形斑。四肢有黑色斑纹。腹部白色。雄性具单个咽下声囊。

生境与习性 栖息于山区靠近水源的环境中。

分布 华中、华南、华东和西南。马来半岛、苏门答腊和印度。在井冈山地区广泛分布。

种群状况 常见种。

Identification Body small, SVL slightly longer than 20 mm; head small; snout pointed; vomerine teeth absent; tympanum indistinct; loreal region nearly vertical; heels overlapping; tibio-tarsal articulation forward reaching the eye; weak disks on tips of digits; toes with a mere rudiment of webs; skin smooth. Dorsum light-grey or light brown, with a fine light-colored vertebral line beginning from tip of snout, extending to anal area, where two arc-shaped black markings are arranged on both sides; limbs with transverse black bands; venter white. Single subgular vocal sac in males.

Habitat and Behavior Inhabits mountainous areas near water.

Distribution C, S, E and SW China. Malay Peninsula, Sumatra and India. Recorded from overall the Jinggangshan Region.

Status of Population Common species.

1.38a

1.38b

1.38c

第 2 章

井冈山地区
爬行动物区系

Chapter 2

Reptile Fauna of the Jinggangshan Region

　　基于 2010-2015 年在井冈山地区密集开展的两栖爬行动物调查和广泛采样，共记录 2 目 16 科 47 属 67 种，其中龟鳖目 3 科 3 种，有鳞目蜥蜴亚目 4 科 13 种，有鳞目蛇亚目 9 科 51 种，包括新种井冈攀蜥及江西省新分布纪录北部湾蜓蜥。其中，中国特有爬行动物 18 种；CITES 附录 I、附录 II、附录 III 各 1 种；被 IUCN 列为极危（CR）1 种，濒危（EN）2 种，易危（VU）1 种。

　　本书收录了井冈山地区记录的爬行动物 62 种。其余 5 种爬行动物，棕黑腹链蛇、乌龟、红纹滞卵蛇、灰鼠蛇和赤练华游蛇均为文献记录物种，本次调查没有记录，故均未予收录。

Based on continual herpetological field surveys coupled with extensive sampling in the Jinggangshan Region conducted from 2010 to 2015, have resulted in discovery of 67 species of reptiles, belonged to 47 genera, 16 families and two orders. Among recorded species, three species belong to three families in Testudines, 13 species belong to four families in Lacertilia, and 51 species belong to five families in Serpentes, including newly described *Japalura jinggangensis* and *Sphenomorphus tonkinensis* as first distribution records for Jiangxi. 18 species are endemic to China; three are listed in CITES App. I, App. II and App.III, respectively; and one in the IUCN Red List as Critically Endangered (CR), two as Endangered (EN), and one as Vulnerable (VU).

We selected 62 reptile species from the Jinggangshan Region for this atlas. The remaining five species were excluded for the following reasons: *Amphiesma sauteri*, *Mauremys reevesii*, *Oocatochus rufodorsatus*, *Ptyas korros* and *Sinonatrix annularis* were only obtained from the literatures and not detected in our surveys.

龟鳖目 TESTUDINES
平胸龟科 Platysternidae

2.1 平胸龟 *Platysternon megacephalum* Gray, 1831

识别特征 头大，不能缩入龟壳内；腹甲扁平，背甲亦扁平，背中央有一显著的纵走隆起脊棱；上颌钩曲如鹰嘴，故名平胸龟、大头龟、鹰嘴龟。5趾型附肢，具爪，指间具蹼；第5趾不发达，无爪。尾长，亦不能缩入壳内；尾鳞方形，环形排列。头背部棕红色或橄榄绿色，盾片具放射状细纹；腹甲黄绿色或黄色。

生境与习性 栖于山溪中，夜间活动。肉食性，以小鱼、螺、蛙等为食。

分布 安徽、福建、广东、广西、贵州、海南、香港、湖南、江苏、江西、云南和浙江。越南北部、老挝、柬埔寨、泰国北部和缅甸南部。

种群状况 IUCN 红色名录：濒危。中国物种红色名录：濒危。CITES 附录 I。

Identification Head large, cannot be retracted into shell; carapace flattened with a significant longitudinal ridge along the dorsal midline; maxillary hooked such as the olecranon. The limbs pentadactyl, digits clawed and webbed, the fifth toe is not developed without claw. Tail long, also cannot be retracted into shell; tail scales square, circularly arrayed. Brown-red or olive-green above, dorsal plastron with radial lines; plastron yellowish-green or yellow.

Habitat and Behavior Nocturnal and carnivorous, takes small fish, snails and frogs for food; inhabits mountain streams.

Distribution Anhui, Fujian, Guangdong, Guangxi, Guizhou, Hainan, Hong Kong, Hunan, Jiangsu, Jiangxi, Yunnan, Zhejiang. N Vietnam, Laos, Cambodia, N Thailand, S Myanmar.

Status of Population IUCN Red List: EN. China Species Red List: EN. CITES App. I.

龟鳖目 TESTUDINES
鳖科 Trionychidae

2.2 中华鳖 *Pelodiscus sinensis* (Wiegmann, 1835)

识别特征 吻长，管状，约等于眼径；2鼻孔开口于吻端；颈部皮肤松软，基部无疣粒；体表被革质皮，散布疣粒；四肢平扁，蹼发达；尾短，雄性尾露出裙边，雌性尾不露出裙边。头尾均能缩入壳内。体背面橄榄绿色，腹面黄色，有对称灰绿色斑块。

生境与习性 见于井冈山农田水沟。

分布 除青海、西藏和新疆外，中国各省均有分布记录。国外分布于远东地区（俄罗斯）、朝鲜、韩国、日本和越南等。

种群状况 IUCN 红色名录：易危。中国物种红色名录：易危。

Identification Snout tubular, long, approximately equal to the diameter of the eye; two nostrils at the tip of snout; skin of neck loose and soft, without tubercles at the base of neck; surface of body covered with leathery skin, scattered tubercles; limbs are compressed with developed webs; tail short, visible from above in the male, not

2.1a

2.1b

2.2

visible from above in female; head and tail can be retracted into shell; olive-green above, ventral surface yellow with symmetrical gray green blotches.

Habitat and Behavior　Inhabits farmland ditches in Mt. Jinggang.

Distribution　Except Qinghai, Xizang and Xinjiang, all provinces in China. Far East (Russian), D. P. R. Korea, R. O. Korea, Japan, Vietnam.

Status of Population　IUCN Red List: VU. China Species Red List: VU.

有鳞目 SQUAMATA
蜥蜴亚目 LACERTILIA
壁虎科 Gekkonidae

2.3 多疣壁虎 *Gekko japonicus* (Schlegel, 1836)

识别特征　体被颗粒鳞，包括前臂和小腿均有疣粒；尾基部每侧肛疣 2-4 个；雄性肛前孔 4-8 个；趾指间具蹼迹，第 4 趾趾下瓣 6-10 个。体背灰色至灰棕色，颈、躯干和尾背具褐色横斑。

生境与习性　栖息于建筑物及岩石缝隙，夜出性动物，主要以昆虫为食。卵生。

分布　安徽、江苏、上海、浙江、福建、江西、湖南、湖北、广西、贵州和四川。国外见于日本及朝鲜半岛。见于井冈山大船、茨坪等地。

种群状况　常见种。

Identification　The head, body and limbs covered with uniform granular scales interspersed with tubercles, including the forearm and shank; 2-4 postcloacal tubercles laterally on each side at the base of the tail; 4-8 precloacal pores in male; digitals with rudiment of webs; 6-10 subdigital lamellae on the fourth toe. Dorsal surface of nape, body and tail grey or grey-brown with transverse brown bands.

Habitat and Behavior　Inhabits building gaps and rock crevices. Nocturnal. Insectivore. Oviparous.

Distribution　Anhui, Jiangsu, Shanghai, Zhejiang, Fujian, Jiangxi, Hunan, Hubei, Guangxi, Guizhou and Sichuan. Japan, Korean Peninsula. Recorded from Dachuan, Ciping in Mt. Jinggang.

Status of Population　Common species.

2.4 蹼趾壁虎 *Gekko subpalmatus* (Günther, 1864)

识别特征　体被均一粒鳞，无疣粒；尾基部每侧肛疣 1 个；雄性肛前孔 5-11 个；趾指间蹼发达，可达趾指的 1/3。第 4 趾趾下瓣 7-10 个。体背深棕色，颈、躯干和尾有镶黑色边的浅色横斑，四肢有模糊的深色横斑纹；体腹面黄色，有散布的深棕色斑点。

生境与习性　生活在亚热带地区，栖息于建筑物及岩石缝隙，夜出性动物，主要以昆虫为食。卵生。

分布　中国特有种。分布于东南、华南及西南。见于井冈山茨坪、湘洲等地。

种群状况　常见种。

Identification　The head, body and limbs covered with uniform granular scales without tubercles; single postcloacal tubercles laterally on each side at the base of the tail; 5-10 precloacal pores in

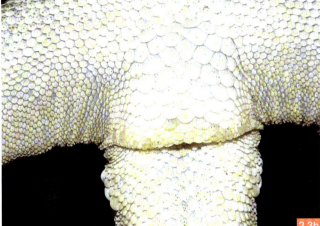

male; digitals with developed of webs, at most one-third webbed; 7-10 subdigital lamellae on the fourth toe. Dorsal surface of nape, body and tail dark-brown with dark-edged transverse bright bands; ventral surface yellow, scattered dark-brown spots.

Habitat and Behavior　Inhabits building gaps and the rock crevices in subtropical regions. Nocturnal. Insectivore. Oviparous.

Distribution　Endemic to China. S, SE and SW China. Recorded from Ciping and Xiangzhou in Mt. Jinggang.

Status of Population　Common species.

蜥蜴亚目 LACERTILIA
鬣蜥科 Agamidae

2.5 丽棘蜥 *Acanthosaura lepidogaster* (Cuvier, 1829)

识别特征　位于背中脊的背鬣鳞小，锯齿状。眼后棘 1 枚，不发达。鼓膜上方有颈棘 1 枚。颈鬣鳞基部扁宽，与低矮的背鬣鳞不连接，有一个小的间隔。尾长，约为头体长的 1.5 倍；活体颜色变异较多，肩背中央有一菱形深色斑，肘部有一镶深色边的醒目白斑，尾背有深色横斑。

生境与习性　栖于山区林下，夜晚常伏于树枝或草秆上休息。卵生。

分布　江西、福建、广东、海南、广西、云南和贵州。越南、柬埔寨和泰国等。在井冈山见于茨坪、八面山、河西垄和湘洲等地。

种群状况　常见种。

Identification　A compressed median dorsal ridge on back surmounted by a low serrate dorsal crest; small spine behind end of supraciliary edge and a small nuchal spine somewhat above tympanum; a nuchal crest of compressed scales, broad at their bases, separated from low dorsal crest by a narrow diastema. Tail long, approximately 1.5 times as long as the length of tip of snout to vent. More color variation in life, a large diamond-shaped dark-brown median spot widened on shoulders; a very distinct dark-edged cream-spot on elbow; dark bands surround the tail.

Habitat and Behavior　Inhabits mountain forests. Rests usually lying on branches or straw at night. Oviparous.

Distribution　Jiangxi, Fujian, Guangdong, Hainan, Guangxi, Yunnan and Guizhou. Vietnam, Cambodia, Thailand. Recorded from Ciping, Bamianshan, Hexilong and Xiangzhou in Mt. Jinggang.

Status of Population　Common species.

2.6 井冈攀蜥 *Japalura jinggangensis* Wang sp.nov.

识别特征　体型中等，头体长 71.9 mm。尾相对较短，尾长是头体长的 1.69 倍。后肢较短，其长度为头体长 66%，贴体前伸第 4 趾（不含爪）仅抵达口角。鼓膜隐蔽。鬣鳞 31 枚，包括 6 枚颈鬣。有喉褶。体侧没有浅色纵斑纹，四肢间有 4 个镶黑边的波浪形棕色横斑带。

生境与习性　栖息于海拔 1210 m 亚热带湿润常绿阔叶林。

分布　模式标本：SYS r000988，成年雄性，2014 年 7 月采集

于井冈山荆竹山。

种群状况 仅有 1 号标本。

Identification Body size moderate, SVL 71.9 mm; tail relatively short, tail length 169% of SVL; hindlimbs relatively short, hindlimb length 66% of SVL, toe IV (not including claw) just reaching the angle of the mouth when hindlimb adpressed; tympanum concealed; a serrated middorsal ridge composed of 31 dorsal crest-like scales, including six nuchal crest-like scales; transverse gular fold present; dorsum of body lacks longitudinal stripe, and decorated with four transverse wavy brown bands edged by black borders between fore and hind limb insertions.

Habitat and Behavior Inhabits moist subtropical evergreen broadleaved forests at elevation of 1210 m a.s.l.

Distribution Holotype: SYS r000988, adult male, collected from the Jingzhushan in Mt. Jinggang on July 2014.

Status of Population Only one has been found.

蜥蜴亚目 LACERTILIA
石龙子科 Scincidae

2.7 宁波滑蜥 *Scincella modesta* (Günther, 1864)

识别特征 体纤细、四肢短、前后肢贴体相向时不相遇。头顶被大型对称鳞；无股窝或鼠蹊窝，亦无肛前孔；有眼睑窗。背鳞为体侧鳞 2 倍，环体中段鳞 26-30 行；第 4 趾趾下瓣 10-16 枚；背古铜色或黄褐色，散布不规则黑色斑点或线纹；自吻端经鼻孔、眼上方、颈侧至尾末端有黑褐色纵纹，该纹较窄，上缘清晰波浪状，下缘模糊。头腹面白色，散布不规则黑色斑点，体腹面浅黄色，尾腹面橘红色。

生境与习性 栖息于森林地面，常见于落叶堆中。日行性陆栖动物。以昆虫为食。卵生。

分布 中国特有种。辽宁、河北、上海、江苏、浙江、安徽、江西、湖南、湖北、四川、福建、广东和香港。见于井冈山梨坪等地。

种群状况 常见种。

Identification Body slender; limbs short and not meeting when adpressed; dorsal surface of head covered with large symmetrical scales; no femoral pores, inguinal pores and precloacal pores; lower eyelid with transparent window; scale size on dorsum of body twice as large as those on flank; midbody scales in 26-30 rows; 10-16 subdigital lamellae on the fourth toe. Bronze or yellowish-brown with scattered black spots or linear markings above; a narrow dark lateral stripe on each side extends from tip of snout to tail; the upper edge of the dark lateral stripe is jagged, the lower edge is blurred; ventral head white with scattered black spots; ventral body yellowish; ventral tail red-orange.

Habitat and Behavior Inhabits forest floor. More frequently found in deciduous leaves. Diurnal and terrestrial. Insectivore. Oviparous.

Distribution Endemic to China. Liaoning, Hebei, Shanghai, Jiangsu, Zhejiang, Anhui, Jiangxi, Hunan, Hubei, Sichuan, Fujian, Guangdong and Hong Kong. Recorded from Liping in Mt. Jinggang.

Status of Population Common species.

2.8 印度蜓蜥 *Sphenomorphus indicus* (Gray, 1853)

识别特征　头顶被大型对称鳞；无股窝或鼠蹊窝，亦无肛前孔。下眼睑被鳞，无眼睑窗；没有上鼻鳞；额鼻鳞与额鳞相接，前额鳞一般不相接；眶上鳞4枚；第4趾趾下瓣16-20枚，趾背鳞2行。背古铜色；体两侧各有一条黑色纵带，其上有一条不连续的浅色线纹；体腹面白色。

生境与习性　栖息在低地常绿阔叶林至山区森林。多白天活动，陆栖种。卵胎生。

分布　河南、台湾、甘肃及长江以南的广大地区。印度、不丹、孟加拉国、马来半岛、缅甸、老挝、柬埔寨、泰国和越南等地。在井冈山地区广泛分布。

种群状况　分布广，种群数量大，常见种。

Identification　Dorsal surface of head covered with large symmetrical scales; no femoral pores, inguinal pores and precloacal pores; lower eyelid scaly; no supranasals; frontonasal in contact with frontal; two prefrontals separated; supraoculars four; 16-20 subdigital lamellae on the fourth toe. Bronze above; outer scale rows on dorsum lighter than back, bordering lateral broad black stripes on flanks, extending from eye to tail; the upper edge of the dark lateral stripe smooth; lower flanks whitish with little or no pigment; chin, venter, and underside of tail immaculate whitish.

Habitat and Behavior　Inhabits lowland evergreen forests to submontane forests. More frequently found in forests and often encountered along forest paths. Diurnal and terrestrial. Ovoviviparous.

Distribution　Henan, Taiwan, Gansu and south of the Yangtze River. Maynmar, Thailand, Vietnam, Laos, Cambodia, Malay Peninsula, India, Bhutan, Bangladesh. Widespread in the Jinggangshan Region.

Status of Population　Common species.

2.9 股鳞蜓蜥 *Sphenomorphus incognitus* (Thompson, 1912)

识别特征　头顶被大型对称鳞；无股窝或鼠蹊窝，亦无肛前孔。下眼睑被鳞，无眼睑窗；没有上鼻鳞；额鼻鳞与额鳞相接，前额鳞一般不相接；眶上鳞4枚；股后外侧有一团大鳞；第4趾趾下瓣17-22枚。背深褐色，具密集黑色斑点；幼体尾橘红色；体两侧各有一条上缘锯齿状的黑色纵带，杂浅黄色斑点；体腹面浅黄色，有黑色斑点。

生境与习性　栖息在丘陵、山地阴湿灌丛间，常见于路边。多中午外出活动。以昆虫、蜘蛛等为食。卵胎生。

分布　台湾、福建、江西、广东、海南、广西、云南和湖北等地。在井冈山地区广泛分布。

种群状况　常见种。

Identification　Dorsal surface of head covered with large symmetrical scales; no femoral pores, inguinal pores and precloacal pores; lower eyelid scaly; no supranasals; frontonasal in contact with frontal; usually two prefrontals separated; supraoculars four; one

2.8a

2.8b

2.8c

2.9a

patch of enlarged scales on the rear of the thigh; 17-22 subdigital lamellae on the fourth toe. Bronze in color speckled with light and dark spots above; tail scarlet in juveniles; a lateral broad dark stripe on each side of body extending from eye to tail, and with yellowish spots; upper edge of the dark lateral stripe jagged; ventral surface yellowish with black spots.

Habitat and Behavior　Inhabits lowland evergreen forests to submontane forests. More frequently found in riparian forests and often seen basking on stream banks. Diurnal and terrestrial. Ovoviviparous.

Distribution　Taiwan, Fujian, Jiangxi, Guangdong, Hainan, Guangxi, Yunnan and Hubei. Widely recorded from the Jinggangshan Region.

Status of Population　Common species.

2.10 北部湾蜓蜥 *Sphenomorphus tonkinensis* Nguyen, Schmitz, Nguyen, Orlov, Böhme and Ziegler, 2011

识别特征　小型蜓蜥，最大头体长 52.5 mm。下眼睑被鳞，无眼睑窗；前额鳞一对，大多个体彼此相接；眶上鳞 4 枚；额鳞与前 2 枚眶上鳞相接；通体背鳞光滑无棱；环体中段鳞分别为 32-38；第 4 趾基部背鳞 3 行，趾下瓣 15-19 枚。体背棕褐色，自颈至尾有黑色斑点，通常排列成一条直线；体侧深色纵纹自肩部起向后破碎成为一块块的深色块斑。

生境与习性　栖息于海拔 190-650 m 的亚热带至热带常绿阔叶林地面。多白天活动，陆栖种。卵生。

分布　江西、广东、广西和海南。越南北部。见于井冈山湘洲。

种群状况　罕见种。

Identification　A small-sized skink with maximum snout to vent length 52.5 mm; lower eyelid scaly, no transparent window; two prefrontals in contact with each other, rarely separated; supraoculars four; frontal in contact with the first two supraoculars; dorsal scales smooth without keel; midbody scales in 32-38 rows; scales on dorsal surface of base of toe IV in three rows; 15-19 lamellae under fourth toe. Brown above, usually with a discontinuous dark vertebral stripe or large dark spots arranged in a line; upper lateral body with black stripe, interspersed by small light spots from behind the neck.

Habitat and Behavior　Inhabits evergreen broad-leaved forest floors. More frequently found in moist leaf-litter and soil surface from tropical and subtropical mountainous forests at altitudes between 190-650 m a.s.l. Diurnal and terrestrial. Oviparous.

Distribution　Jiangxi, Guangdong, Guangxi and Hainan. N Vietnam. Only recorded from Xiangzhou in Mt. Jinggang.

Status of Population　Rare species.

2.11 蓝尾石龙子 *Plestiodon elegans* Boulenger, 1887

识别特征　身体粗壮，四肢和尾发达；吻短圆；没有后鼻鳞；上唇鳞 7 枚，下唇鳞 7 枚；额鳞与前 3 枚眶上鳞相接；环体鳞 26-28 行；股后有一团大鳞。幼体和青年个体体背黑色，背具 5

条黄色纵纹，尾蓝色；老年个体背面橄榄色，褐色侧纵纹显著，尾蓝色减少甚至消失。

生境与习性 栖息于森林地面，常见于落叶堆中和山溪附近区域。日行性陆栖动物。以昆虫为食。卵生。

分布 北京、天津、上海、江苏、浙江、安徽、江西、福建、台湾、河南、湖北、湖南、广东、香港、广西、贵州、云南、四川和陕西。国外见于越南。广泛分布于井冈山地区。

种群状况 常见种。

2.11a

Identification Body robust, limbs and tail well developed; snout short and rounded; no postnasal; supralabials 7, infralabials 7; frontal in contact with first three supraoculars; midbody scale rows 26-28; one patch of enlarged scales on the rear of the thigh. Dorsal surface bright black with five longitudinal yellowish stripes, tail bright blue in young individuals; black fades to olive-brown, blue reduced or even lost in old individuals.

2.11b

Habitat and Behavior Inhabits the forest floor. More frequently found in deciduous leaves and the vicinity of montane streams. Diurnal and terrestrial. Insectivore. Oviparous.

Distribution Beijing, Tianjin, Shanghai, Jiangsu, Zhejiang, Anhui, Jiangxi, Fujian, Taiwan, Henan, Hubei, Hunan, Guangdong, Hong Kong, Guangxi, Guizhou, Yunnan, Sichuan and Shaanxi.Vietnam. Widely recorded from the Jinggangshan Region.

Status of Population Common species.

2.11c

2.12 中国石龙子 *Plestiodon chinensis* (Gray, 1838)

识别特征 身体粗壮，四肢和尾发达；吻短圆；没有后鼻鳞；上唇鳞 7 枚，少数 9 枚；下唇鳞 7 枚；额鳞与前 2 枚眶上鳞相接；环体鳞 24 行，少数 22 行或 26 行。幼体背黑色，有 1 条不分叉的浅黄线纹起于顶间鳞，2 条背侧线起于最后一枚眶上鳞，尾蓝色。亮丽的幼年色斑模式随着年龄增长而消失，头变成红棕色，背橄榄色或橄榄棕色，有红色或橘红色斑块出现在体侧，喉乳白色有灰色鳞缘，腹表面其余部分乳白色或黄色。

生境与习性 栖息于低海拔地区，包括农田和城市绿地。日行性动物。卵生。

分布 台湾、香港、澳门、福建、浙江、江苏、安徽、江西、湖南、广东、广西、海南、云南和贵州。国外见于越南。广泛分布于井冈山地区。

种群状况 常见种。

2.12a

Identification Body robust; limbs and tail well developed; snout short, rounded; no postnasal; supralabials 7, rarely 9; infralabials 7; frontal in contact with first two supraoculars; midbody scale rows 24, rarely 22 or 26. In young individuals, dorsal surface black-brown, with a mid-dorsal light yellow stripe not forked, beginning from interparietal; two dorsolateral light-colored stripes beginning from the last supraocular on each side, respectively, tail blue. In old individuals, the light-colored stripes lost, head becoming reddish-brown and dorsum olive or brownish-olive, with scattered red or orange blotches on flanks; throat cream with grey scale edged; rest of ventral surface cream white or yellow.

2.12b

Habitat and Behavior　Inhabits lowlands including agricultural areas, green belts in cities. Diurnal. Oviparous.

Distribution　Taiwan, Hong Kong, Macao, Fujian, Zhejiang, Jiangsu, Anhui, Jiangxi, Hunan, Guangdong, Guangxi, Hainan, Yunnan and Guizhou. Vietnam. Widespread in the Jinggangshan Region.

Status of Population　Common species.

2.12c

蜥蜴亚目 LACERTILIA
蜥蜴科 Lacertidae

2.13 古氏草蜥 *Takydromus kuehnei* (Van Denburgh, 1909)

识别特征　体纤细；四肢纤弱；尾长，圆柱形，约为头体长 2 倍。头三角形，吻长且钝圆。前鼻鳞 1 枚，与对侧的彼此相接；2 枚后鼻鳞；头顶被大型对称鳞。有围领。颈背在顶鳞后为颗粒鳞；躯干背部有 6 行大棱鳞，中间 2 行大鳞中间有一行断续小棱鳞；躯干部腹鳞为覆瓦状大鳞，除外侧 2 行起棱外，中央 4 行平滑。尾被起棱大鳞，排列成环，不分节。成体鼠蹊窝每侧 3-5 个。背棕色，腹面乳白色，体侧自鼻孔经眼下缘、经耳孔、前肢上缘有一条上缘黑色的白色纵纹。

生境与习性　栖息于亚热带、热带低地森林。白天常见于地面和草地上。晚上栖于五节芒等植物的顶端。卵生。

分布　台湾、福建、江西、广东、海南、广西和贵州。国外见于越南。井冈山见于茨坪等地。

种群状况　常见种。

2.13a

Identification　Body slender; limbs long and slender; tail long and cylindrical, approximately twice as long as length from the tip of snout to vent; a single prenasal, in contact with contralateral one; two superposed postnasals; head triangular, snout elongated and rounded; dorsum of head covered with large symmetrical scales; neck with collar; granular scale on nape behind parietals; dorsum of trunk with keeled large scales in six rows and an intermittent small scale row between middle two large scale rows; ventral surface of trunk large scales in six rows, outer two rows keeled, rest four rows smooth; tail not segmented and covered with large keeled scales, which arranged in transverse rows; 3-5 inguinal pores in adults. Brown above, venter whitish-cream; a longitudinal pale stripe edged with black on each side, beginning from nostril, across subocular region, lower part of ear opening, extends to above forelimb insertion.

2.13b

Habitat and Behavior　Inhabits subtropical and tropical forests in lowlands. Frequently found on the forest floor and grass at the daytime; often sleep at top of *Miscanthus floridulus* and other plants at night. Diurnal. Oviparous.

Distribution　Taiwan, Fujian, Jiangxi, Guangdong, Hainan, Guangxi and Guizhou. Vietnam. Recorded from Ciping in Mt. Jinggang.

Status of Population　Common species.

2.13c

2.14 崇安草蜥 *Takydromus sylvaticus* (Pope, 1928)

识别特征　体纤细；四肢长而纤弱；尾长，圆柱形，为头体长 3.0-3.7 倍。头长为头宽 2 倍；吻窄长。头顶被大型对称鳞。左右前鼻鳞相接，将吻鳞与额鼻鳞分隔，或左右前鼻鳞分离，吻鳞与额鼻鳞相接；前额鳞通常彼此相接；眶上鳞 4 枚，前后两枚较小。颏片 4-5 对。有弱围领。体背鳞较体侧鳞略大，起强棱，平铺，排列不呈明显纵行；体侧为颗粒鳞；腹鳞 6 行，为覆瓦状大鳞，雄性均起强棱，雌性除外侧 2 行起棱外，中央 4 行平滑或有弱棱。尾被起棱大鳞，排列成环，不分节。鼠蹊窝每侧 3 个。幼体背棕色，略带绿色，腹面黄绿色，体侧有一条黄白色腹侧纵线，始自鼻鳞，经眼下缘、经耳孔、前肢上缘向后延伸，尾棕红色；成年个体头、体、四肢和尾背墨绿色，染棕红色，腹黄绿色，仅有一条腹侧纵纹；老年个体背翠绿色，除腹侧纵纹外，出现一条浅色背侧纵纹。

生境与习性　栖息于中亚热带低地森林，多在地面活动，常见于近水林缘和灌丛地面。日行性动物，白天多见于空旷地面晒太阳。于植物根部土壤中过冬。卵生。

分布　中国特有种。分布于福建（邵武、南平）、浙江（泰顺）、江西（贵溪、婺源、龙南、井冈山）、广东（仁化）、湖南（桂东）。井冈山见于双溪口和荆竹山。

种群状况　过去被认为是罕见物种，作者近年调查证实该种广泛分布于东南地区。

Identification　Body slender; limbs long and slender; tail long and cylindrical, 3.0-3.7 times as long as length from the tip of snout to vent; head twice as long as broad; snout acutely pointed; dorsum of head covered with large symmetrical scales; right and left prenasals in contact with each other behind rostral, that separated rostral and frontonasal; or rostral in contact with frontonasal, that separate right and left prenasals; usually prefrontals in contact with each other; supraoculars 4, the most anterior and posterior ones significantly small; 4-5 pairs of chin-shields; neck with a weak collar; scales on back of body slightly large, strongly keeled, juxtaposed, not arranged in distinct longitudinal rows, those on sides granular; ventral scales of body large, arranged in six longitudinal rows, strongly keeled in males, the medial four rows smooth or weakly keeled and outermost two rows keeled in females; tail not segmented and covered with large keeled scales, which arranged in transverse rows; three inguinal pores. In juveniles, dorsal surface of head, body, limbs brown, tinging green; tail red-brown; venter greenish-yellow; a ventrolateral longitudinal yellowish-white stripe edged with black dorsally, beginning from nasal, along the lower margin of the orbit and ear opening, upper region forelimb insertion, posteriorly extending. In adults, dorsal surface of head, body, limbs and tail green, tinging brown; with a same ventrolateral stripe white. In older individuals, dorsal surface jade-green, faded brown, with same ventrolateral white stripe, and newly added dorsolateral white stripe.

Habitat and Behavior　Inhabits mid-subtropical forests in lowlands. Often seen basking on the forest floor. It is dormant in the soil at the roots of plants in winter. Diurnal. Oviparous.

2.14a

2.14b

2.14c

Distribution　Endemic to China. Fujian (Shaowu, Nanping), Zhejiang (Taishun), Jiangxi (Guixi, Wuyuan, Longnan and Mt. Jinggang), Guangdong (Renhua), Hunan (Guidong). Recorded from Shuangxikou and Jingzhushan in Mt. Jinggang.

Status of Population　Previously considered a rare species, but recently confirmed to be widely distributed in SE China.

2.15a

2.15 北草蜥 *Takydromus septentrionalis* (Günther, 1864)

识别特征　尾甚长，为头体长 2-3 倍。头顶被大型对称鳞。背部大棱鳞 6 行，腹部 8 行棱鳞；鼠蹊孔 1 对。背绿色或棕色，腹面白色或灰棕色。雄性背鳞外侧有一条起于顶鳞后缘至尾部的绿色纵纹，体侧有不规则深色斑。

生境与习性　栖息于山地草丛中。日行性食虫动物。卵生。

分布　中国秦岭以南各省。在井冈山地区广泛分布。

种群状况　常见种。

2.15b

Identification　Tail long, 2-3 times as long as length from the tip of snout to vent; dorsum of head covered with large symmetrical scales; dorsum with large keeled scales in six rows; venter with large keeled in eight rows; one pair of inguinal pores. Green or brown above, ventral surface white or grey-brown; a green longitudinal stripe from the trailing edge of parietal to the tail along outermost dorsal scale row on each side in males; flanks with irregular dark spots.

Habitat and Behavior　Inhabits mountain grass and brushwood. Diurnal. Oviparous.

Distribution　Widespread in south of Qinling Mountains. Widely recorded from the Jinggangshan Region.

Status of Population　Common species.

2.15c

有鳞目 SQUAMATA
蛇亚目 SERPENTES
盲蛇科 Typhlopidae

2.16 钩盲蛇 *Ramphotyphlops braminus* (Daudin, 1803)

识别特征　小型蛇类，最大体长不足 20 cm。体纤细、圆筒状。头和颈部区分不明显。吻端圆。外鼻孔侧置。上唇鳞 4 枚。眼不清晰，隐于眼鳞下。头部鳞显著大于体背鳞。通身鳞大小相似，覆瓦状排列。环体鳞 20 行。鼻鳞分裂为 2，背视可见。尾端有刺。背黑色或深棕色，腹色浅，吻和尾尖苍白。

生境与习性　见于人类活动区，如小片林地，也见于山区林下。夜出性穴居动物，白天偶尔也会钻出地面活动。以昆虫为食。卵生。

分布　江西、浙江、福建、台湾、广东、香港、海南、广西、云南、贵州、重庆和湖北。西亚、南亚、东南亚、澳大利亚、非洲和墨西哥。在井冈山见于茨坪、大井、河西垄等地。

种群状况　常见种。

Identification　A small-sized snake with maximum length from the tip of snout to tip of tail 20 mm ; body slender and cylindrical; nostrils lateral; supralabials 4; eye indistinct and hidden beneath the ocular

2.15d

scale; head scales significantly greater than dorsal scales; body scales uniformed and imbricated; midbody scales in 20 rows; nasal scale completely divided and visible from above; caudal spine present. Dark purplish-brown to almost black, paler snout and caudal spine.

Habitat and Behavior　Encountered in human dwellings and lightly forested areas; also found in mountain forests. Nocturnal and subfossorial, or active on the ground surface, especially after rain. Insectivore. Oviparous.

Distribution　Jiangxi, Zhejiang, Fujian, Taiwan, Guangdong, Hong Kong, Hainan, Guangxi, Yunnan, Guizhou, Chongqing and Hubei. W, S and SE Asia, Australia, Africa and Mexico. Recorded from Ciping, Dajing and Hexilong in Mt. Jinggang.

Status of Population　Common species.

2.16a

2.16b

2.16c

蛇亚目 SERPENTES
闪鳞蛇科 Xenopeltidae

2.17 海南闪鳞蛇 *Xenopeltis hainanensis* Hu and Zhao, 1972

识别特征　身体粗壮，圆柱形。吻圆。顶间鳞在 4 枚顶鳞之间。眶后鳞单枚；上唇鳞 7 枚，3-2-3 排列，少数 6 枚或 8 枚。下唇鳞 6-7 枚。上颌齿 22-24 枚。眼小。尾短，末端尖且硬。背鳞光滑无棱。环体鳞 15 行。腹鳞 152-163，尾下鳞 16-19 对，肛鳞对分。背蓝褐色，闪金属光泽。腹面躯干和尾基部蓝褐色，其余部分灰白色。

生境与习性　栖息于低山森林。以两栖类和爬行类为食。卵生。

分布　华南和华东地区。越南北部。

种群状况　常见种。

Identification　Body robust, cylindrical; snout rounded; interparietal surrounded by 4 parietals; single postocular; supralabials 7 (3-2-3), rarely 6 or 8; infralabials 6-7; maxillary teeth 22-24; eye small; tail short, tip sharp and hard; dorsal scales smooth, not keeled; midbody scales in 15 rows; ventrals 152-163; subcaudals 16-19, paired; anal divided. Dorsum iridescent bluish-brown; ventral surface of trunk and base of tail greyish-white; rest venter bluish-brown.

Habitat and Behavior　Inhabits submontane forests. Diet consists of amphibians and reptiles. Oviparous.

Distribution　E and S China. N Vietnam.

Status of Population　Common species.

2.17a

2.17b

蛇亚目 SERPENTES
闪皮蛇科 Xenodermatidae

2.18 井冈脊蛇 *Achalinus jinggangensis* (Zong and Ma, 1983)

识别特征　体细长。头长椭圆形，与颈部区分不显著；吻鳞三角形，背视可见其上缘。无颊鳞，前额鳞转入颊区，入眶，并与上唇鳞相接。无眶前鳞和眶后鳞。前颞鳞入眶；第三级颞鳞上枚最大，左右两侧三级颞鳞在顶鳞后被 1-2 枚小鳞分隔。上唇鳞 6 枚，第 4、第 5 枚入眶，第 6 枚最大。下唇鳞 6 枚。颌

2.17c

片 2 对, 其后为腹鳞。背鳞 23 行, 起棱, 仅外侧行平滑而且扩大。肛鳞完整不对分; 尾下鳞单行。体背棕黑色, 有金属光泽。腹面黑色, 鳞缘色淡。

生境与习性 栖息于海拔 550-1600 m 山区森林和村落。夜行性, 穴居, 食虫。卵生。

分布 井冈山地区特有种。井冈山见于茨坪、大井、荆竹山等地, 以及比邻的桃源洞自然保护区。

种群状况 目前仅发现 11 个个体。IUCN 红色名录: 极危; 中国物种红色名录: 极危。

Identification Body slender; head oval, elongate and indistinct from neck; rostral triangular, barely visible from above; loreal absent; prefrontal bending in to the loreal region, in contact with supralabials and entering the eye; preocular and postocular absent; anterior temporals in contact with the eye; upper third temporal very large, separated with each other by 1-2 small scales; supralabials 6, fourth and fifth entering the eye, sixth much elongate; infralabials 6; chin-shields two pairs, immediately followed by the ventrals; dorsal scales in 23 rows, keeled, the outer scale row expanded and smooth; anal entire; subcaudals in single row, not paired. Uniform iridescent brownish-black above; uniform black beneath, ventral scale edges pale.

Habitat and Behavior Inhabits forests and villages at elevations between 550-1600 m a.s.l. Nocturnal and subfossorial. Insectivore. Oviparous.

Distribution Endemic to the Jinggangshan Region, where it recorded from Ciping, Dajing and Jingzhushan in Mt. Jinggang and Taoyuandong Nature Reserve.

Status of Population Only 11 individuals have been found. IUCN Red List: CR. China Species Red List: CR.

2.19 棕脊蛇 *Achalinus rufescens* Boulenger, 1888

识别特征 头窄长; 吻鳞小, 高宽相等, 背视不可见。额鳞约为顶鳞的一半。前颞鳞入眶。每侧有 3 枚盾状鳞与顶鳞相接, 第 3 枚非常大, 彼此相接或被 1 枚小鳞分隔。上唇鳞 6 枚, 第一枚非常小, 第 4、第 5 枚入眶, 第 6 枚最大。下唇鳞 5 枚, 少数 6 枚。颔片 2 对。背鳞 23 行, 起强棱或仅外侧行平滑。腹鳞 136-165 枚; 肛鳞完整不对分; 尾下鳞单行 82 枚。体背均一橄榄绿色或棕色, 有的个体有 1 条清晰的黑色脊线延伸至尾尖。腹面均一微黄色, 鳞缘白色。

生境与习性 栖息于平原、丘陵、山区。夜行性, 穴居, 食虫。卵生。

分布 浙江、江西、广东、香港、海南和广西。越南北部。井冈山见于湘洲、白银湖。

种群状况 常见种。

Identification Head narrow, elongate; rostral small, as deep as broad, not visible from above; frontal hardly half as long as the parietals; anterior temporals in contact with the eye; two shields bordering the parietals on each side, the third very large in contact with each other or separated from its fellow by a small scale; supralabials six, first very small, fourth and fifth entering the eye, sixth much elongate; infralabials five, rarely six; two pairs of chin-shields; dorsal scales strongly keeled or outer scale row smooth, in

23 rows; ventrals 136-165; anal entire; subcaudals 82, not paired. Uniform olive-green or brown above; sometime a significant longitudinal black vertebral line extends to tip of tail; uniform yellowish beneath; scale edges white.

Habitat and Behavior　Inhabits glasslands, hills and mountains. Nocturnal and subfossorial. Insectivore. Oviparous.

Distribution　Zhejiang, Jiangxi, Guangdong, Hong Kong, Hainan and Guangxi. N Vietnam. Recorded from Xiangzhou and Baiyinhu in Mt. Jinggang.

Status of Population　Common species.

蛇亚目 SERPENTES
钝头蛇科 Pareatidae

2.20 台湾钝头蛇 *Pareas formosensis* (Van Denhurgh, 1909)

识别特征　身体细长，稍侧扁。头短，与颈部区分显著。吻短钝。眼大，瞳孔直立。颊鳞单枚，不入眶。前额鳞入眶。眶前鳞单枚或 2 枚；眶后鳞单枚（少数缺或 2 枚）。眶下鳞单枚，细长，分隔上唇鳞与眼眶。上唇鳞通常 6 枚，下唇鳞通常 7 枚。背鳞光滑，通体 15 行。腹鳞163-193，肛鳞完整单枚。尾下鳞对分，56-85 对。上体浅棕色，身体和尾有不规则横斑。头顶深棕色，有黑色斑点，并镶有黑色斑纹，向后延伸至颈侧。有黑色条纹自眶后鳞至嘴角。下体浅黄色。虹膜橘红色。

生境与习性　栖息于山区森林和农田。夜行性树栖蛇类。卵生。

分布　关于台湾钝头和中国钝头蛇 *Pareas chinensis* (Barbour, 1912) 的分类问题一直存有争议。江耀明（2004）认为中国钝头蛇是台湾钝头的次定同物异名。Guo *et al.* (2011) 基于采集自四川（宝兴和天全）和台湾样品的分子证据认为台湾钝头蛇和中国钝头蛇均为有效种。目前普遍采纳的分类观点是台湾钝头蛇只分布于台湾，中国钝头蛇分布于中国南方各省，包括福建、江西、广东、广西、贵州、四川、云南、西藏和香港（Uetz and Hošek, 2017）。但基于形态和分子证据的研究，我们认为中国东南（江西、福建、广东、广西）的居群更近于台湾钝头蛇，而非中国钝头蛇。在此，我们暂时仍将其按台湾钝头蛇处理。

种群状况　常见种。

Identification　Body slender, slightly laterally compressed; head short, distinct from neck; snout short, truncated; eye large, pupil vertical; loreal single, not touching the eye; prefrontal touching the eye; preocular single or two; postocular single (rarely lacking or two); subocular single, elongated, separated supralabials from orbit; supralabials usually six; infralabials usually seven; dorsal scales smooth, in 15 rows throughout; ventrals 163-192; anal entire; subcaudals 56-85, paired. Light brown above with irregular black transverse bars on body and tail; top of head dark-brown with black spots and bordered with black stripes, posteriorly extending to sides of neck; black streak from postocular to angle of jaw on each side; light yellow below; iris orange.

Habitat and Behavior　Inhabits montane forests and farmland. Nocturnal and arboreal. Oviparous.

2.20a

2.20b

Distribution　The taxonomy of *Pareas formosensis* (Van Denhurgh, 1909) and *Pareas chinensis* (Barbour, 1912) is still an ongoing controversy. Jiang (2004) suggested that latter should be a Junior synonym of former. Guo *et al.* (2011) suggested that *P. chinensis* is a valid species and does not form a complex with *P. formosensis* based on molecular data from Sichuan (Baoxing and Tianquan) and Taiwan. At present, the widely accepted opinion is that *P. formosensis* is only found in Taiwan, *P. chinensis* is distributed in southern China (Fujian, Jiangxi, Guandong, Guangxi, Guizhou, Sichuan, Yunnan, Hong Kong) (Uetz and Hošek, 2017). However, our results of morphological and molecular studies indicated that populations from southeastern China (Jiangxi, Fujian, Guangdong and Guangxi) is closer to *P. formosensis*, and rather than *P. chinensis*. Herein, the population of Jinggangshan Region is still treated as *P. formosensis*.

Status of Population　Common species.

2.21 福建钝头蛇 *Pareas stanleyi* (Boulenger, 1914)

识别特征　身体细长。头略长，与颈部区分显著。吻短钝。眼大，瞳孔直立。颊鳞单枚，入眶。前额鳞入眶。没有眶前鳞；眶后鳞单枚。眶下鳞单枚，细长，分隔上唇鳞与眼眶。上唇鳞通常 7-8 枚，下唇鳞通常 7-8 枚。背鳞通体 15 行，除两侧最外行平滑外，其余具强棱。腹鳞 151-160，肛鳞完整单枚。尾下鳞对分，48-60 对。上体浅棕色，身体和尾有断续黑色横斑带。头顶自鼻间鳞之后到颈部有一大黑斑块，该黑斑自枕部向后分叉，有一黑线自眼眶向后达颈部。下体黄白色，散有稀疏黑斑点。虹膜浅棕色。

生境与习性　栖息于山区森林。夜行性树栖蛇类。卵生。

分布　江西、浙江、福建和贵州。在井冈山见于白银湖等地。

种群状况　较罕见。中国物种红色名录：易危。

Identification　Body slender; head slightly long, distinct from neck; snout short, truncated; eye large, pupil vertical; loreal single, entering the eye; prefrontal entering the eye; no preocular; postocular single; subocular single, elongated, separated supralabials from orbit; supralabials usually 7-8; infralabials usually 7-8; dorsal scales in 15 rows throughout, strongly keeled, the outer row smooth; ventrals 151-160; anal entire; subcaudals 48-60, paired. Light-brown above with intermittent black transverse bars on body and tail; top of head with large black shadings from posterior border of two internasals to nape of the neck, bifurcated into two black stripes posteriorly; a black stripe beginning from posterior border of orbit extending to sides of neck; yellowish-white with sparse black speckle below; iris light-brown.

Habitat and Behavior　Inhabits montane forests. Nocturnal and arboreal. Oviparous.

Distribution　Jiangxi, Zhejiang, Fujian and Guizhou. Recorded from Baiyinhu in Mt. Jinggang.

Status of Population　Rare species. China Species Red List: VU.

蛇亚目 SERPENTES
蝰科 Viperidae

2.22 白头蝰 *Azemiops kharini* Orlov, Ryabov and Nguyen, 2013

识别特征　具有管状毒牙的毒蛇。身体圆筒形，中等大小。头椭圆形，与颈部区分显著。吻钝。外鼻孔大，侧向。眼小，圆形，瞳孔直立。颊鳞单枚，很小。无颊窝。眶前鳞 3 枚，眶后鳞 2 枚或 3 枚。上唇鳞 6 枚，第 3 枚入眶。下唇鳞 8 枚，偶有 7 或 9 枚。背鳞光滑，通常 17-17-15 行排列，偶有通身 15 行者。腹鳞 170-197 枚。肛鳞完整单枚。尾下鳞 39-52，对分，偶有不对分的。头背橘红色，从前额鳞至枕部区域浅橘红色，有成对棕红色斑；身体和尾背面深棕色，有黑色斑点和镶黑边的橘红色横斑纹，这些斑纹是交错的，偶有连在一起的。腹面灰白色。

生境与习性　栖息于山地和丘陵。常见于农业区和民居附近。夜晚或晨昏活动。以小型哺乳动物为食。

分布　江西、安徽、浙江、福建、广东、广西、云南、贵州、四川、陕西和甘肃。国外见于缅甸和越南北部。井冈山见于荆竹山。

种群状况　常见种。

Identification　Solenoglyphous venomous snakes. Body cylindrical, moderate-sized; head elliptical, distinct from neck; snout blunt; nostril large, laterally; eye rounded, slightly small, pupil vertical; loreal single, small; loreal pit absent; preoculars 3; postoculars 2 or 3; supralabials 6, third entering the eye; infralabials 8, rarely 7 or 9; dorsal scales smooth, usually arranged in 17-17-15 rows, rarely in 15 rows throughout; ventrals 170-197; anal entire;

subcaudals 39-52, usually paired, rarely not divided. Orange-red above head; the region from frontal to occiput light orange with paired red-brown marks; dorsal surface of body and tail dark-brown with black spots, and paired orange cross-bars with black edges that usually alternating, rarely meeting with one another laterally; grey white below.

Habitat and Behavior Inhabits hills and mountainous areas. More frequently found in agricultural areas and in the vicinity of human settlements. Active at night or dusk. Diet consists of small mammals.

Distribution Jiangxi, Anhui, Zhejiang, Fujian, Guangdong, Guangxi, Yunnan, Guizhou, Sichuan, Shaanxi and Gansu. N Vietnam, Myanmar. Recorded from Jingzhushan in Mt. Jinggang.

Status of Population Common species.

2.22c

2.23 尖吻蝮 *Deinagkistrodon acutus* (Günther, 1888)

识别特征 具管牙毒蛇。体粗壮。头三角形,与颈部区分显著。头顶鳞大,对称排列。吻端上翘。上唇鳞7,少数6枚或8枚,第2枚高大于长,构成颊窝前缘。第3、第4枚最大,入眶。眼大,瞳孔直立。眶前鳞2枚。眶下鳞单枚,长显著大于高。眶后鳞2枚。下颌鳞大。背鳞起棱,鳞尖具二疣。体中段鳞21行,少数23行。腹鳞158-172。肛鳞完整。尾下鳞对分,43-63对,或不对分。尾尖尖长。上背灰或棕色,每侧有一系列近三角形大斑,相对或交错排列。有腹面镶浅黄色边的深棕色眼后斑纹。尾末段浅黄色,尾尖黑色。腹面乳白色,杂以灰色和清晰黑斑。

生境与习性 栖息于低地和山麓的森林、农耕地和居民区附近。夜行性陆栖动物。食物包括小型哺乳动物、鸟类和蛙类。卵生。

分布 江西、安徽、浙江、江苏、福建、广东、广西、贵州、湖南、湖北和台湾。越南北部。广泛分布于井冈山地区。

种群状况 常见种。

2.23a

2.23b

Identification Solenoglyphous snakes. Body stout; head triangular, distinct from neck; scales on the top of head large, symmetrical; snout upturned with rostral raised upwards; supralabials 7, rarely 6 or 8, second higher than long, forming anterior margin of loreal pit; third and fourth largest, entering the orbit; infralabials 10-11, rarely 9 or 12; eye large, pupil vertical; preoculars 2; subocular single, significantly longer than high; postoculars 2; lower temporals enlarged; dorsal scales keeled, scale tips bituberculate; midbody scale rows 21, rarely 23; ventrals 158-172; anal entire; subcaudals 43-63, paired or undivided; tail-tip acute. Grey or brown above, with series of pairs of large, subtriangular dark marks on each side that meet or alternate; the presence of dark-brown postocular stripe edged yellowish ventrally; tail end pale yellow and tail-tip black; cream below, mottled grey and distinct black spots.

2.23c

Habitat and Behavior Inhabits forests, agricultural areas and areas around human settlements from lowlands to submontane limits. Nocturnal and terrestrial. Diet consists of small mammals, birds and frogs. Oviparous.

Distribution Jiangxi, Anhui, Zhejiang, Jiangsu, Fujian, Guangdong, Guangxi, Guizhou, Hunan, Hubei and Taiwan. N Vietnam. Widely occurs in the Jinggangshan Region.

2.23d

Status of Population　Common species.

2.24 短尾蝮 *Gloydius brevicaudus* (Stejneger, 1907)

识别特征　具管牙毒蛇。身体粗壮。尾短。头略呈三角形，与颈部区分显著。吻短钝。头顶鳞大，对称排列。枕鳞起棱。鼻间鳞窄长，彼此相接。吻棱清晰。鼻鳞大，对分，鼻孔大，呈圆形。颊鳞单枚，近方形。有颊窝。眶前鳞 2 枚。眶上鳞单枚。眶后鳞 3（2）枚，最下一枚弯至眼下。上唇鳞 7 枚，偶尔 6 枚或 8 枚；第二枚最小，不入颊窝；第三枚最大，入眶。下唇鳞 9-12 枚。背鳞强棱，鳞式 21（23）-21-17（偶见 15）。肛鳞完整。尾下鳞对分，偶尔有几枚不对分。体背浅棕色，有 2 排大马蹄形深色斑，通常有一条纵走红色脊线。尾末段浅色，尾端黑色。头顶有 2 个大黑棕色侧斑，开始于眶上鳞和额鳞后缘，延伸至颈背。有一条白色线开始于上眶前鳞后缘，穿过眼睛延展至颈部。其下有一个暗棕色宽线开始于颊窝，穿过眼延伸至颈部。唇白色，有红色小点；腹面灰白色，密布棕色或深棕色斑点。

生境与习性　栖息于平原丘陵。常见于鼠洞、沟渠、路边和民居。卵胎生。

分布　广泛分布于中国，从四川、甘肃至浙江、从广东至吉林。延伸至朝鲜半岛。井冈山见于茨坪。

种群状况　常见种。

Identification　Solenoglyphous snakes. Body stout; tail short; head slightly triangular, flattened, distinct from neck; snout short, truncate; scales on the top of head large, symmetrical; occipital scales keeled; internasals narrow and long, in contact with each other; canthus rostralis distinct; nasal large, divided, with a large rounded nostril; loreal single, square-like; loreal pit present; preoculars 2; supraoculars single; postoculars 3 (2), the most lower one curved to below eye; supralabials 7, rarely 6 or 8; second smallest, no entering the loreal pit; third largest, entering the orbit; infralabials 9-12; dorsal scales strongly keeled, in 21 (23)-21-17 (rarely 15); anal entire; subcaudals paired, rarely several subcaudals undivided. Light brown above, with two rows of large horseshoe-shaped dark blotches, and usually with a longitudinal reddish vertebral line; tail end is lightly colored and tip black; top of head with two black-brown lateral blotches beginning from posterior edge of supraoculars and frontal and extended to nape of the neck; white stripe beginning from the posterior edge of upper-preoculars, through eye to the neck; broad, dark-brown stripe below it beginning from loreal pit, through eye, posteriorly extended to the neck; lips white with reddish specks; grey-white below, with densely arranged brown or dark-brown spots.

Habitat and Behavior　Inhabits plains and hills. More frequently found in mouse holes, ditches, roadsides and the vicinity of human settlements. Nocturnal and diurnal. Diet consists of rodents, rats, birds, lizards and frogs. Ovoviviparous.

Distribution　Broadly distributed in China from Sichuan and Gansu to Zhejiang, and from Guangdong to Jilin. Korean Peninsula. Recorded from Ciping in Mt. Jinggang.

Status of Population　Common species.

2.24a

2.24b

2.25 山烙铁头 *Ovophis monticola orientalis* (Günther, 1864)

识别特征　具管牙毒蛇。体粗壮。头三角形，宽且头顶平坦，与颈部区分显著。眼小。吻短而平钝，长于眼径 2 倍。头顶鳞小，不等大，稍覆瓦状排列，前部平滑，后部有弱棱。眶上鳞通常大而单枚，其间在一条直线上排布有 5-10 枚鳞。鼻间鳞大，彼此相接，偶被一枚小鳞分隔。鼻鳞较大，不完全对分。鼻孔大，圆形，位于鼻鳞中部。有颊窝。8-10 枚上唇鳞，第 1 枚与鼻鳞完全分离，第 2 枚有时和颊窝前鳞分离。第 4 枚最大。在唇和眼间有 2-4 行小鳞，眶下鳞通常碎裂成小鳞。下唇鳞 9-12 枚。背鳞平滑或有弱棱。中段鳞 21-25 行。肛鳞完整。尾下鳞对分。背面棕红色，体背有较大方形不规则排列的棕黑色斑或斑纹，体侧有小斑。头侧深棕色，有不规则白色花纹。通常有一条从眼至颌角的浅色纹。尾背深棕色，有一系列白斑点。腹面发白，点缀棕色形成网状花纹。

生境与习性　栖息于山地常绿阔叶林。夜行性陆栖蛇类，见于灌丛、草地，也常进入鸡舍或民居内寻找食物。食物包括啮齿动物、蝙蝠、鸟类及鸟卵、蜥蜴和蛙类。卵生。

分布　广泛分布从四川和云南，向东至台湾。越南北部。井冈山见于大坝里、荆竹山。

种群状况　常见种。

Identification Solenoglyphous snakes. Body stout; head triangular, broad and flattened, distinct from neck; eye small; snout short and truncate, more than twice as long as diameter of eye; scales on the top of head small, unequal, feebly imbricate, anterior ones smooth, posterior weakly keeled; supraoculars usually large and entire, 5-10 scales on a line between two supraocalars; internasals large, in contact with each other, rarely separated by single scale; nasal significantly larger, incompletely divided; nostril large, rounded, and on the middle of nasal; loreal pit present; 8-10 supralabials, the first completely separated from the nasal, second sometimes separated from the scale formed the anterior border of the loreal pit; fourth supralabials largest; 2-4 series of small scales between the eye and labials, suboculars being usually broken up into small scales; infralabials 9-12; dorsal scales smooth or feebly keeled, midbody scale rows 21-25; anal entire; subcaudals paired. Brown-red above, with large, squarish, irregularly placed black-brown blotches upon the back, and smaller ones upon the sides; sides of head dark-brown, with irregular white pattern; usually a light streak from the eye to the angle of the jaw; dorsal surface of tail dark-brown, with a series of white spots; whitish below, dusted brown and forming a reticular pattern.

Habitat and Behavior Inhabits montane and submontane evergreen forests. Nocturnal, terrestrial, associated with bushes and grassland. May enter chicken coops and human dwellings in search of food. Diet consists of rodents, rats, birds and their eggs, lizards and frogs. Oviparous.

Distribution Broadly distributed in China from Sichuan and Yunnan to Taiwan. N Vietnam. Recorded from Dabali and Jingzhushan in Mt. Jinggang.

Status of Population Common species

2.25a

2.25b

2.25c

2.26 原矛头蝮 *Protobothrops mucrosquamatus* (Cantor, 1839)

识别特征 具管牙类毒蛇。体长。头长，三角形，与颈部区分显著。吻延长，2-3 倍于眼直径。头顶被不等小鳞，后部鳞有钝棱。眶上鳞窄长，完整单枚，两鳞间直线相隔11-18枚鳞。鼻间鳞稍小，彼此被2-6枚小鳞分隔。鼻鳞单枚，对分。鼻孔小，圆形，在单枚鼻鳞上。有颊窝。上唇鳞9-10枚，偶见8枚、12枚或13枚。第1枚上唇鳞与鼻鳞分开。下唇鳞14-15枚，偶见12枚或16枚。背鳞起强棱，常见鳞式25 (25-29)-25-19。尾下鳞对分。肛鳞完整。背棕色或红棕色，有一系列大型不规则斑，该斑镶宽黑边和黄色窄边。两胁有小的深色斑。有眶后深色纹。头顶有"V"形背侧暗纹。腹面发白，散布浅棕色，有近方形斑块。

生境与习性 栖息于热带、亚热带常绿阔叶林。夜行性陆栖蛇类。食物包括蛙类、蜥蜴、蛇类、鸟类和啮齿动物，也会进入鸡舍或民居寻找食物。卵生。

分布 江西、安徽、江苏、浙江、福建、广东、海南、广西、云南、四川、甘肃、陕西、湖南、香港和台湾。缅甸、老挝、越南北部、印度东北部和孟加拉国。井冈山地区广泛分布。

种群状况 常见种。

2.26a

2.26b

Identification　Solenoglyphous snakes. Body slender; head triangular, elongated, and significantly distinct from neck; snout elongated, 2-3 times as long as eye diameter; scales on the top of head small, unequal, obtusely keeled on the posterior part; supraoculars long and narrow, entire, 11-18 scales on a line between supraoculars; internasals rather small, separated from one another by 2-6 small scales; nasal single, no divided; nostril in the single nasal, small, rounded; loreal pit present; suprlabials 9-10, rarely 8, 12 or 13, first supralabial separated from the nasal; infralabials 14-15, rarely 12 or 16; dorsal scales strongly keeled, mostly in 25 (25-29)-25-19; subcaudals paired; anal entire. Dorsal surface brown or reddish-brown, with a series of large irregular spots edged with wide black and narrow yellow; a series of small dark-brown spots on flanks; the presence of dark postocular streak; V-shaped dorsolateral dark streak on top of head; whitish below, with light brown speckles.

Habitat and Behavior　Inhabits tropical and subtropical evergreen forests. Nocturnal, terrestrial. Diet consists of frogs, skinks, snakes, birds and rodents. May enter chicken coops and human habitations in search of food. Oviparous.

Distribution　Jiangxi, Anhui, Jiangsu, Zhejiang, Fujian, Guangdong, Hainan, Guangxi, Yunnan, Sichuan, Gansu, Shaansi, Hunan, Hong Kong and Taiwan. Myanmar, Laos, N Vietnam, NE India, Bangladesh. Widely recorded from the Jinggangshan Region.

Status of Population　Common species.

2.27 竹叶青 *Viridovipera stejnegeri* (Schmidt, 1925)

识别特征　具管牙毒蛇。体适中，稍平扁。头三角形，显著宽于颈部。吻鳞三角形，背视稍可见。头顶被大小不等小鳞，光滑，稍覆瓦状排列。眶上鳞每侧单枚，窄长，两枚之间直线相距 9-15 枚鳞。鼻鳞单枚，不对分；鼻间鳞比其相邻的鳞稍大，与另外一枚相隔 1-4 枚鳞。有颊窝。上唇鳞 8-12 枚，第 1 枚与鼻鳞分离；下唇鳞 10-14 枚。背鳞起棱，最外侧 3 行平滑，通常鳞式 21 (23)-21-15。尾下鳞对分。肛鳞单枚完整。头体背亮绿色，尾背棕红色，腹面浅绿色。腹侧线在雄性下半部分橘红色或红色，上半部分白色；雌性白色，有时其下有模糊红色。虹膜橘红色。

生境与习性　栖息于山地森林近溪流环境。夜行性地栖和树栖蛇类，常见于矮树和灌丛，也常见于林下地面。以啮齿类、鸟类、蜥蜴、蛙类为食。卵胎生。

分布　江西、安徽、江苏、浙江、福建、广东、海南、广西、云南、贵州、重庆、四川、湖南、吉林（长白山）、香港和台湾。老挝、越南北部、泰国和缅甸。井冈山地区广泛分布。

种群状况　常见种。

Identification　Solenoglyphous snakes. Body moderate, slightly compressed; head triangular, significantly broader than neck; rostral triangular, barely visible from above; scales on the top of head small, unequal, subimbricate, smooth; supraoculars long, narrow, entire; nasal single, undivided; 9-15 scales between midsection of supraoculars in a transverse line; internasal slightly larger than adjacent scales, separated from one another by 1-4 scales; loreal pit present; supralabials 8-12, first supralabial separated from nasal; infralabials 10-14; dorsal scales keeled, outer three rows smooth, mostly arranged in 21(23)-21-15 rows; subcaudals paired; anal entire. Dorsal surface of body and head bright green; dorsal surface of tail brown-red; pale green below; ventrolateral stripe orange or red ventrally, white dorsally in males; white in females, sometimes with faint red marks below; iris orange-red.

Habitat and Behavior　Inhabits montane forests near streams. Nocturnal, terrestrial and arboreal, associated with bushes and low trees. More frequently found on the forest floor. Diet consists of rodents, birds, lizards and frogs. Ovoviviparous.

Distribution　Jiangxi, Anhui, Jiangsu, Zhejiang, Fujian, Guangdong, Hainan, Guangxi, Yunnan, Guizhou, Chongqing, Sichuan, Hunan and Jilin (Changbai Mountain), Hong Kong and Taiwan. Laos, N Vietnam, Thailand, Myanmar. Widely recorded from the Jinggangshan Region.

Status of Population　Common species.

2.27a

2.27b

蛇亚目 SERPENTES
水蛇科 Homalopsidae

2.28 中国水蛇 *Myrrophis chinensis* (Gray, 1842)

识别特征   具后沟牙毒蛇。体粗壮。头颈区分稍显。吻鳞宽钝，背视可见。鼻孔背侧位。眼小，瞳孔小，圆形。尾短。左右鼻鳞相接。鼻间鳞小，单枚，不与颊鳞相接。颊鳞单枚。眶前鳞单枚。眶后鳞 2 枚，少数 1 枚。上唇鳞 7 枚，仅第 4 枚入眶。背鳞平滑，通常 23-23-21 行。腹鳞 138-154 枚。肛鳞对分。尾下鳞对分，40-51 对。体背灰棕色，散布有黑斑，有时在颈背形成黑线。上唇鳞下部、下唇鳞黄白色。背鳞外侧 2-3 行棕红色。每一腹鳞前部分暗灰色，后部分黄白色。

生境与习性   栖息于平原、丘陵、山脚的溪流、池塘和水田。夜行性水生蛇类。以小型鱼类和蝌蚪为食。卵胎生。

分布   江西、安徽、江苏、浙江、福建、广东、海南、广西、湖南、湖北、香港和台湾。越南北部。井冈山见于下井、西坪等地。

种群状况   常见种。

Identification   Opisthoglyphous snakes. Body robust; head slightly distinct from neck; rostral broad and blunt, rounded, visible from above; nostril dorsolateral; eye small, pupil small and rounded; tail short; left nasal in contact with right one; internasal single, not in contact with loreal; loreal single; preocular single; postoculars 2, rarely single; supralabials 7, only fourth entering the eye; dorsal scales smooth, mostly in 23-23-21 rows; ventrals 138-154; anal divided; subcaudals 40-51, paired. Grey-brown above with scattered black spots, sometimes formed a vertebral line on the nape; low part of supralabials and infralabials yellowish-white; outer 2-3 dorsal scale rows reddish-brown; anterior part of each ventral dark-grey, posterior part yellow-white.

Habitat and Behavior   Inhabits streams, ponds and water fields in plains, hills and submontane regions. Nocturnal and aquatic. Diet consists of small fish, tadpoles. Ovoviviparous.

Distribution   Jiangxi, Anhui, Jiangsu, Zhejiang, Fujian, Guangdong, Hainan, Guangxi, Hunan, Hubei, Hong Kong and Taiwan. N Vietnam. Recorded from Xiajing and Xiping in Mt. Jinggang.

Status of Population   Common species.

2.29 铅色水蛇 *Hypsiscopus plumbea* (Bole, 1827)

识别特征   具后沟牙毒蛇。小型蛇类，通常全长小于 450 mm。头颈区分不显。尾短。吻鳞宽圆，背视稍可见。眼中等大，瞳孔圆形。鼻孔位于鼻鳞内，有鼻瓣，位于吻的背侧面。鼻间鳞单枚，不与吻鳞相接。左右 2 鼻鳞相接。上唇鳞 8 枚，第 4、第 5 枚入眶。背鳞光滑，19-19-17 行。腹鳞 124-132 行。肛鳞对分。尾下鳞对分，31-42 对。体背橄榄色或绿棕色，通常有暗色鳞缘。外侧 2-3 行黄色。上唇鳞下部分和下唇鳞及腹面微黄色或白色，尾腹面中央有黑色线或系列黑色斑。

生境与习性   栖息于平原、丘陵和低山区域的水库、池塘等静水水域。夜行性水生动物。以鱼类、蛙类和蝌蚪为食。卵胎生。

分布   江西、江苏、浙江、福建、广东、海南、广西、云南、

2.28a

2.28b

2.29a

2.29b

香港和台湾。缅甸、泰国、越南、马来西亚和印度尼西亚。井
冈山见于下井等地。

种群状况　常见种。

Identification　Opisthoglyphous snakes. Body small, total length
usually less than 450 mm; head indistinct from neck; tail short;
rostral broadly rounded, barely visible from above; eye moderate,
pupil rounded; internasal single, not touching the rostral; nostril
with valves, dorsolaterally placed in nasal; two nasals touching each
other; supralabials 8, usually 4th and 5th touching the eye; dorsal
scales smooth, in 19-19-17 rows; ventrals 124-132; anal divided;
subcaudals 31-42, paired. Olive or green-brown above, usually with
dark scale margins; outer 2-3 scale rows yellow; infralabials, lower
part of supralabials and below body yellowish or whitish; ventral
surface of tail with a median black line or series of dots.

Habitat and Behavior　Inhabits reservoirs, ponds and lentic
fields in plains, hills and submontane regions. Nocturnal and aquatic.
Diet consists of fish, frogs and tadpoles. Ovoviviparous.

Distribution　Jiangxi, Jiangsu, Zhejiang, Fujian, Guangdong,
Hainan, Guangxi, Yunnan, Hong Kong and Taiwan. Myanmar,
Thailand, Vietnam, Malaysia, Indonesia. Recorded from Xiajing in
Mt. Jinggang.

Status of Population　Common species.

蛇亚目 SERPENTES
鳗形蛇科 Lamprophiidae

2.30 紫砂蛇 *Psammodynastes pulverulentus* (Boie, 1827)

识别特征　具后沟牙毒蛇。中等体型，略粗壮。吻短，侧视截
形。头短而高，顶平，与颈部区分显著。唇肿胀。眼中等大，
瞳孔直立椭圆形。鼻孔在单一鼻鳞上。吻鳞宽稍大于高。鼻间
鳞比前额鳞显著小。颊鳞不入眶，单枚，宽高几相等，有时横
裂为二，少数缺失。眶前鳞单枚，少数 2 枚。眶后鳞 2 枚，少
数 3 枚。上唇鳞通常 8 枚，下唇鳞通常 8 枚或 7 枚。背鳞光滑，
17-17-15（偶有 13）排列。肛鳞完整。尾下鳞对分，45-69 对。
体色多变。上体深棕色、黑色或微红、微黄色，头顶有几条镶
浅色边的深色纵纹，身体和尾有镶黑边的浅色色斑。腹面微黄色，
有深棕色纵纹。

生境与习性　栖息于平原、低地和山地森林。常见于潮湿的森
林地面，植物茎或大叶片上或溪流附近石缝中。以青蛙、蜥蜴、
蛇为食。卵胎生。

分布　江西、福建、广东、海南、广西、云南、贵州、湖南、西藏、
香港和台湾。广泛分布于东南亚和南亚国家。井冈山见于湘洲。

种群状况　常见种。

Identification　Opisthoglyphous snakes. Body moderate, slightly
stout; snout short, truncated in profile; head short, high, crown flat,
distinct from neck; lip swollen; eye moderate, pupil vertical ellipse;
nostril in single nasal; rostral a little broader than deep; internasals
much smaller than prefrontals; loreal single, about as long as
deep, not touching the eye, sometimes transversely divided, rarely

lacking; preocular single, rarely two; postoculars two, rarely three;
supralabials usually eight; infralabials eight or seven; dorsal scales
smooth, in 17-17-15 (rarely 13); anal entire; ventrals 141-169;
subcaudals 45-69, paired. Color variable; dark-brown or blackish,
reddish, yellowish above, with several longitudinal dark streaks
edged light on top of head and light marks edged black on body
and tail; yellowish below, with longitudinal dark-brown streaks.

Habitat and Behavior　Inhabits plains and lowland forests to submontane forests. More frequently found in dark, damp forest floor, plant stems and large leaves or in stones near streams. Diet consists of frogs, skinks and small snakes. Ovoviviparous.

Distribution　Jiangxi, Fujian, Guangdong, Hainan, Guangxi, Yunnan, Guizhou, Hunan, Xizang, Hong Kong and Taiwan. Widely distributed in S and SE Asia. Recorded from Xiangzhou in Mt. Jinggang.

Status of Population　Common species.

蛇亚目 SERPENTES
眼镜蛇科 Elapidae

2.31 银环蛇 *Bungarus multicinctus* Blyth, 1861

识别特征　前沟牙毒蛇。体细长。头卵圆形，与颈部区分不显。眼小，圆形。无颊鳞。眶前鳞单枚。眶后鳞 2 枚，少数单枚。上唇鳞 7 枚，少数 6 枚或 8 枚；下唇鳞 7 枚，少数 6 枚或 8 枚。背鳞光滑，通体 15 行。肛鳞完整。尾下鳞完整，少数前 3 枚对分。上体黑色，有窄横向白斑带。腹面黄白色或灰白色，有散布黑斑。

生境与习性　栖息于平原丘陵，常见于近水区域。食物包括鱼类、蛙类、蜥蜴类、蛇类和小型哺乳动物。夜行性蛇类。卵生。

分布　江西、安徽、浙江、江苏、福建、广东、香港、海南、广西、云南、四川、贵州、湖南、湖北和台湾。越南、老挝和缅甸。井冈山地区广泛分布。

种群状况　常见种。

Identification　Proteroglyphous snakes. Body slender; head elliptic, indistinct from neck; eye small, rounded; loreal absent; preocular single, postocular 2, rarely single; supralabials 7, rarely 6 or 8; infralabials 7, rarely 6 or 8; dorsal scales smooth, in 15 rows throughout; anal entire; subcaudals entire, rarely first three paired. Black above, with narrow transverse white bands; yellowish-white or grey-white below, with scattered black spots.

Habitat and Behavior　Inhabits plains and hills. More frequently found in the vicinity of water. Diet consists of fish, frogs, lizards, snakes and small mammals. Nocturnal. Oviparous.

Distribution　Jiangxi, Anhui, Zhejiang, Jiangsu, Fujian, Guangdong, Hong Kong, Hainan, Guangxi, Yunnan, Sichuan, Guizhou, Hunan, Hubei and Taiwan. Myanmar, Vietnam, Laos. Widely recorded from the Jinggangshan Region.

Status of Population　Common species.

2.32 中国丽纹蛇 *Sinomicrurus macclellandi* (Reinhardt, 1844)

识别特征　前沟牙毒蛇。体细长，圆筒形。头短，与颈部区别不显著。鼻鳞单枚，鼻孔位于鼻鳞中央。没有颊鳞。眶前鳞单枚；眶后鳞 2 枚。上唇鳞 7 枚，第 3 和第 4 枚入眶。下唇鳞 6 枚，有时 5 枚或 7 枚。背鳞光滑，通体 13 行。肛鳞对分。尾下鳞对分。体背棕红色，有镶黄色边的横向黑带纹。头背黑色，有 2 个横带，一个黄白色，在吻端，穿过鼻间鳞延展至鼻鳞和第一上唇鳞及第 2 上唇鳞前缘；另一个在眼后，乳白色。腹面黄白色，有黑

2.31a

2.31b

2.32a

2.32b

带或近方形黑斑。

生境与习性　栖息于低地和中海拔丘陵亚热带常绿阔叶林。夜行性陆栖掘洞蛇类，躲藏在松软泥土中或植被里。食物包括其他蛇、蜥蜴，卵生。

分布　江西、安徽、江苏、浙江、福建、广东、香港、海南、广西、云南、贵州、湖南、四川、甘肃、西藏和台湾。缅甸北部、越南北部、泰国北部、老挝、印度、尼泊尔和孟加拉国。井冈山见于茨坪。

种群状况　常见种。

Identification　Proteroglyphous snakes. Body slender, cylindrical; head short, indistinct from neck; nasal single; nostril on the middle of nasal; loreal absent; preocular single; postoculars 2; supralabials 7, third and fourth entering the eye; infralabials 6, sometimes 5 or 7; dorsal scales smooth, in 13 rows throughout; anal divided; subcaudals paired. Reddish-brown above, with transverse black bands bordered with yellow edges; dorsal surface of head black, with two transverse bands, one pale yellowish on tip of snout, through internasals and rostral, extended to nasals, first supralabians and anterior edge of second supralabial; another cream behind eye; yellowish-cream below, with black bands or square-like marks.

Habitat and Behavior　Inhabits lowlands and mid-hills in subtropical evergreen forests. Nocturnal, terrestrial and subfossorial, sheltering under loose soil or in vegetation. Diet consists of other snakes and lizards. Oviparous.

Distribution　Jiangxi, Anhui, Jiangsu, Zhejiang, Fujian, Guangdong, Hong Kong, Hainan, Guangxi, Yunnan, Guizhou, Hunan, Sichuan, Gansu, Xizang and Taiwan. N Myanmar, N Vietnam, N Thailand, Laos, India, Nepal, Bangladesh. Found from Ciping in Mt. Jinggang.

Status of Population　Common species.

2.33 舟山眼镜蛇 *Naja atra* Cantor, 1842

识别特征　前沟牙毒蛇。体长可达 2000 mm。头稍清晰区别于颈部。眼中等，没有颊鳞。眶前鳞单枚，通常与鼻间鳞相接。眶后鳞 3 枚，少数 2 枚。上唇鳞 7 枚，第 3 枚最高，入眶，第 4 枚也入眶。下唇鳞 8-9 枚，少数 7 枚或 10 枚。第 4 和第 5 枚大，通常有 1 枚三角形小鳞在这二鳞之间的唇边缘上。背鳞光滑斜列，中段鳞 19 枚或 21 枚。肛鳞完整，少数分分。尾下鳞对分。生活时体色变异较大。通常上背浅灰色、黄褐色、灰黑色至亮黑色，有或没有成对的浅黄色成对不等距离的横斑，在幼体该斑显著。头侧浅色。颈有眼镜状斑纹扩展至浅色喉部。喉部通常有一对黑斑。腹面白色至灰色、深灰色杂以白色或黑色斑。

生境与习性　栖息于森林、灌丛、草地、红树林、开阔地，甚至人口稠密地区。日行性和夜行性蛇类。食物包括鱼类、蛙类、蜥蜴、蛇类、鸟类和鸟卵、小型哺乳动物。卵生。

分布　四川、福建、广东、广西、贵州、湖南、湖北、浙江、海南、台湾和香港。老挝北部、越南北部。

种群状况　IUCN 红色名录：易危。

Identification　Proteroglyphous snakes. Body elongated with total length up to 2000 mm; head slightly distinct from neck; eye moderate, loreal absent; preocular single, usually in contact with the internasal; postoculars 3, rarely 2; supralabials 7, third highest, entering the eye; fourth also entering the eye; infralabials 8-9, rarely 7 or 10; fourth and fifth largest, usually with a small triangular scale between them on the oral margin; dorsal scales smooth and glossy, obliquely arranged; midbody scale rows 19 or 21, rarely 20; anal usually entire, rarely divided; subcaudals paired. More color variation in life, usually light grey, tan, grey-black to iridescent black above, with or without narrow transversal double pale yellowish bands at irregular intervals which are especially prominent in juveniles; sides of head are lightly colored; the hood is a spectacle-like mark on the expanded neck and linked to light throat area; the throat area usually with a pair of black lateral spots; white to grey, dark-grey mottled with white, or blackish below.

Habitat and Behavior Inhabits forests, shrubs, grasslands, mangroves, open fields and heavily populated regions. Diurnal and nocturnal. Diet consists of fish, frogs, skinks, snakes, birds and their eggs, and small mammals. Oviparous.

Distribution Sichuan, Fujian, Guangdong, Guangxi, Guizhou, Hunan, Hubei, Zhejiang, Hainan, Taiwan and Hong Kong. N Laos, N Vietnam.

Status of Population IUCN Red List: VU.

2.34a

蛇亚目 SERPENTES
游蛇科 Colubridae

2.34 锈链腹链蛇 *Hebius craspedogaster* (Boulenger, 1899)

识别特征 身体细长；头颈区分显著。鼻鳞对分。颊鳞单枚，少数2枚。眶前鳞1枚，眶后鳞3枚，少数4枚。上唇鳞8枚（少数7枚或9枚），2-3-3或3-2-3排列。下唇鳞通常10枚。眼大，瞳孔圆形。背鳞19-19-17，起强棱，少数2外侧行平滑。肛鳞对分；尾下鳞对分。头背锈红色，向后直到尾尖为暗棕色，散布锈红色斑；体侧各有一条锈红色带有黄色点斑的纵线。唇白色或黄色，鳞缝有黑斑。从最后一枚上唇鳞至颈侧有一镶黑边的白色或黄色斜斑。腹面白色或黄色，腹鳞和尾下鳞外侧边缘有一长黑斑，在腹面形成一对链状黑线。

生境与习性 生活在山区森林，常见于近水区域，或路边、草丛。白天活动，以蛙、蟾蜍、蝌蚪为食。可能卵生。

分布 安徽、重庆、福建、甘肃、广东、广西、贵州、河南、湖北、湖南、江苏、陕西、山西、四川和浙江。越南北部。井冈山见于茨坪、下井、荆竹山和八面山等地。

种群状况 常见种。

2.34b

Identification Body slendar; head distinct from neck; nasal divided; loreal single, rarely 2; preocular single, postoculars 3, rarely 4; supralabials 8 (rarely 7 or 9), 2-3-3 or 3-2-3; infralabials usually 10; eye large, pupil rounded; dorsal scale rows 19-19-17, strongly keeled, rarely two outer rows smooth; anal divided; subcaudals, paired. Dorsal surface of head rusty-red, posteriorly dark-brown until tip of tail, scattered rusty-red spots; two rusty-red longitudinal stripes with yellow spots along flanks; labials white or yellow with black bars on sutures; short oblique white or yellow streak with black-edge from last supralabial to the side of the neck; ventral surface white or yellow with elongated black spot at outer extremities of ventrals, forming longitudinal line on each side of venter and tail.

Habitat and Behavior Inhabits montane forests, generally in the vicinity of water bodies, or roads or in thickets. Diet consists of frogs and tadpoles. Diurnal. May be oviparous.

Distribution Anhui, Chongqing, Fujian, Gansu, Guangdong, Guangxi, Guizhou, Henan, Hubei, Hunan, Jingsu, Shanxi, Shanxi,

2.34c

Sichuan, Zhejiang. N Vietnam. Recorded from Ciping, Xiajing, Jingzhushan and Bamianshan in Mt. Jinggang.

Status of Population Common species.

2.35 草腹链蛇 *Amphiesma stolatum* (Linnaeus, 1758)

识别特征 体相对粗壮。头颈区分明显。鼻鳞对分。鼻间鳞较短。颊鳞单枚，小，方形。眶前鳞单枚，眶后鳞3枚，较少2枚或4枚。上唇鳞8枚（较少9枚），2-3-3 或 3-2-3 排列。下唇鳞10枚。眼大，瞳孔圆形。背鳞式 19-19-17，起强棱，两外侧行鳞平滑。肛鳞对分。尾下鳞对分。微棕色或橄榄褐色，或多或少有清晰黑斑或黑色网纹斑，与两条黄色纵带相交，后部更加清晰。眶前鳞和眶后鳞黄色。腹面白色，通常在每个腹鳞两端有一个黑斑点。

生境与习性 栖息于平原、丘陵和山地。

分布 华南和华东地区。南亚和东南亚。井冈山见于湘洲等地。

种群状况 常见种。

Identification Body relatively robust; head distinct from neck; nasal divided; internasals short; loreal single, small and square; preocular single, postoculars 3, rarely 2 or 4; supralabials 8 (rarely 9), 2-3-3 or 3-2-3; infralabials 10; eye large, pupil rounded; dorsal scale rows 19-19-17, strongly keeled, two outer rows smooth; anal divided; subcaudals paired. Greenish or brownish-olive, with more or less distinct black spots or reticulated cross-bars intersected by two yellow longitudinal stripes, which are more marked posteriorly; preoculars and postoculars yellowish; lower surfaces white, usually with a black spot at outer extremities of each ventral scale.

Habitat and Behavior Inhabits glasslands, hills and mountains.

Distribution S and E China. S and SE Asia. Recorded from Xiangzhou in Mt. Jinggang.

Status of Population Common species.

2.36 绞花林蛇 *Boiga kraepelini* Stejneger, 1902

识别特征 具后沟牙毒蛇。头大，颈部较细，头颈区分明显。体纤长，侧扁。尾细长。眼大，瞳孔直立。无颊窝。颞部鳞小，不成列。背鳞平滑，鳞式 23（21-25）-21（19-23）-17（15-17）；脊鳞稍扩大；尾下鳞对分；肛鳞多对分，少数完整不对分。通体背面棕红色，头背有不显的深色的"V"形斑纹；体尾有粗大镶黄边的深棕色横斑和体侧的暗色斑。腹面白色，密布棕褐色或浅紫色斑点。

生境与习性 栖息于山区和丘陵，常见于溪边灌木上，亦见于溪流中岩石上。多夜间活动，以鸟类、鸟卵、爬行动物为食。卵生。

分布 江西、浙江、安徽、福建、湖南、广东、海南、广西、贵州、四川和台湾。越南北部。广泛分布于井冈山地区。

种群状况 常见种。

Identification Opisthoglyphous snakes. Head large and distinct from neck; neck narrow; body slender, laterally compressed; tail long and slender; eye large, pupil vertical; temporal scales small, not arranged in rows; dorsal scales smooth, arranged in 23 (21-25) - 21 (19-23) - 17 (15-17) rows; vertebral scale row slightly enlarged;

subcaudals paired; anal divided, rarely entire. Dorsum brownish-red with yellowish-edge dark cross-bars and lateral fait spots; dorsum of head with indistinct V-shaped dark streak; lower surfaces pale, densely with brown or pale purple spots.

Habitat and Behavior Inhabits mountains and hills. More frequently found in riparian shrubs and on rocks in streams. Nocturnal and arboreal, associated with trees. Diet consists of birds and their eggs, and reptiles. Oviparous.

Distribution Taiwan, Jiangxi, Zhejiang, Anhui, Fujian, Hunan, Guangdong, Hainan, Guangxi, Guizhou and Sichuan. N Vietnam. Widely recorded from the Jinggangshan Region.

Status of Population Common species.

2.36c

2.37 尖尾两头蛇 *Calamaria pavimentata* Duméril, Bibron and Duméril, 1854

识别特征 身体纤细，圆柱形。头短，与颈部区分不显著。吻鳞宽显著大于高，背视可见部分长度是前额鳞缝的 1/2-2/3。尾短，末端尖。没有颊鳞和鼻间鳞。前额鳞前与吻鳞相接，侧与前 2 枚上唇鳞相接。额鳞长大于宽。眶上鳞、眶前鳞和眶后鳞均为单枚。上唇鳞 4 枚，第 2、第 3 枚入眶。下唇鳞 5 或 6 枚，少数 7 枚。眼小，瞳孔圆形。上颌齿 8-11 枚。背鳞光滑，通体 13 行；肛鳞单枚；尾下鳞对分。背红棕色，有暗色纵线纹。颈侧有黑色围领，其前后有黄斑；有 2 个黄斑在尾基部，2 个黄斑接近尾尖。腹面黄色，腹部和尾腹面有一条暗色中线。

生境与习性 栖息于低山林地。食物有蚯蚓、昆虫幼虫等。卵生。

分布 江西、浙江、福建、广东、海南、广西、贵州、四川和台湾。国外见于越南北部、印度（阿萨姆邦）、马来半岛和爪哇（印尼）。井冈山见于香洲、茨坪、龙潭和水口等地。

种群状况 常见种，但非常隐蔽。

2.37a

Identification Body slender, cylindrical; head short, indistinct from neck; rostral much broader than high, its portion visible from above 1/2-2/3 as long as the interprefrontal suture; tail short, gradually tapering to a point; loreal and internasal absent; prefrontals anteriorly in contact with rostral, laterally in contact with first two supralabials; frontal longer than broad; supraocular, preocular and postocular single; supralabials 4, 2nd and 3rd touching the eye; infralabials 5 or 6, rarely 7; eye small, pupil rounded; maxillary teeth 8-11; dorsal scales smooth in 13 rows throughout; anal entire; subcaudals paired. Reddish-brown above, with narrow longitudinal dark lines; solid black collar on the nape, anteriorly and posteriorly with yellow markings; two yellow spots at the base of tail and two near the tip of the tail; uniform yellow beneath, with a dark median line along the belly and tail.

Habitat and Behavior Inhabits hilly forests. Diet consists of earthworm and worms. Oviparous.

Distribution Jiangxi, Zhejiang, Fujian, Guangdong, Hainan, Guangxi, Guizhou, Sichuan and Taiwan. N Vietnam, India (Assam), Malay Peninsula, Java. Recorded from Xiangzhou, Cipng, Longtan and Shuikou in Mt. Jinggang.

Status of Population Very cryptic common species.

2.37b

2.38 钝尾两头蛇 *Calamaria septentrionalis* Boulenger, 1890

识别特征　外形很像尖尾两头蛇，与其区别在于吻鳞背视刚刚可见；吻短而宽圆。尾短钝，与身体等粗。尾下鳞对分。背面黑棕色或蓝棕色，鳞有许多小白点，形成网状。有一个背中部断开的黄色围领。尾基部有 2 个黄斑。腹面珊瑚红色，尾正中有一条黑线。

生境与习性　栖息于低地林中。吃蚯蚓或昆虫幼虫。卵生。

分布　江西、安徽、江苏、浙江、福建、广东、海南、广西、贵州、四川、湖南、湖北、河南、香港和台湾。越南北部。井冈山见于罗浮、茨坪、荆竹山、龙潭和水口等地。

种群状况　常见种，但很隐蔽。

Identification　Most similar to *C. pavimentata*, from which it differs by rostral only just visible from above; snout shorter and more broadly rounded than latter; tail short and blunt, as thick as body; frontal as broad as long; anal entire; subcaudals paired. Blackish-brown or bluish-brown above, scales with numerous small pale dots, forming networks; a yellow nuchal collar interrupted on the middle; two yellow spots at the base of tail; uniform coral-red beneath, with a black line along the middle of the tail.

Habitat and Behavior　Inhabits lowland forests. Diet consists of earthworm and worms. Oviparous.

Distribution　Jiangxi, Anhui, Jiangsu, Zhejiang, Fujian, Guangdong, Hainan, Guangxi, Guizhou, Sichuan, Hunan, Hubei, Henan, Hong Kong and Taiwan. N Vietnam. Occurs at Luofu, Ciping, Jingzhushan, Longtan and Shuikou in Mt. Jinggang.

Status of Population　Very cryptic common species.

2.39 翠青蛇 *Cyclophiops major* (Günther, 1858)

识别特征　身体适度粗壮。头颈区分显著。眼大，瞳孔圆形。尾适度长。上唇鳞 8 枚，3-2-3 排列，个别 7 枚或 9 枚。下唇鳞 6 枚。颊鳞单枚。眶前鳞单枚，眶后鳞 2 枚。颞鳞 1+2。背鳞光滑，通体 15 行，雄性荐背鳞起弱棱。肛鳞对分。尾下鳞对分。背亮绿色，幼体有黑色斑点。腹面和上唇鳞下部、下唇鳞浅黄绿色，或乳黄色。

生境与习性　栖息于丘陵和山地森林。日行性陆栖动物，有时在夜间活动。食物包括蚯蚓和昆虫幼虫。卵生。

分布　甘肃、陕西、河南、四川、重庆、湖北、湖南、江西、江苏、浙江、安徽、上海、浙江、福建、广东、海南、广西、贵州、台湾和香港。越南北部。广泛分布于井冈山地区。

种群状况　常见种。

Identification　Body moderately robust; head distinct from neck; eye large, pupil rounded; tail moderately long; supralabials 8, 3-2-3, rarely 7 or 9; infralabials 6; loreal single; preocular single; postoculars 2; temporals 1+2; dorsal scales smooth in 15 rows throughout, sacral dorsum covered weakly keeled scales in males; anal divided; subcaudals paired. Bright green above, scattered black spots in juvenile; greenish-yellow or cream beneath and low portion of supralabials, infralabials.

Habitat and Behavior   Inhabits the hills and montane forests. Diurnal and sometimes active at night, and terrestrial. Diet consists of earthworm and worms. Oviparous.

Distribution   Gansu, Shaanxi, Henan, Sichuan, Chongqing, Hubei, Hunan, Jiangxi, Jiangsu, Zhejiang, Anhui, Shanghai, Zhejiang, Fujian, Guangdong, Hainan, Guangxi, Guizhou, Taiwan and Hong Kong. N Vietnam. Widely recorded from the Jinggangshan Region.

Status of Population   Common species.

2.39c

2.40 黄链蛇 *Lycodon flavozonatum* (Pope, 1928)

识别特征   体适度延长，稍平扁。头短，与颈部区分不甚明显。眼中等大，瞳孔竖椭圆形。鼻间鳞显著短于前额鳞。颊鳞单枚，不入眶。眶前鳞1枚或2枚。眶后鳞2枚。颞鳞2+3（2）。上唇鳞8枚，2-3-3，少数7枚。下唇鳞10枚。背鳞17-17-15行排列，中间几行起弱棱。腹鳞有清晰的侧棱。肛鳞单枚完整。尾下鳞对分。上颌齿12-13枚，最后3枚扩大。背面黑色，身体部分有50-96条黄色横斑带，尾部有13-28条，这些斑带在体侧分叉，并有黑色斑点。腹面白色，有黑斑。尾下鳞暗色。头黑色有黄色斑纹，其中之一宽斜带从眼至嘴角，另一个"V"形始于顶鳞后缘。

生境与习性   栖息于山区森林，常见于溪流、水沟和草丛附近。夜间活动。以蜥蜴、小蛇为食。

分布   江西、安徽、福建、广东、海南、广西、云南、贵州和湖南。缅甸、越南。井冈山地区广泛分布。

种群状况   常见种。

2.40a

Identification   Body moderately elongated, slightly compressed; head short, indistinct from the neck; eye moderate, pupil vertical elliptical; internasals distinctly shorter than prefrontals; loreal single, not entering the eye; preoculars single or 2; postoculars 2; temporals 2+3(2); supralabials 8, 2-3-3, rarely 7; infralabials 10; dorsal scales in 17-17-15 rows, a few mid-dorsal scale rows feebly keeled, rest smooth; ventrals with a distinct lateral keel; anal entire; subcaudals paired; maxillary teeth 12-13, the posterior-most three enlarged. Black above, with 50-96 narrow yellow cross-bands on body and 13-28 on tail, that bifurcate on the sides, enclosing dark spots; white below with black spots; subcaudals dark; head black with yellow stripes, one of which wide oblique band from the eye to the angle of the mouth, another V-shaped marking starting from the hinder margin of the parietals.

2.40b

Habitat and Behavior   Inhabits montane forests. More frequently found in riparian shrubs. Nocturnal. Diet consists of skinks and snakes.

Distribution   Jiangxi, Anhui, Fujian, Guangdong, Hainan, Guangxi, Yunnan, Guizhou and Hunan. Myanmar, Vietnam. Widely recorded from the Jinggangshan Region.

Status of Population   Common species.

2.40c

2.41 赤链蛇 *Lycodon rufozonatum* (Cantor, 1842)

识别特征   体长。身体稍平扁。头和颈部区分不显著。眼中等大，瞳孔直立椭圆形。鼻间鳞显著短于前额鳞。颊鳞单枚，窄长入眶。眶前鳞单枚，少数2枚；眶后鳞2枚。上唇鳞8枚（2-3-3，少

2.41a

数 3-2-3）。下唇鳞 8-10 枚。背鳞除中间几行有弱棱外均平滑，中部背鳞 17 行或 19 行。肛鳞单枚，完整。尾下鳞对分。色彩模式与黄链蛇很像，只是黄色被珊瑚红色所替换。

生境与习性 栖息于平原、丘陵、山区的田野和村舍附近。夜行性。以鱼类、蛙类、蜥蜴、蛇、鸟类和鼠类为食。卵生。

分布 在中国除宁夏、新疆、西藏、青海外所有各省。国外见于远东地区（俄罗斯）、朝鲜半岛、日本和越南。广泛分布于井冈山地区。

种群状况 常见种。

Identification Body elongated body slightly compressed; head indistinct from the neck; eye moderate, pupil vertical elliptical; internasals distinct shortor than prefrontals; loreal single, narrow and elongated, entering the eye; preocular single, rarely 2; postoculars 2; supralabials 8 (2-3-3, rarely 3-2-3), or 7 (2-2-3); infralabials 8-10; dorsal scales smooth, except a few mid-dorsal rows weakly keeled, in 17 or 19 rows at midbody; anal entire; subcaudals paired; maxillary teeth 12-13, the posterior-most three enlarged. It is similar to *L. flavozonatum* in color pattern, but yellow is replaced by coral-red.

Habitat and Behavior Inhabits plains, hills, montane regions, agricultural fields and vicinity of cottages. Nocturnal. Diet consists of fish, skinks, snakes, birds and rodents. Oviparous.

Distribution All provinces in China except Ningxia, Xinjiang, Xizang and Qinghai. Far East (Russia), Korean Peninsula, Japan, Vietnam. Widely recorded from the Jinggangshan Region.

Status of Population Common species.

2.42 黑背白环蛇 *Lycodon ruhstrati* (Fischer, 1886)

识别特征 身体细长。头颈区分不显。吻鳞小，三角形，背视刚刚可见。鼻间鳞短于前额鳞，不与颊鳞相接。额鳞长度勉强接近顶鳞的 1/2。颊鳞单枚，长是高的 2 倍，入眶或不入眶。眶前鳞单枚，不与额鳞相接。眶后鳞 2 枚。上唇鳞 8 枚，少数 9 枚，通常第 3 至第 5 枚入眶。下唇鳞 9 或 10 枚。背鳞 17-17-15 行，光滑，少数个体中间几行有弱棱。肛鳞单枚。尾下鳞对分。背黑色，有灰白色或白色斑带。头前部分黑色，后部分灰白色。上下唇白色。腹面通常乳白色，在有些个体黑色。

生境与习性 栖息于山区林地。有日行也有夜行个体，树栖和陆栖。食物包括石龙子和草蜥等蜥蜴类动物。卵生。

分布 江西、福建、广东、广西、贵州、云南、四川和湖南。越南、缅甸、老挝、柬埔寨、泰国、斯里兰卡、尼泊尔、印度、菲律宾和印度尼西亚。广泛分布于井冈山地区。

种群状况 常见种。

Identification Body slender and elongated; head indistinct from neck; rostral small, triangular, just visible from above; internasal shorter than prefrontal, not touching loreal; frontal hardly half as long as the parietals; loreal single, twice as long as high, touching or not the eye; preocular single, not touching frontal; postoculars 2; supralabials 8, rarely 9, usually third to fifth contact orbit; infralabials 9 or 10; dorsal scales in 17-17-15 rows, smooth, rarely a few mid-dorsal rows feebly keeled; anal entire; subcaudals paired. Black above, with pale grey or white crossbars; anterior part of

head black, posterior part pale grey or white; supra- and infralabials white; typically cream below, or black in some individuals.

Habitat and Behavior Inhabits montane forests. Nocturnal and diurnal, and arboreal as well as terrestrial. Diet consists of skinks and lacertids. Oviparous.

Distribution Jiangxi, Fujian, Guangdong, Guangxi, Guizhou, Yunnan, Sichuan and Hunan. Laos, Vietnam, Myanmar, Cambodia, Thailand, Sri Lanka, Nepal, India, the Philippines, Indonesia. Widely recorded from the Jinggangshan Region.

Status of Population Common species.

2.43 玉斑锦蛇 *Euprepiophis mandarinus* (Cantor, 1842)

识别特征 体型较大。头颈区分稍显。眼中等大，瞳孔圆形。吻平钝。颊鳞1枚或缺失。眶前鳞1枚。眶后鳞1-3枚。上唇鳞6-8枚。下唇鳞8-10枚。背鳞平滑，鳞式23（21-25）-23（21）-19（17-21）。尾下鳞对分。肛鳞对分。头背亮黄色，有3条黑色横斑带，第1条在两第1上唇鳞之间；第二条在两眼之间，眼下分叉；第3条"V"形，尖端在额鳞，穿过口角至喉部。颈背、体背和尾背灰色或灰棕色，背鳞中心区棕红色；背有一排大的黄色菱形斑，镶有宽的黑色边和窄黄色边。腹面乳白色，有时有大黑斑。

生境与习性 栖息于平原、丘陵和山地。常见于开放森林的多石区域，林下灌丛，农耕地及近水区域。陆栖蛇类。以蜥蜴及其卵、啮齿类、鼩鼱类为食。卵生。

分布 除黑龙江、吉林、内蒙古、宁夏、青海、新疆和海南外，广泛分布于中国各省区。越南北部、缅甸和印度东北部。井冈山见于大坝里等地。

种群状况 中国物种红色名录：易危。

Identification Body size large. head slightly distinct from the neck; eye moderate, pupil rounded; snout obtuse; loreal single or absent; preocular single; postoculars 1-3; supralabials 6-8; infralabials 8-10; dorsal scales smooth, 23(21-25)-23(21)-19(17-21); subcaudals paired; anal divided. Dorsum of head bright yellow with three black cross-bands, first covered between two first supralabial; second covered between two eyes and bifurcated below the eyes; third V-shaped, apex on frontal, through angle of the mouth reaching throat; dorsal surface of neck, body and tail grey to greyish-brown and dorsal scales with brownish-red centers; dorsal surface with large, rhombic yellow marks edged with broad black border and narrow yellow border; cream below, sometimes with large black blotches.

Habitat and Behavior Inhabits plains, hills and montane regions. It is more frequently found in open forests with rocky substrate, scrubland, agricultural fields and the vicinity of water. Terrestrial. Diet consists of skinks and their eggs, rodents and shrews. Oviparous.

Distribution It is widely distributed in China, excluding Heilongjiang, Jilin, Neimenggu, Ningxia, Qinghai, Xinjiang and Hainan. N Vietnam, Myanmar, NE India. Recorded from Dabali in Mt. Jinggang.

Status of Population China Species Red List: VU.

2.43a

2.43b

2.44 王锦蛇 *Elaphe carinata* (Günther, 1864)

识别特征 体型较大。成年和幼体色斑差异较大。头体背黑黄相间，头背面有"王"字形黑纹。背鳞除最外侧1-2行平滑外，均起强棱。体中段背鳞21行以上；腹鳞大多超过200枚。

生境与习性 栖息于平原、丘陵和山地。以蛙类、蛇类、蜥蜴和鸟类及其卵、啮齿类等为食。卵生。

分布 在中国广泛分布在从河北、天津到两广北部，从台湾到陕甘、川滇等广大地区。国外见于越南。

种群状况 中国物种红色名录：易危。

Identification Body size large; color patterns are significantly different among juveniles and adults; dorsal surface black, mixed with yellow; presence of a 王 -shaped black decorative pattern on top of head; dorsal scales strong keeled, the most outer 1-2 rows smooth, at last 21 dorsal scale rows at midbody; ventrals usually more than 200.

Habitat and Behavior It inhabits plains, hills and montane regions. Diet consists of frogs, snakes, skinks, birds and their eggs, and rodents. Oviparous.

Distribution　It is widely distributed in China from Hebei and Tianjin to northern Guangdong and northern Guangxi, from Taiwan to Shaanxi, Gansu, Sichuan and Yunnan. Vietnam.

Status of Population　China Species Red List: VU.

2.45 黑眉锦蛇 *Orthriophis taeniurus* (Cope, 1861)

识别特征　体细长。头颈区分显著。眼中等大,瞳孔圆形。尾细长。颊鳞单枚。眶前鳞单枚或 2 枚,通常在眶前鳞下有 1 枚小鳞。上唇鳞 6-9 枚,下唇鳞 9-13 枚。背鳞 25(23)-23(21, 25)-19(17),在体中段除最外一行平滑外,均起弱棱。腹鳞外侧有弱侧棱。尾下鳞对分。肛鳞对分。背面灰棕色,头的每侧有一条黑色带,在眼后最宽。体前段背部有一个由蝴蝶型黑斑组成的脊列,在黑斑侧方有黑色小斑。体后段背部有一条 3-4 行鳞宽的浅色脊带,两侧各有一条 5-6 行鳞宽的黑带。腹面黄色或乳黄色。

生境与习性　栖息于平原、丘陵和山地。以蛙类、鸟类和啮齿动物为食。陆栖蛇类。卵生。

分布　除宁夏、新疆、青海、黑龙江和吉林外,见于中国各省。越南、泰国、缅甸、印度、日本、朝鲜半岛和俄罗斯滨海地区。井冈山见于茨坪、湘洲等地。

种群状况　常见种。

Identification　Body slender and elongated; head distinct from neck; eye moderate, pupil rounded; tail long and slender; loreal single; preocular single (or 2), usually presence of a small scale below preocular; supralabials 6-9; infralabials 9-13; dorsal scales in 25(23)-23(21, 25)-19(17), weakly keeled at midbody, but outer most row smooth; ventrals with a feebly lateral keel; subcaudals paired; anal divided. Greyish-brown above; a black stripe on each side of the head, broadest behind the eye; anterior part of dorsal body with a series of large black butterfly-shaped vertebral spots, and smaller diamond-shaped ones on the sides; posterior part of body with a pale grey vertebral stripe in three or four scale rows, and a broad black stripe in five or six scale rows on each side; yellow or cream below.

Habitat and Behavior　Inhabits plains, hills and montane regions. Diet consists of frogs, birds and rodents. Terrestrial. Oviparous.

Distribution　All provinces in China excluding Ningxia, Xinjiang, Qinghai, Heilongjiang and Jilin. Vietnam, Thailand, Myanmar, India, Korean Peninsula, Japan, coastal Russia. Recorded from Ciping and Xiangzhou in Mt. Jinggang.

Status of Population　Common species.

2.46 紫灰锦蛇黑线亚种 *Oreocryptophis porphyracea nigrofasciata* (Cantor, 1839)

识别特征　体中等长。头颈区分不甚显著。眼小,瞳孔圆形。颊鳞单枚或缺失。眶前鳞单枚。眶后鳞 2 枚,少数 1 枚。颞鳞 1+2 或 1+1。上唇鳞 8 枚,少数 7 枚或 9 枚。下唇鳞 11 枚。背鳞光滑,19(18, 17)-19(17)-17(15) 行。肛鳞对分。尾下鳞对分,通常少于 70 对。背棕红色,有镶黑边的暗色宽斑。头背面有 3 条纵纹,一条位于头中线,由吻鳞到顶鳞后缘;另 2 条位于侧面,

2.44a

2.44b

2.45a

2.45b

始于眼，向后延伸至尾末端。腹面纯苍白色。

生境与习性　栖息于山地森林、农田和村落。以小型哺乳动物如鼠类、鼩鼱为食。卵生。

分布　江西、安徽、江苏、浙江、福建、广东、广西、贵州、湖南、香港和台湾。井冈山见于茨坪、湘洲、水口、荆竹山等。

种群状况　常见种。

Identification　Body moderately elongated; head slightly distinct from neck; eye small, pupil rounded; loreal single or absent; preocular single; postoculars 2, rarely 1; temporals 1+2 or 1+1; supralabials 8, rarely 7 or 9; infralabials 11; dorsal scales in 19(18, 17)-19(17)-17(15) rows and smooth; anal divided; subcaudals usually less than 70, paired. Reddish-brown above with broad dark-brown cross-bars with black edges; dorsal surface of head with three longitudinal black stripes, one down the middle of the head from rostral to hinder edge of parietal scale, and other two on each side, usually starting from the eye and extending to end of the tail; uniform pale below.

Habitat and Behavior　Inhabits montane forests, farmlands and villages. Diet consists of small mammals such as voles and shrews. Oviparous.

Distribution　Jiangxi, Anhui, Jiangsu, Zhejiang, Fujiang, Guangdong, Guangxi, Guizhou, Hunan, Hong Kong and Taiwan. Recorded from Ciping, Xiangzhou, Shuikou and Jingzhushan in Mt. Jinggang.

Status of Population　Common species.

2.46a

2.46b

2.47 灰腹绿锦蛇 *Rhadinophis frenatum* (Gray, 1853)

识别特征　体细长。头稍大，与颈部明显可分。眼中等大，瞳孔小，圆形。前额鳞侧向与上唇鳞相接，没有颊鳞。眶前鳞单枚或缺失。上唇鳞8枚，少数9枚或7枚。下唇鳞9-11枚。背鳞19 (20, 21) - 19 (17) - 15 (13, 14)，除最外侧1-3行光滑外，均起弱棱。腹鳞具侧棱；肛鳞对分。尾下鳞对分。成体背草绿色，头两侧各有一条黑色条纹从鼻鳞开始经瞳孔向后延伸至颈侧；背鳞间皮肤黑色。上唇鳞和下唇鳞黄绿色；腹面微黄色。在幼体，背灰色；头顶鳞缘黑色；背鳞间皮肤黑色。

生境与习性　栖息于丘陵山地森林。以鼠类、鸟类和蜥蜴等为食。日行性地栖蛇类。卵生。

分布　江西、安徽、浙江、福建、广东、广西、贵州、四川和河南。印度和越南。井冈山见于五指峰。

种群状况　常见种。

Identification　Body slender and elongated; head slightly large, distinct from neck; eye moderate, pupil small and rounded; prefrontal in contact with supralabials laterally; loreal absent; preocular single or absent; supralabials 8, rarely 9 or 7; infralabials 9-11; dorsal scales in 19 (20, 21) - 19 (17) - 15 (13, 14) rows, feebly keeled, but outer most 1-3 scale rows smooth; ventrals with two lateral keel; anal divided; subcaudals paired. In adults, grass green above, with a black stripe on each side of head starting from nasal, through pupil and extending to side of the neck; skin between dorsal scales black; supralabials and infralabials greenish-yellow; yellowish below. In juveniles, grey above; scales on top of head with black edges; skin between dorsal scales black.

2.47a

2.47b

Habitat and Behavior Inhabits hilly and montane forests. Diet consists of voles, birds and skinks. Diurnal and terrestrial. Oviparous.

Distribution Jiangxi, Anhui, Zhejiang, Fujian, Guangdong, Guangxi, Guizhou, Sichuan and Henan. India and Vietnam. Recorded from Wuzhifeng in Mt. Jinggang.

Status of Population Common species.

2.48 颈棱蛇 *Macropisthodon rudis* Boulenger, 1906

识别特征 体中等长，粗壮。头大，略呈三角形，与颈部区分显著。吻鳞三角形，背视刚可见。前额鳞 2 枚。顶鳞短，与额鳞等长。鼻鳞大，完全对分为 2。颊鳞单枚，上与鼻间鳞及前额鳞相接，下与上唇鳞相接。眼大，瞳孔圆形。眶前鳞 3-4 枚，眶后鳞 3-4 枚，眶下鳞 3 枚，上唇鳞 7-8 枚，不入眶。颞鳞和背鳞起强棱。背鳞 23-23-19 行。肛鳞对分。尾下鳞对分。头背深褐色，侧面有一条黑纹始于吻鳞、过眼、向后与颈部第一个黑棕色横斑相连。黑纹之下区域橘红色。背其余部分棕黄色，颈背有宽黑棕色横斑，向后有成对的圆形或椭圆形大斑。头腹面浅黄色，其余部分浅黄棕色，散布黑色斑纹。幼体头背浅棕黄色。

生境与习性 栖息于山地。以蛙类和蜥蜴为食。卵胎生。

分布 中国特有种。分布于江西、安徽、浙江、福建、广东、广西、云南、贵州、湖南、河南和台湾。广泛分布于井冈山地区。

种群状况 常见种。

Identification Body moderate, robust; head large, slightly triangular, distinct from neck; rostral triangular, just only visible from above; prefrontals 2; parietals short, as long as the frontal; nasal large, completed divided into two; loreal single, touching the internasal and prefrontal dorsally, and the supralabials ventrally; eye large, pupil rounded; preoculars 3-4; postoculars 3-4; suboculars 3; supralabials 7-8, not touching the eye; temporals and dorsal scales strongly keeled; dorsal scales in 23-23-19 rows; anal divided; subcaudals paired. Dorsum of head dark brown; a blackish brown stripe on each side of head, starting from tip of rostral, through eye, connecting with first transverse band on the neck; the region below the stripe orange; rest part yellowish-brown; nape of neck with broad dark brown transverse bands, posteriorly with paired rounded or elliptical marks; ventral surface of head yellowish, posteriorly yellow-brown, with scattered black blotches. In juveniles, dorsal surface of head yellowish-brown.

Habitat and Behavior Inhabits montane regions. Diet consists of frogs and skinks. Ovoviviparous.

Distribution Endemic to China. Jiangxi, Anhui, Zhejiang, Fujiang, Guangdong, Guangxi, Yunnan, Guizhou, Hunan, Henan and Taiwan. Widely recorded from the Jinggangshan Region.

Status of Population Common species.

2.49 台湾小头蛇 *Oligodon formosanus* (Günther, 1872)

识别特征 体粗壮，近圆筒形。头小、短，与颈部区分不显著。吻鳞三角形，背视可见。眼中等大，瞳孔圆形。尾短。上颌齿 10-11 枚。颊鳞单枚，少数个体 2 枚。有下眶前鳞；眶前鳞单枚

或2枚。上唇鳞8枚（罕见7枚），3-2-3排列；下唇鳞6-9枚。背鳞光滑，19-19-17行。肛鳞完整。尾下鳞对分。背棕灰色或红棕色。身体和尾背网状，有不规则细黑色横斑；有一条橘红色脊纹。头背表面有一条暗棕色斑纹从额鳞穿过眼到达上唇鳞；一个尖端向前的"V"形斑纹从额鳞延伸到颈部；2个饰有黑色和橘红色边的暗斑分别位于顶鳞上。头腹面和身体前腹部苍白色，有褐色或橘红色侧斑；后腹部和尾腹面粉红色。

生境与习性　栖息于平原到山地。噬吃爬行动物的卵，包括它自己的卵。卵生。

分布　福建、广东、广西、江西、江苏、贵州、海南、浙江和台湾。越南。井冈山见于湘洲、茨坪。

种群状况　常见种。

Identification　Body robust, subcylindrical; head small, short, indistinct from neck; rostral triangular, visible from above; eye moderate, pupil rounded; maxillary teeth 10-11; tail short; loreal single, rarely 2; presubocular present; preocular single or 2; supralabials 8 (rarely 7), 3-2-3; infralabials 6-9; dorsal scales smooth, in 19-19-17 rows; anal entire; subcaudals paired. Greyish-brown or reddish-brown above; dorsal body and tail reticulated with irregular thin black cross bars; vertebral stripe orange; dorsal surface of head with a dark-brown stripe passing from the front, through eye, to the supralabials; the presence of a pointed forward dark-brown V-shaped mark that extends from frontal to neck; two dark blotches edged with black and orange on parietals, respectively; ventral surface of head and anterior venter of body pale white, with black and orange lateral spots; posterior venter and ventral surface of tail pink.

Habitat and Behavior　Inhabits plains to mountains. Preys almost exclusively on reptile eggs, included its own. Oviparous.

Distribution　Fujian, Guangdong, Guangxi, Jiangxi, Jiangsu, Guizhou, Hainan, Zhejiang and Taiwan. Vietnam. Recorded from Xiangzhou and Ciping in Mt. Jinggang.

Status of Population　Common species.

2.50 中国小头蛇 *Oligodon chinensis* (Günther, 1888)

识别特征　身体圆柱形。头小，三角形，与颈部区分不显。吻鳞三角形，大部分背视可见。眼中等大，瞳孔圆形。额鳞几乎等于顶鳞。眶上鳞单枚，大。颊鳞单枚，少数2枚。眶前鳞单枚，少数2枚或3枚。眶后鳞2枚，少数单枚。上唇鳞6-8枚，下唇鳞7-9枚，通常8枚。背鳞光滑，17-17-15行。肛鳞完整单枚。尾下鳞对分。背棕色或灰棕色，有一个深棕色有黑色和浅黄色镶边的横纹，覆盖鼻间鳞后部、前额鳞和额鳞及眶上鳞前部，穿过眼，延伸至上唇鳞；一个"V"形斑纹，尖端在额鳞后部，向后延伸至颈部腹侧；2个圆形深棕色斑在颞部；身体和尾部有14-19个深棕色横斑带。腹前部腹鳞浅黄色，有黑色斑点，向后几乎全为黑色。

生境与习性　栖息于平原和山区。吃爬行动物卵。卵生。

分布　江西、安徽、江苏、浙江、福建、广东、广西、云南、贵州、湖南、湖北和河南。越南北部。井冈山见于湘洲、大井、茨坪和白银湖。

2.49b

2.50a

2.50b

种群状况　常见种。

Identification　Body cylindrical; head small, triangular, indistinct from neck; rostral triangular, large part visible from above; eye moderate, pupil rounded; frontal approximately equal to parietals; supraocular single, large; loreal single, rarely 2; preocular single, rarely 2 or 3; postoculars 2, rarely single; supralabials 6-8, usually 8; infralabials 7-9, usually 8; dorsal scales smooth, in 17-17-15 rows; anal entire; subcaudals paired. Brown or greyish-brown above; a dark-brown transverse stripe edged with black and yellowish, covered posterior part of internasals, prefrontals, anterior part of frontal and supraocular, throught eye, extended to supralabials; a dark-brown V-shaped marking edged by black and yellowish, apex in posterior part of frontal, posteriorly extended to ventrolateral side of neck; two dark-brown rounded spots in temporal regions; 14-19 dark-brown transverse bands on body and tail; ventrals of anterior part yellowish, with black spots, posteriorly almost completely black.

Habitat and Behavior　Inhabit plains and mountains. Preys almost exclusively on reptile eggs. Oviparous.

Distribution　Jiangxi, Anhui, Jiangsu, Zhejiang, Fujian, Guangdong, Guangxi, Yunnan, Guizhou, Hunan, Hubei and Henan. Northern Vietnam. Recorded from Xiangzhou, Dajing, Ciping and Baiyinhu in Mt. Jinggang.

Status of Population　Common species.

2.51 饰纹小头蛇 *Oligodon ornatus* Van Denhurgh, 1909

识别特征　身体圆柱形。头小，与颈部区分不显。吻鳞三角形，背视可见大部分。眼中等大，瞳孔圆形。前额鳞转向颊部，与上唇鳞相接。没有颊鳞。眶前鳞单枚。眶后鳞 2 枚，少数 1 枚。上唇鳞 6 枚。下唇鳞 6 枚。背鳞平滑，均为 15 行。肛鳞对分。尾下鳞对分。体背棕色或黄棕色，头颈部有 3 个镶黑边的 V 形横纹，体尾有 4 条黑色纵纹和（7-8）+（2-3）个镶黑边的波浪形横斑。腹面黄白色，正中有一条粉红色纵线，其两侧为方形黑斑。

生境与习性　栖息于山地森林。

分布　中国特有种。分布于江西、浙江、福建、台湾、广西、湖南和四川。

种群状况　常见种。

Identification　Body cylindrical; head small, indistinct from neck; rostral triangular, large part visible from above; eye moderate, pupil rounded; prefrontal bending in to the loreal region, in contact with supralabials; loreal absent; preocular single; postoculars 2, rarely single; supralabials 6; infralabials 6; dorsal scales smooth in 15 rows throughout; anal divided; subcaudals paired. Brown or yellow-brown above, with three dark-brown V-shaped transverse stripes edged with black on dorsum of head and neck, with 4 longitudinal black stripes and （7-8）+（2-3）dark-brown wavy transverse bands edged with black on body and tail; ventral surface yellowish-white, with a pink stripe in middle region, bordered by black spots at two sides.

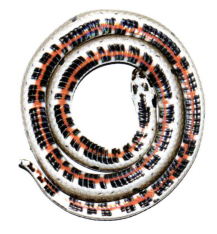

Habitat and Behavior　Inhabits montane forests.

Distribution　Endemic to China. Jiangxi, Zhejiang, Fujian, Taiwan, Guangxi, Hunan and Sichuan.

Status of Population　Common species.

2.52 挂墩后棱蛇 *Opisthotropis kuatunensis* Pope, 1928

识别特征　体圆柱形。头宽平扁，与颈部区分不显。上颌齿小，几等大。鼻孔接近鼻鳞上缘。鼻间鳞长弯向外侧。前额鳞单枚。额鳞大，长稍大于宽，显著短于顶鳞。眶上鳞对分或完整。颊

鳞长大于高。眶前鳞2枚，少数1枚或3枚。眶后鳞2枚，少数3枚。上唇鳞14（少数13枚），前6枚（少数5枚）完整，其后上唇鳞横裂。下唇鳞小，更加不规则。前额片皱褶状，几倍于后额片。背鳞通身19行，有沟和起强棱。肛鳞对分。尾下鳞对分。背橄榄棕色，有3条昏暗的黑色纵纹，每一纵纹1行鳞宽。腹部和每侧1-3行鳞均匀浅色。尾腹面除肛后区域外，有暗色云斑。

生境与习性 栖息于山区溪流。常见于水中和岩石下。以蚯蚓为食。夜行性水生动物。卵生。

分布 中国特有种。江西、福建、浙江、香港、广东和广西。井冈山见于白银湖。

种群状况 少见。

Identification Body cylindrical; head broad, depressed, indistinct from neck; maxillary small, subequal; nostrils near upper edge of the nasals which are not in contact; internasals long, curved outward; prefrontal single; frontal large, a little longer than broad, much shorter than parietals; supraocular divided or entire; loreal longer than deep; preoculars 2, rarely 1 or 3; postoculars 2, rarely 3; supralabials 14 (rarely 13), first 6 supralabials (rarely 5) entire; those following are divided horizontally; infralabials small, even more irregular; the very finely rugose anterior chin-shields are several times the size of the narrow posterior ones; dorsal scales in 19 rows, striated and strongly keeled throughout; anal divided; subcaudals paired. Uniform olive-brown above, extends down on either side to the third row of scales and is marked only by three very obscure, black, longitudinal lines, each as wide as a scale; the belly and first three rows of scales on either side are uniformly light; the ventral surface of the tail is clouded except for just behind the anus.

Habitat and Behavior Inhabits montane streams. More frequently found in water or under stones. Preys on earthworms. Nocturnal and aquatic. Oviparous.

Distribution Endemic to China. Jiangxi, Fujian, Zhejiang, Hong Kong, Guangdong and Guangxi. Recorded from Baiyinhu in Mt. Jinggang.

Status of Population Rare species.

2.53 山溪后棱蛇 *Opisthotropis latouchii* (Boulenger, 1899)

识别特征 体圆筒形。头宽平扁，与颈部区分不显。眼小，瞳孔圆形。吻鳞长方形，宽大于高，背视刚刚可见。鼻孔背置，在单一鼻鳞上。鼻间鳞小。前额鳞单枚，宽2.5倍于长。额鳞大，长宽几相等。颊鳞单枚，入眶。眶前鳞无，少数单枚。眶后鳞2枚。上唇鳞8-10枚，下唇鳞7-10枚。通体背鳞17行，中间7-15行起弱棱。肛鳞对分。尾下鳞对分。上体黑色，头背有橄榄绿色云状斑，身体和尾背有数条纵走黄绿色线纹。腹面浅黄色。

生境与习性 栖息于山区溪流和水稻田。常见于水中、石下和腐烂植物间。以环节动物动物为食。夜行性。卵生。

分布 中国特有种。江西、安徽、浙江、福建、广东、广西、贵州、湖南和重庆。井冈山地区广泛分布。

种群状况 常见种。

Identification　Body cylindrical; head broad, depressed, indistinct from neck; eye small, pupil rounded; rostral rectangular, broader than high, just visible from above; nostril in the single nasal, directed upward; internasals small; prefrontal single, 2.5 times as broad as long; frontal large, as broad as long; loreal single, touching the eye; preocular absent, rarely single; postoculars 2; supralabials 8-10, infralabials 7-10; dorsal scales in 17 rows throughout, 7-15 middle rows feebly keeled; anal divided; subcaudals paired. Black above; dorsum of head with olive-green blotches; dorsal body and tail with several longitudinal yellowish-green lines; pale yellowish below.

Habitat and Behavior　Inhabits montane streams and paddy fields. More frequently found in water, under stones and in decaying plants. Preys on annelids. Nocturnal. Oviparous.

Distribution　Endemic to China. Jiangxi, Anhui, Zhejiang, Fujian, Guangdong, Guangxi, Guizhou, Hunan and Chongqing. Widely recorded from the Jinggangshan Region.

Status of Population　Common species.

2.54 崇安斜鳞蛇 *Pseudoxenodon karlschmidti* Pope, 1928

识别特征　与本属其他种类相比，头较窄，背面无黄色，通常颈部有一个清晰的倒"V"形白线。体型中等。头与颈部稍可区分。眼大、瞳孔圆形。吻宽大于高，背视刚可见。鼻间鳞远短于前额鳞。颊鳞单枚。眶前鳞单枚。眶后鳞 3 枚。上唇鳞 8 枚，少数 7 枚；下唇鳞 9-10 枚。背鳞起棱，中段环体鳞 19 行。肛鳞对分；尾下鳞对分。体色变异较大。上体黑灰色，通常颈背有一个可见或模糊的倒"V"形黑斑，其前缘镶有清晰的窄白边。在体和尾背中线有若干的浅色横斑。唇及头腹面纯白色；上唇鳞鳞缝黑色；体腹面灰白色，至少前 20 枚腹鳞有长方形大黑斑，向后直到尾部小黑斑逐渐增多；小黑斑密布在腹鳞两侧尖部、形成一条黑线。

生境与习性　栖息于山地森林、茶园和农田。以蛙类为食。卵生。

分布　中国特有种。江西、福建、广东、海南、广西、贵州和湖南。记录于井冈山西坪。

种群状况　稀见物种。

Identification　From congeners differs by lack of yellow in the dorsal pattern, having a narrower head, usually having a distinct reverse V-shaped white line. Body moderate; head slightly distinct from neck; eye large; pupil rounded; rostral broader than deep, just visible from above; internasals much shorter than prefrontals; loreal single; preocular single; postoculars 3; supralabials 8, rarely 7; infralabials 9-10; dorsal scales keeled, midbody scales in 19 rows; anal divided; subcaudals paired. Blackish grey above; the presence of a visible or barely visible reverse V-shaped black marking anteriorly bordered by a distinct narrow white line on nape of neck; the middle of the dorsum of body and tail with several transverse light colored bands; labials and ventral surface of the head immaculate white; supralabial sutures black; ventral surface of body and tail grey-white, with several oblong-shaped large black patches at least on first twenty ventrals, followed ventrals speckled

with black more and more profusely toward the tail; laterally the speckles are concentrated along the tips of the ventrals to form a line.

Habitat and Behavior　Inhabits montane forests, tea plantations and farmland. Preys on frogs. Oviparous.

Distribution　Endemic to China. Jiangxi, Fujian, Guangdong, Hainan, Guangxi, Guizhou and Hunan. Recorded from the Xiping in Mt. Jinggang.

Status of Population　Rare species.

2.55 大眼斜鳞蛇 *Pseudoxenodon macrops fukienensis* (Blyth, 1854)

识别特征　中等体型。头颈稍可区分。眼大，瞳孔圆形。吻宽大于高，背视稍可见。鼻间鳞短于前额鳞。颊鳞单枚。眶前鳞单枚。眶后鳞 3 枚。上唇鳞 8 枚，少数 7 枚；下唇鳞 8-11 枚。背鳞起棱，最外行平滑，环体中段鳞 19 行；肛鳞对分；尾下鳞对分。上体棕色，有 3 个斑纹，第一个从前额鳞过眼至嘴角，第二个倒 "V" 形，覆盖整个顶鳞，延伸至腹鳞，第三个在颈部，倒 "V" 形，无白色镶边，其尖端几近顶鳞，向两侧延伸至腹鳞。在体尾背中部有若干浅色横斑，均被 4 个黑斑环绕。唇和头腹面纯白色，上唇鳞鳞缝黑色。腹鳞有深色斑点，但前 20 枚腹鳞几乎没有斑点。

生境与习性　栖息于山地森林。日行性。以蛙类为食。卵生。

分布　江西、福建和湖南。记录于井冈山茅坪。

种群状况　常见种。

Identification　Body moderate; head slightly distinct from neck; eye large; pupil rounded; rostral broader than deep, just visible from above; internasals shorter than prefrontals; loreal single; preocular single; postoculars 3; supralabials 8, rarely 7; infralabials 8-11; dorsal scales keeled, but outermost row smooth; midbody scales in 19 rows; anal divided; subcaudals paired. Brown above with three black marks, first covered prefrontals, through eyes, extended to angle of jaw on each side; second reverse V-shaped, covered parietals and posteriorly extended to ventrals, third reverse V-shaped without white border, across the neck whose apex reached almost to the parietals, and laterally extended to ventrals; the middle of the dorsum of body and tail with several transverse light bands surrounded by four black spots; supralabial sutures black; labials and ventral surface of head immaculate white; the ventrum has the usual dark speckling which is almost lacking on the first twenty ventrals.

Habitat and Behavior　Inhabits montane forests. Diurnal. Preys on frogs. Oviparous.

Distribution　Jiangxi, Fujian and Hunan. Recorded from the Maoping in Mt. Jinggang.

Status of Population　Common species.

2.56 纹尾斜鳞蛇 *Pseudoxenodon stejnegeri* (Harbor, 1908)

识别特征　体型中等。头颈区分稍显。眼和瞳孔均大，圆形。吻鳞背视稍可见。颊鳞单枚。眶前鳞单枚，偶尔 2 枚。眶后鳞 3 枚或 4 枚，偶见 2 枚。上唇鳞 8 枚，少数 7 枚。下唇鳞 8-10 枚。背鳞起棱，最外一行鳞平滑，鳞式 19 (17-21)-17-15(14)。肛鳞对分。尾下鳞对分。头背灰棕色，颈背黄绿色，有 2 个黑色大而宽的斜斑带。向后身体背面灰棕色，有一系列菱形的浅棕色镶黑色边的斑纹，再向后直至尾尖形成一条背中线纹。两眼后各有一条粗黑斑纹延至嘴角。唇浅色，前 5 个鳞缝黑色。腹面灰白色，有黑色方形斑。

生境与习性　栖于山地森林。捕食蛙类。

分布　江西、安徽、浙江、福建、广西、贵州、四川和台湾。

2.55a

2.55b

2.55c

2.56a

井冈山见于茨坪和河西垄等地。

种群状况　常见种。

Identification　Body moderate; head slightly distinct from neck; eye and pupil large, rounded; rostral just visible from above; loreal single; preocular single, rarely 2; postoculars 3 or 4, rarely 2; supralabials 8, rarely 7; infralabials 8-10; dorsal scales keeled, most outer rows smooth, arranged in 19 (17-21)-17-15(14); anal divided; subcaudals paired. Color variable; brown-grey above the head, nape of neck green-yellowish with two wide, moderately long oblique black stripes; posteriorly brown-grey above body with series of transverse diamond light brown marks bordered with black edges, posteriorly until the tip of tail forming a dorsal middle line; a wide black stripe extended to the angle of jaw behind eye on each side; lip light brown; anterior five sutures of supralabials black; iris brown-red, pupil black; grey-white below, with square black spots.

Habitat and Behavior　Inhabits montane forests. Preys on frogs.

Distribution　Jiangxi, Anhui, Zhejiang, Fujian, Guangxi, Guizhou, Sichuan and Taiwan. Recorded from Ciping, Hexilong in Mt. Jinggang.

Status of Population　Common species.

2.57 虎斑颈槽蛇 *Rhabdophis tigrinus* (Boie, 1826)

识别特征　体中等。头颈区分明显。颈沟清晰。上颌齿 22-23 枚，最后 2 枚强壮而延长，与前面的颌齿有齿间隙相隔。颊鳞 1 枚或 2 枚。眶前鳞 1-2 枚。眶后鳞 2-4 枚。上唇鳞 7 (2-2-3) 或 8 (2-3-3) 枚。下唇鳞 8-10 枚。背鳞起棱，或最外侧一行光滑，鳞式 19-19-17 (15)。肛鳞对分。尾下鳞对分。上背橄榄绿色，身体和尾背有一系列大黑横斑。身体前部两个横斑之间染红色。颈背有大黑斑，其后染红色。唇浅绿，有两个黑色斜斑，一条在眼下，一条在眼后。上唇鳞鳞缝黑色。腹面黄绿色。

生境与习性　栖息于山区。常见于水田和杂草丛生的潮湿多水环境。日行性陆栖蛇类。以甲虫、蛙类、蟾蜍和其他蛇类为食。卵生。

分布　广泛分布于华东地区。俄罗斯、朝鲜半岛和日本。井冈山见于茨坪、八面山和新城区等地。

种群状况　常见种。

Identification　Body moderate; head distinct from neck; nuchal groove distinct; maxillary teeth 22-23, the posterior-most two teeth strongly enlarged, and separated by diastemata from other; loreal single or two; preoculars 1-2; postoculars 2-4; supralabials 7 (2-2-3) or 8 (2-3-3); infralabials 8-10; dorsal scales keeled or most outer row smooth, arranged in 19-19-17 (15) rows; anal divided; subcaudals paired. Greenish-olive above, with a series of large black cross-bars on body and tail; anterior region of body between cross-bars reddish; nape of neck with large black marks, posteriorly reddish; lips light-green with two black oblique bands, one below the eye, another behind the eye; sutures of supralabials black; yellowish-green below.

Habitat and Behavior　Inhabits mountainous area. More frequently found in watery fields, weedy and watery environments. Diurnal and terrestrial. Diet consists of beetles, frogs, toads and

other snakes. Oviparous.

Distribution　Widely distributed in E China. Russia, Korean Peninsula, Japan. Recorded from the Ciping, Bamianshan and Xinchengqu in Mt. Jinggang.

Status of Population　Common species.

2.58 黑头剑蛇 *Sibynophis chinensis* (Günther, 1889)

识别特征　体细长圆柱状。头颈区分不显。尾甚长。鼻孔大，在鼻鳞中央。颊鳞小。眶前鳞单枚。眶后鳞 2 枚。背鳞 17 行，平滑。肛鳞对分。尾下鳞对分。头背灰棕色，有 2 条黑色横斑，1 条在眼后，另 1 条在枕部。颈背有镶浅黄色后缘的黑色大斑。体和尾背棕色，有一条清晰的黑色脊线，通常还有 2 条不太清晰的背侧纵线。唇白色。腹面浅黄色，通常每一个腹鳞有外侧黑斑点，形成 2 条纵线。

生境与习性　栖息于低山森林。夜行性陆栖动物。食物有蜥蜴和蛙类。卵生。

分布　江西、安徽、浙江、江苏、福建、广东、香港、海南、广西、云南、四川、甘肃、陕西、贵州和湖南。越南北部和老挝。井冈山见于八面山、河西垄。

种群状况　常见种。

Identification　Body slender and cylindrical; head indistinct from neck; tail significantly longer; nostril large, rounded, on the middle of nasal; loreal small; preocular single; postoculars 2; dorsal scales smooth, in 17 rows throughout; anal divided; subcaudals paired. Dorsal surface of head grey-brown, with two black transverse bars, one behind the eyes, other across occiput; a large black mark posteriorly bordered with light yellow on the nape of neck; posteriorly brown above the body and tail, with a distinct black vertebral line, usually with two less distinct dorsolateral lines; lip white; light yellow below, usually each ventral scale with an outer black spot and forming two longitudinal lines.

Habitat and Behavior　Inhabits submontane forests. Nocturnal and terrestrial. Diet consists of skinks and frogs. Oviparous.

Distribution　Jiangxi, Anhui, Zhejiang, Jiangsu, Fujian, Guangdong, Hong Kong, Hainan, Guangxi, Yunnan, Sichuan, Gansu, Shaanxi, Guizhou and Hunan. N Vietnam, Laos. Recorded from Bamianshan and Hexilong in Mt. Jinggang.

Status of Population　Common species.

2.59 环纹华游蛇 *Sinonatrix aequifasciata* (Barbour, 1908)

识别特征　体强壮而长。吻长。头三角形，头颈区分显著。眼中等大。鼻孔侧向。鼻鳞单枚，长方形，宽2倍于高。颊鳞单枚。眶前鳞单枚，少数2枚。眶后鳞2-4枚。通常有小眶下鳞1-3枚。上唇鳞9枚或8枚，不入眶或1-2枚入眶。下唇鳞8-11枚。背鳞19-19-17行，中央13-17行起强棱。肛鳞对分。尾下鳞对分。头背红棕色。唇灰白色。体尾背灰色或棕色，有若干个独特的暗横带，该斑带中心浅色，向体侧延展并形成"X"形。腹面白色，有不完整的黑斑带。

生境与习性　栖息于低地到山区森林内大型多石溪流。以鱼类和蛙类为食。卵生。

分布　江西、浙江、福建、广东、香港、海南、广西、云南、重庆、贵州和湖南。越南、缅甸和老挝。记录于整个井冈山地区。

种群状况　常见种。

Identification　Body stout and elongated; snout elongated; head triangular and distinct from neck; eye moderate; nostril lateral; nasal entire and rectangular, twice as wide as high; loreal single; preocular single, rarely 2; postoculars 2-4; usually small suboculars 1-3; supraliabials 9 or 8, not enter or 1-2 supralabials enter orbit; infralabials 8-11; dorsal scales in 19-19-17 rows, middle 13-17 rows strongly keeled; anal divided; subcaudals paired. Dorsal surface of head red brown; lip grey-white; grey or brown above body and tail, with distinctive several transverse dark bands with light centers, extended to flanks and becoming X-shaped; white below, with incomplete black bands.

Habitat and Behavior　Inhabits large, rocky streams in lowland

2.58a

2.58b

2.58c

2.59a

or montane forests. Diet consists of fishes and frogs. Oviparous.

Distribution Jiangxi, Zhejiang, Fujian, Guangdong, Hong Kong, Hainan, Guangxi, Yunnan, Chongqing, Guizhou and Hunan. Vietnam, Myanmar, Laos. Recorded from overall the Jinggangshan Region.

Status of Population Common species.

2.60 乌华游蛇 *Sinonatrix percarinata* (Boulenger, 1899)

识别特征 体粗壮而长。吻长。头三角形，与颈部区分显著。鼻孔侧位。鼻鳞对分。颊鳞单枚。眶前鳞单枚，偶尔2枚。眶后鳞3-5枚。通常有小型眶下鳞。上唇鳞8-9枚，有1-2枚入眶。下唇鳞8-11枚。背鳞起棱，19-19-17排列，最外行弱棱或光滑。肛鳞对分。尾下鳞对分。成年上体橄榄棕色，年幼个体橄榄色，有超过36个镶浅色边缘的黑色横带，该斑带在背部分叉。体侧斑带间在幼体桃红色。体腹面发白或发灰色，有不完整暗色斑带。

生境与习性 栖息于山间水库、水凼、溪流和水田。水生及陆生蛇类。以鱼类和蛙类为食。卵生。

分布 江西、安徽、江苏、浙江、福建、广东、香港、海南、广西、云南、四川、甘肃、陕西、河南、湖北、湖南、贵州和台湾。缅甸、泰国、老挝和越南。记录于整个井冈山地区。

种群状况 常见种。

Identification Body stout and elongated; snout elongated; head triangular and distinct from neck; nostril lateral; nasal divided; loreal single; preocular single, rarely 2; postoculars 3-5; usually presence of small suboculars; supraliabials 8-9, 1-2 supralabials enter orbit; infralabials 8-11; dorsal scales keeled and in 19-19-17 rows, the most outer row weakly keeled or smooth; anal divided; subcaudals paired. Above olive-brown in old individuals, olive in juveniles, with up to 36 black transverse bands bordered with light edges which fork dorsally; flanks between bands pinkish in juveniles; whitish or greyish below, with dark incomplete bands.

Habitat and Behavior Inhabits montane reservoirs, pools, streams and paddy fields. Aquatic and terrestrial. Diet consists of fish and frogs. Oviparous.

Distribution Jiangxi, Anhui, Jiangsu, Zhejiang, Fujian, Guangdong, Hong Kong, Hainan, Guangxi, Yunnan, Sichuan, Gansu, Shaanxi, Henan, Hubei, Hunan, Guizhou and Taiwan. Myanmar, Thailand, Laos, Vietnam. Recorded from overall the Jinggangshan Region.

Status of Population Common species.

2.61 乌梢蛇 *Ptyas dhumnades* (Cantor, 1842)

识别特征 体型较大。头颈区分显著。颊鳞单枚。眶前鳞2枚。眶后鳞2枚，少数3枚。上唇鳞8枚。下唇鳞8-10枚。背鳞16-16-14，中央2-6行起棱。肛鳞对分。尾下鳞对分。头背橄榄色，身体前部背面棕色，鳞缘黑色。亚成体身体和尾有4条清晰黑色纵带，这些纵带在成体只在前段清晰。身体前段腹鳞黄色或土黄，向后变成灰黑至棕黑。

生境与习性 栖息于平原丘陵。食物包括鱼类、蛙类、蜥蜴和啮齿动物。卵生。

分布　江西、安徽、江苏、浙江、福建、广东、香港、海南、广西、云南、四川、甘肃、陕西、河南、湖北、湖南、贵州和台湾。记录于整个井冈山地区。

种群状况　常见种。

Identification　Body size large. head distinct from neck; loreal single; preoculars 2; postoculars 2, rarely 3; supralabials 8; infralabials 8-10; dorsal scales in 16-16-14 rows; the 2-6 middorsal scale rows keeled; anal divided, rarely entire; subcaudals paired. Dorsal surface of head olive; dorsum of anterior body brown, with black edges of the scales; body and tail with distinct four longitudinal black stripes in subadults, these stripes distinct only on anterior part of the body in adults; ventrals of anterior part of body yellow or tan, posteriorly becoming grey-black to brown-black.

Habitat and Behavior　Inhabits plains and hills. Diet consists of fish, frogs, skinks and rodents. Oviparous.

Distribution　Jiangxi, Anhui, Jiangsu, Zhejiang, Fujian, Guangdong, Hong Kong, Hainan, Guangxi, Yunnan, Sichuan, Gansu, Shaanxi, Henan, Hubei, Hunan, Guizhou and Taiwan. Recorded from overall the Jinggangshan Region.

Status of Population　Common species.

2.61a

2.61b

2.62 黄斑渔游蛇 *Xenochrophis flavipunctatus* (Hallowell, 1860)

识别特征　体型中等。头颈区分显著。眼大，圆形。颊鳞单枚。眶前鳞单枚。眶后鳞 2-3 枚。上唇鳞 7-9 枚。下唇鳞 9-10 枚。背鳞 19-19-17 行，中间 9-15 行起棱。肛鳞对分。尾下鳞对分。上体橄榄绿色。头后颈背有 "V" 形黑斑。身体和尾有模糊的黑色横斑，在斑纹间染红色。有 2 个显著的黑色斜斑纹，一个在眼下，一个在眼后。腹面灰白色或浅黄色。腹鳞有黑色鳞缘。

生境与习性　栖息于森林、灌丛、草地、红树林、开阔水域和社区。半水栖蛇类。食物包括鱼类、蛙类及其卵、蝌蚪、蜥蜴、昆虫和小型哺乳动物。卵生。

分布　江西、安徽、江苏、浙江、福建、广东、香港、海南、广西、云南、西藏、四川、陕西、湖南、贵州和台湾。中南半岛、东南亚和南亚。

种群状况　常见种。

Identification　Body moderate; head distinct from neck; eye large, rounded; loreal single; preocular single; postoculars 2-3; supralabials 7-9; infralabials 9-10; dorsal scales in 19-19-17 rows, 9-15 middorsal rows keeled; anal divided; subcaudals paired. Olive-green above; nape behind head with V-shaped large black mark; the body and tail with indistinct black cross-bars, and tinged reddish between cross-bars; two conspicuous black oblique bands, one below the eye, other behind the eye; grey-white or yellowish below; ventrals scales edged with black.

Habitat and Behavior　Inhabits forests, shrubs, grasslands, mangroves, open fields and heavily populated regions. Semi-aquatic. Diet consists of fish, frogs and their eggs, tadpoles, skinks, insects

2.62a

and small mammals. Oviparous.

Distribution　Jiangxi, Anhui, Jiangsu, Zhejiang, Fujian, Guangdong, Hong Kong, Hainan, Guangxi, Yunnan, Xizang, Sichuan, Shaanxi, Hunan, Guizhou and Taiwan. Indo-China Peninsula, S and SE Asia.

Status of Population　Common species.

第 3 章
井冈山地区鸟类区系

Chapter 3
Avifauna of the Jinggangshan Region

在 2010-2015 年野外调查的基础上，结合井冈山自然保护区的标本，并有选择地引用了区域内过往正式发表的研究文献，共记录鸟类 17 目 56 科 167 属 290 种。其中，中国特有鸟类 9 种；国家 I 级重点保护野生动物 2 种，国家 II 级重点保护野生动物 38 种；CITES 附录 I 收录物种 3 种，附录 II 收录 31 种；被 IUCN 红色名录列为濒危（EN）2 种、易危（VU）4 种。本书选择性地收录了其中 260 种。

Based on a series of surveys between 2010 and 2015 at Jinggangshan National Nature Reserve and surroundings, examination of specimens held at this reserve and a review of the literature, we recorded 17 orders, 56 families, 167 genera and 290 bird species at the Jinggangshan Region. Among these, nine species are endemic to China; two species are listed in China Key List: I and 38 species are listed in China Key List: II; 58 species are listed in CITES App. I and 31 species in App. II; two species are Endangered (EN) and four species are Vulnerable (VU) on the IUCN Red List. This atlas describes 260 species of birds.

Abbreviations in this book are L = the length from tip of bill to tip of tail; N = northern; S = southern; E = eastern; W = western; C = central; NE = northeastern; SE = southeastern; SW = southwestern; juv. = juvenile; ad. = adult; a.s.l. = above sea level; and Mt. = mount.

鸊鷉目 PODICIPEDIFORMES
鸊鷉科 Podicipedidae

3.1 小鸊鷉 *Tachybaptus ruficollis* (Pallas, 1764)

识别特征　体小而短胖的鸊鷉，约 27 cm。繁殖期头顶及上体深褐色，颊部及前颈栗红色，具明显黄绿色嘴斑，下体灰白色。非繁殖期上体灰褐，下体皮黄。尾短小，呈绒毛状；瓣蹼足，脚在身体遥后方。虹膜黄色；嘴黑色或角质色；脚蓝灰色。

生境与习性　栖息于池塘、湖泊、江河、沼泽等地。有时成小群，也与其他水鸟混群。常潜水取食水生昆虫及其幼虫、鱼虾等。求偶期间相互追逐时常发出重复的高音吱叫声。营浮巢于水生植物上。窝卵数 4-8 枚，通常 4-5 枚。孵卵期 18-24 天。早成鸟，孵出后第 2 日即可下水游泳。

分布　非洲、欧亚大陆、印度、中国、日本、东南亚至新几内亚北部。在井冈山见于茨坪、新城区、荆竹山等地。

种群状况　在井冈山全年常见。

Identification　L ca. 27 cm. A very small, short-rounded grebe. Toes lobed and legs positioned well to the rear. Tail-less "bottom" with fluffy plumage. Breeding adult has chestnut cheek and foreneck, dark crown, nape and upperparts, yellow gape on bill. Non-breeding adult blackish above with pale brown sides of head and body, underparts buff. Eyes pale yellow; bill black or pale horn; tarsi dark grey.

Habitat and Behavior　Inhabits marshes, ponds, lakes and rivers. Sometimes gregarious, and with other waterfowl. Often dives to catch fish, insects and crustaceans. Also dives when alarmed. Gives rapid series of high pitched notes in breeding season. Nest a raft of vegetation. Clutch size is 4-5 eggs usually. Incubation period is 18-24 days. Precocies can swim.

Distribution　Africa, Eurasia, India, China, Japan, SE Asia to N New Guinea. At Ciping, Xinchengqu, and Jingzhushan in Mt. Jinggang.

Status of Population　Common all year round in Mt. Jinggang.

3.2 凤头鸊鷉 *Podiceps cristatus* (Linnaeus, 1758)

识别特征　体大、颈修长的鸊鷉，约 50 cm。繁殖期成鸟上体深褐色，黑色羽冠显著，耳羽和颈侧的领状饰羽栗色和黑色，下体白色，胁部栗褐色。非繁殖期上体灰褐色，下体近白，头部无栗黑色饰羽，眼上方白色（区别于赤颈鸊鷉）。虹膜近红；嘴黄色，下颚基部带红色，嘴峰近黑；脚近黑。

生境与习性　栖息于江河、湖泊、大型池塘等各类水域中。受惊时总潜入水中。以水栖昆虫、鱼虾及部分水生植物为食。繁殖期成做精湛的求偶炫耀，两相对视，身体高高挺起并同时点头，有时嘴上还衔着植物。

分布　古北区、非洲中南部、印度、澳大利亚及新西兰。在井冈山见于新城区。

种群状况　在井冈山迁徙季节可见。

Identification　L ca. 50 cm. A large grebe with a long thin neck and distinct black crown. Breeding adult has prominent black crest, orange and black head plumes, upperparts dark brown, underparts

3.1a

3.1b

3.2a

white with red-brown flanks. Non-breeding adult paler buff-brown without ear-tufts; white above eye (vs Red-necked Grebe). Eyes dark red; bill pinkish-horn; tarsi dark grey.

Habitat and Behavior　Inhabits rivers, large lakes and ponds. Submerges smoothly when alarmed. Often dives to catch fish, insects, crustaceans and some water plants. Breeding pairs perform elaborate mating rituals, distinctive head waving and nodding display.

Distribution　Palearctic, South-central Africa, India, Australia and New Zealand. At Xinchengqu in Mt. Jinggang.

Status of Population　Visible on migration in Mt. Jinggang.

鹳形目 CICONIIFORMES
鹭科 Ardeidae

3.3 白鹭 *Egretta garzetta* (Linnaeus, 1766)

识别特征　中等体型的白色鹭，约 60 cm。全身羽毛白色，繁殖期枕部具两根细长饰羽，背及胸具蓑状羽。与非繁殖期牛背鹭的区别在于体型较大而纤瘦。虹膜黄色；脸部裸露皮肤黄绿色，繁殖期为淡粉色；嘴黑色；腿及脚黑色，趾黄色。

生境与习性　栖息于低海拔的沼泽、稻田、湖泊、滩涂及沿海小溪流。以鱼、蛙、昆虫等为食，兼食植物性食物。单独或成散群活动。常与其他鹭集群营巢于阔叶林或杉林的树冠处。树上栖止时常呈缩头驼背状。窝卵数 3-5 枚。孵化期 21-25 天。

分布　非洲、欧洲、亚洲及大洋洲。在井冈山见于茨坪、新城区、罗浮、下庄、黄坳等地。

种群状况　在井冈山全年常见。

Identification　L ca. 60 cm. Mid-sized white egret. All white, breeding adult has two elongated nape plumes, and elongated, erectile, lacy plumes on breast and lower back. Taller and slimmer than Eastern Cattle Egret in non-breeding season. Eyes yellow; bill black; lores yellow to greyish-green (non-breeding), pink or orange (breeding); long tarsi generally all black, toes yellow.

Habitat and Behavior　Inhabits marsh, wet rice fields, lakes, streams and shallow coastal lagoons. Feeds on fish, frogs, insects and some plants. Usually forages alone or in loose group. Nests at the top of trees or bushes with other herons. Stands crouched when roosting. Clutch size is 3-5 eggs. Incubation period is 21-25 days.

Distribution　Africa, Europe, Asia and Oceania. At Ciping, Xinchengqu, Luofu, Xiazhuang and Huang'ao in Mt. Jinggang.

Status of Population　Common all year round in Mt. Jinggang.

3.4 苍鹭 *Ardea cinerea* Linnaeus, 1758

识别特征　体型较大的鹭，约 92 cm。上体青灰色，头、颈、下体白色，有黑色过眼纹及延长的枕羽，飞羽、翼角及两道胸斑黑色，前颈中部具 2-3 列黑色纵纹。幼鸟的头及颈灰色较重。虹膜黄色；嘴黄绿色；脚橘红至褐色。

生境与习性　栖息于河流、湖泊、沼泽及海岸滩涂等。常单个或成对活动于浅水处，颈缩至两肩间，腿亦常缩起一只于腹下。以鱼、虾、蛙、昆虫等为食，有时也寻食鼠类等。性机警，飞行沉重而缓慢。多集群或混群营巢在水域附近的岩壁、树上或

芦苇丛中。窝卵数 3-5 枚。雌雄共同孵卵，孵化期 24-28 天，幼鹭 40-50 天离巢。

分布 非洲、欧亚大陆、朝鲜半岛、日本至菲律宾及巽他群岛。在井冈山见于新城区、宁冈等地。

种群状况 在井冈山夏季和迁徙季常见。

Identification L ca. 92 cm. A large, pale grey heron. Adult has white head and neck with black eyestripe and nape plumes, 2-3 black streaks on foreneck, black flight feathers and wrist, and some black on sides of belly. Juvenile duskier, and no nape plumes. Eyes yellow; bill yellow-green; long tarsi pinkish-orange to greyish-flesh.

Habitat and Behavior Inhabits rivers, lakes, marshes and coastal mudflats. Solitary or in pairs in shallows with neck semi-retracted in the shoulder and one leg retracted under belly. Feeds on fish, shrimp, frogs, insects and mice. Flight heavy and slow. Breeds colonially in stick nests in trees, rocks, and reeds near water. Clutch size is 3-5 eggs. Both parents incubate. Incubation period is 24-28 days. Brooding period is 40-50 days.

Distribution Africa, Eurasia, Korean Peninsula and Japan to the Philippines and Sunda Islands. At Xinchengqu and Ninggang in Mt. Jinggang.

Status of Population Common in summer and on migration in Mt. Jinggang.

3.4a

3.5 草鹭 *Ardea purpurea* Linnaeus, 1766

识别特征 体型较大的鹭，约 80 cm。大体栗褐色，头顶蓝黑色，繁殖期枕部具两枚黑灰色饰羽，颈棕色，颈侧黑色纵纹延至胸部。背及覆羽、尾灰色，飞羽黑色，胁部及大腿栗色。幼鸟褐色较重。虹膜黄色；嘴褐色，嘴峰近黑；脚黄褐色。

生境与习性 栖息于沼泽、湖泊、稻田等地。飞行时振翅缓慢而沉重。常单独或结 3-5 只小群活动于水边，以鱼、虾、蛙、蜥蜴及昆虫等为食。有时会站着不动，静静等候鱼类和其他动物性食物。集群营巢在树上或芦苇、杂草丛中。窝卵数 3-6 枚，雌雄共同孵卵，孵化期 24-28 天，幼鹭约 42 天离巢。

分布 非洲、欧亚大陆至菲律宾、苏拉威西岛及巽他群岛。在井冈山见于新城区等地。

种群状况 在井冈山夏季常见。

3.4b

Identification L ca. 80 cm. Mostly greyish-rufous heron with black crown and nape, prominent black and chestnut stripes from face to chest, mantle, wing coverts and tail dark grey, flanks and thighs chestnut. Breeding adult has two nape plumes. Juv. browner overall. Eyes yellow; bill dark culmen and yellowish sides; tarsi yellowish-brown.

Habitat and Behavior Inhabits marshes, lakes and rice fields. Forages alone or in small group. Feeds on fish, shrimp, frogs, lizards and insects. Usually static to waiting for prey. Breeds mostly in small colonies in reed beds or trees. Clutch size is 3-6 eggs. Both parents incubate. Incubation period is 24-28 days. Brooding period is about 42 days.

Distribution Africa, Eurasia to the Philippines, Sulawesi and Sunda Islands. At Xinchengqu in Mt. Jinggang.

Status of Population Common in summer in Mt. Jinggang.

3.5

3.6 大白鹭 *Ardea alba* Linnaeus, 1758

识别特征　体大的白色鹭，约 95 cm。嘴较厚重，颈长且具特别的扭结。嘴角有一黑线直达眼后。繁殖期脸颊裸露皮肤蓝绿色，仅下背着生蓑羽。非繁殖期脸颊裸露皮肤黄色。虹膜黄色；嘴黑色（繁殖期）或黄色，嘴端有时为深色（非繁殖期）；脚全黑（非繁殖期）或腿部裸露皮肤红色而跗跖和趾黑色（繁殖期）。

生境与习性　栖息于稻田、湖泊、河流、海滨及沼泽地等。常单只或小群活动，在浅水处涉水觅食，边走边啄食。以鱼、蛙、田螺、水生昆虫等为食。常与白鹭、池鹭等混群筑巢于高大树木上或芦苇丛中。窝卵数 3-6 枚，多为 4 枚，孵卵期 25-26 天，幼鹭 30-42 天离巢。

分布　全世界。在井冈山见于新城区、大井等地。

种群状况　在井冈山秋冬季常见。

Identification　L ca. 95 cm. A large white egret with large stout bill and long neck (with characteristic kink). Black gape line extends rear of eye. Breeding adult has blue-green facial skin, long and filamentous dorsal plumes extending beyond tail. Non-breeding adult has orange-yellow facial skin. Eyes yellow; bill all black (breeding), or yellow with dark tip (non-breeding); legs all black (breeding), or tibia dark red, tarsi and toes black (non-breeding).

Habitat and Behavior　Inhabits rice fields, lakes, rivers, coasts and marshes. Forages alone or in small groups in shallows by slow stalking. Feeds on fish, frogs, mollusks and insects. Breeds in mixed colonies in flimsy stick nests in trees and reed beds. Clutch size is 3-6 eggs, mostly 4 eggs. Incubation period is 25-26 days. Nesting period of offspring is 30-42 days.

Distribution　Cosmopolitan. At Xinchengqu and Dajing in Mt. Jinggang.

Status of Population　Common in autumn and winter in Mt. Jinggang.

3.7 池鹭 *Ardeola bacchus* (Bonaparte, 1855)

识别特征　体型略小的鹭，约 47cm。繁殖期头及颈深栗色，胸深绛紫色，从肩披至尾的蓑羽蓝黑色，余部白色。非繁殖期大体灰褐色，具褐色纵纹，飞行时双翼及下体白而背部深褐色。虹膜金黄色，眼先裸部黄绿色；嘴黄色，嘴端黑色；腿至趾黄绿色。

生境与习性　栖息于池塘、湖泊、沼泽及稻田等水域及附近的树上。单独或成分散小群进食，以动物性食物为主。每晚三两成群飞回栖处，飞行时振翼缓慢，翼显短。常与夜鹭、白鹭、牛背鹭等组成巢群，在竹林、杉林等林木的顶部营巢。窝卵数 4-5 枚，孵化期 18-22 天，幼鹭 33-36 天离巢。

分布　孟加拉至中国及东南亚。越冬至马来半岛、中南半岛及大巽他群岛。迷鸟至日本。广泛分布于井冈山农田地区。

种群状况　在井冈山全年常见。

Identification　L ca. 47 cm. Smallish brown heron with white wing coverts and tail. Breeding adult has chestnut head, neck and breast with long thin chestnut nape plumes, and dark blue mantle with filamentous dorsal plumes. Non-breeding adult and juvenile have brown mantle and heavily grey-brown streaks from head to chest. In

3.6a

3.6b

3.7a

flight, broad white wings and tail contrast dark back. Eyes yellow, bare facial skin yellow-green; bill yellow with black tip (breeding), or duller with dark upper mandible (non-breeding); tarsi yellow.

Habitat and Behavior Inhabits ponds, lakes, marshes, and rice fields. Solitary or small loose flocks feed at wetlands. Feeds on animal food. Breeds in stick nests on tree tops, including pine and bamboo, in colonies with other ardeids. Clutch size is 4-5 eggs. Incubation period is 18-22 days. Brooding period is 33-36 days.

Distribution Bangladesh to China and SE Asia. Winters to Malay Peninsula, Indo-China Peninsula and Greater Sunda Islands. Wanders to Japan. Widespread in farmland at Mt. Jinggang.

Status of Population Common all year round in Mt. Jinggang.

3.7b

3.8 绿鹭 *Butorides striata* (Linnaeus, 1758)

识别特征 体小腿短的深灰色鹭，约 43 cm。成鸟头顶羽冠黑色并具绿色金属光泽，嘴基部的黑线延至脸颊。上背灰色，两翼及尾青蓝色并具绿色光泽，覆羽羽缘白色。颏白，腹部粉灰。雌鸟较雄鸟略小，更显褐色。幼鸟具褐色纵纹。虹膜黄色，眼周黄绿色；嘴黑色；脚偏绿。

生境与习性 性孤僻，栖息于山间溪流、湖泊、滩涂，也栖于灌丛、红树林等有浓密覆盖的地方。常单个或 2-3 只结小群活动，在溪边或水中岩石边注视水流伺机捕食。食物主要为鱼、蛙类、螺类及昆虫等。在近水的阔叶林或灌木林的树冠隐蔽处筑巢。窝卵数 3-5 枚，雌雄共同孵卵，孵卵期 20-22 天。

分布 美洲、非洲、马达加斯加、印度、中国、东北亚及东南亚、巽他群岛、菲律宾、新几内亚、澳大利亚。在井冈山见于新城区、茅坪、宁冈等地。

种群状况 在井冈山夏季常见。

Identification L ca. 43 cm. Dark grey heron with short leg. Adult has black crown and long thin black nape plumes, black moustachial stripe. Wing coverts dark grey with white at edge and green gloss. Tail short and blackish-grey. Female is smaller and browner than male. Juvenile also browner, with heavily streaked and spotted neck and underparts. Eyes yellow, eyering greenish-yellow; bill black; tarsi yellow.

Habitat and Behavior Inhabits streams, ponds, wet fields, marshes and mangroves, usually near dense cover. Shy and usually solitary, habitually forages from same posts by waiting for prey to pass. Feeds on fish, frogs, mollusks and insects. Breeds solitarily in waterside trees or bushes, and is secretive near nest. Clutch size is 3-5 eggs. Both parents incubate. Incubation period is 20-22 days.

Distribution America, Africa, Madagascar, Indian, China, NE Asia, SE Asia, Sunda Islands, the Philippines, New Guinea, Australia. At Xinchengqu, Ninggang and Maoping in Mt. Jinggang.

Status of Population Common in summer in Mt. Jinggang.

3.8a

3.8b

3.9 牛背鹭 *Bubulcus coromandus* (Boddaert, 1783)

识别特征 体型略小的白色或粉色鹭，约 50 cm。繁殖期大体白色，头、颈、胸披着橙黄色的饰羽，背上着红棕色蓑羽。非繁殖期体羽纯白，仅部分鸟额部沾橙黄。喙、颈较白鹭为短。虹膜黄色；嘴橙黄；脚暗黄至近黑。

生境与习性　栖息于稻田、牧场、水塘、农田及沼泽地等。常成对或结小群，多跟在家畜周围捕食被惊扰起来的昆虫，也吃鱼、虾等食物。繁殖期常与白鹭、池鹭等混群营巢于近水的大树、竹林或杉林。窝卵数 3-6 枚，孵卵期 18-22 天。

分布　除南极之外的几乎所有大陆。广泛分布于井冈山农田地区。

种群状况　在井冈山全年常见。

Identification　L ca. 50 cm. A small stocky white or pink egret. Breeding adult has golden-orange head, neck and breast, and long fine pale orange dorsal plumes. Other feathers are white. Non-breeding adult is all white. Eyes yellow to red; bill orange-yellow; tarsi pinkish-orange in breeding season, dark grey in winter.

Habitat and Behavior　Inhabits rice fields, meadows, ponds, farmlands and marshes. Forages as a pair or in small group. Follows or perches on large domestic animals, grabbing small disturbed prey. Also catches fish and shrimp. Breeds in stick nests in trees in single species or mixed colonies. Clutch size is 3-6 eggs. Incubation period is 18-22 days.

Distribution　All continent except Antarctic. Widespread in farmland at Mt. Jinggang.

Status of Population　Common all year round in Mt. Jinggang.

3.10 黑鳽 *Dupetor flavicollis* (Latham, 1790)

识别特征　中等体型、嘴长的黑色鹭，约 54 cm。成年雄鸟通体近黑色，黄色的颈侧与喉、颈、上胸的黄色纵纹十分明显。雌鸟褐色较浓，下体白色较多。幼鸟更为暗淡，翼上覆羽具浅色羽缘。虹膜黄色或浅绿色；嘴黄褐色；脚黄色至黑褐色。

生境与习性　栖息于芦苇丛、沼泽、滩涂、红树林及林间溪流。性隐秘。白天喜栖于森林或植物茂密缠结的沼泽地，夜晚飞至其他地点觅食鱼、虾、泥鳅及水生昆虫等。在芦苇丛中、树上或竹林中营巢。窝卵数 3-6 枚。雌雄共同孵卵。

分布　印度、中国南部、东南亚、菲律宾及印度尼西亚至澳大利亚。在井冈山见于宁冈等地。

种群状况　在井冈山夏季常见。

Identification　L ca. 54 cm. Mid-sized dark bittern with long bill. Adult male almost black except pale throat and yellowish streaks on foreneck to breast. Adult female dark brown above with pale neck stripes, underparts more buff. Juvenile duller than female, wing coverts have more marked pale edgings. Eyes yellow or pale green; bill yellow-brown; tarsi dull yellow to dark.

Habitat and Behavior　Inhabits reed beds, marshes, mudflats, mangroves and streams in forests. Shy. Feeds stealthily, mostly at night and in dim light. Feeds on fish, shrimp and insects. Breeds in reed beds, bamboo and trees. Clutch size is 3-6 eggs. Both parents incubate.

Distribution　India, Southern China, SE Asia, the Philippines and Indonesia to Australia. At Ninggang in Mt. Jinggang.

Status of Population　Common in summer in Mt. Jinggang.

3.11 黄苇鳽 *Ixobrychus sinensis* (Gmelin, JF, 1789)

识别特征　体型较小的浅褐色鳽，约 32 cm。成鸟顶冠黑色，上体淡黄褐色，下体皮黄色，飞行时黑色的飞羽和尾与皮黄色的覆羽和身体成强烈对比。幼鸟似成鸟，但褐色较浓，全身满

布黑褐色或黄褐色纵纹。虹膜黄色；眼周裸露皮肤黄绿色；嘴淡黄色，嘴峰和先端褐色；脚黄绿色。

生境与习性 栖息于湖泊、水库附近的稻田、芦苇丛、沼泽草地及滩涂中。以鱼、虾、蛙类及水生昆虫为食。常见沿水面掠飞，停歇在芦苇茎上，颈僵直不动。在水边的苇丛或灌丛中营巢，也筑巢于树上或竹林上。窝卵数4-7枚，孵化期14-15天，幼鹭14-15天离巢。

分布 印度、东亚至菲律宾、密克罗尼西亚及苏门答腊。冬季至印度尼西亚及新几内亚。在井冈山见于罗浮、新城区等地。

种群状况 在井冈山夏季和迁徙季常见。

Identification L ca. 32 cm. Small brown bittern. Adult has black crown, pale tawny-brown hind neck and mantle, paler foreneck and buff belly. In flight, pale buff wing coverts and body contrast strongly with dark flight feathers, primary coverts and tail. Juvenile browner with blackish-brown streaks. Eyes yellow, eyering yellow-green; bill orange-yellow with dark culmen and tip; tarsi yellow-green.

Habitat and Behavior Inhabits lakes, rice fields, reed beds and mudflats. Feeds on fish, shrimp, frogs and aquatic insects. Usually low flight over the surface of the water; clambers over reeds, flicking tail and wings; freezes with stretched neck. Breeds on reed nest near water or twig nest higher in trees or bamboo. Clutch size is 4-7 eggs. Incubation period is 14-15 days. Brooding period is 14-15 days.

Distribution India, E Asia to the Philippines, Micronesia and Sumatra. In winter to Indonesia and New Guinea. At Luofu and Xinchengqu in Mt. Jinggang.

Status of Population Common in summer and on migration in Mt. Jinggang.

3.11a

3.12 栗苇鳽 *Ixobrychus cinnamomeus* (Gmelin, JF, 1789)

识别特征 体型略小、嘴显粗短的栗褐色鳽，约41 cm。成年雄鸟上体栗色，下体黄褐，白色的下颊纹延至颈侧，两胁具黑色纵纹，尾下覆羽白色。雌鸟褐色较浓，头顶深褐色，胸腹具黑色纵纹。幼鸟较雌鸟更为深色，上体具浅色点斑，颈侧和胸具浓密黑色纵纹。虹膜黄色；嘴黄色，嘴峰黑色；脚黄绿色。

生境与习性 栖息于低海拔的芦苇丛、沼泽草地及滩涂。性羞怯孤僻，常单独活动。受惊时一跃而起，飞行低，振翼缓慢有力。以小鱼、蛙类和昆虫为食，兼食植物种子。在湿地草丛或芦苇丛中营巢。窝卵数4-6枚。雌雄共同孵卵。

分布 印度、中国、东南亚、苏拉威西岛及巽他群岛。在井冈山见于宁冈等地。

种群状况 在井冈山夏季和迁徙季可见。

Identification L ca. 41 cm. Small chestnut bittern with short heavy bill. Adult male is relatively monochromatic, unmarked rufous above and buffy below with white submoustachial stripe reaching neck, black stripes on flanks, and pale undertail coverts. Female dull brown with dark brown crown and somewhat spotted above, heavy dark streaks on breast and belly. Juvenile is much darker overall, heavily streaked blackish on neck and breast and spotted buff above. Eyes and eyering yellow; facial skin orange-yellow; bill yellow with

3.11b

dark culmen; tarsi yellow-green.

Habitat and Behavior　Inhabits reed beds, swamps, mudflats and mangroves. Shy, usually solitary. When disturbed jumps up and flies low with a slow, powerful rhythm. Breeds in stick nest on reed beds or just above ground. Clutch size is 4-6 eggs. Both parents incubate.

Distribution　India, China, SE Asia, Sulawesi and Sunda Islands. At Ninggang in Mt. Jinggang.

Status of Population　Visible in summer and on migration in Mt. Jinggang.

3.13 海南鸦 *Gorsachius magnificus* (Ogilvie-Grant, 1899)

识别特征　中等体型的深色鸦，约 58 cm。成年雄鸟头部图案显著，上体、顶冠、颊部及颈侧线条深褐色，具粗大的白色过眼纹；喉白，颈侧栗红色；翼上覆羽具白色点斑。下体浅褐色，颈、胸有大量白色斑点。成年雌鸟头顶及上体不及雄性色深，冠羽亦不及雄性明显，背部多浅色纵纹和斑点。虹膜黄色；嘴偏黄，嘴端深色；脚黄绿色。

生境与习性　栖于山间河谷或林中小溪旁（内）的稠密低矮草丛。被赶时飞至林上层。晨昏活动。

分布　中国南部。越南东部有一记录。在井冈山见于湘洲、宁冈等地。

种群状况　在井冈山夏季可见。IUCN 红色名录（2014）：濒危。国家 II 级保护野生动物。

Identification　L ca. 58 cm. Mid-sized distinctive marked night heron. Adult male generally dark brown with black crown, nape plumes, cheeks and ear coverts; lores eye patch and curving postocular stripe, chin and throat all white; neck-sides chestnut; some spots on wing coverts. Underparts brown with heavy white flecks on lower neck and breast. Female has less distinct head pattern, shorter crest and many pale streaks or spots on upperparts. Eyes yellow; bill yellowish with dark tip; tarsi greenish-yellow.

Habitat and Behavior　Inhabits dense herb undergrowth in marshy patches around small streams in forest. Flies up into canopy when flushed. Active at dawn and dusk.

Distribution　Southern China, one record E Vietnam. At Xiangzhou and Ninggang in Mt. Jinggang.

Status of Population　Visible in summer in Mt. Jinggang. IUCN Red List (2014): EN. China Key List: II.

雁形目 ANSERIFORMES
鸭科 Anatidae

3.14 鸳鸯 *Aix galericulata* (Linnaeus, 1758)

识别特征　中等体型的鲜艳鸭类，约 40 cm。雌雄异色。繁殖期雄鸟具暗色羽冠，醒目的白色眼罩延至颈侧，栗红色的颈部饰羽具淡黄色羽轴，上背深色，拢翼时可见棕黄色的"帆"状饰羽，翼镜绿色具白色边缘。雌鸟大体灰褐色，白色眼圈与细长的眼后线相接，胸和胁具浅色纵纹和斑点。雄鸟的非婚羽似雌鸟，但嘴为红色。虹膜深褐，外缘淡黄；雄鸟嘴暗红色，雌

鸟灰色；脚橙黄色。

生境与习性 栖息于山地的河谷、溪流，阔叶林和针阔混交林附近的沼泽、芦苇塘及湖泊等。除繁殖期外常成小群活动。性机警，遇警立即起飞，边飞边叫。睡觉时把头插在翅膀下。杂食性。巢营于树洞或河岸，树洞离地面较高。

分布 东北亚、中国东部及日本。引种其他地区。在井冈山见于茅坪等地。

种群状况 在井冈山罕见。国家Ⅱ级保护野生动物。

Identification L ca. 40 cm. Mid-sized colorful duck. Breeding male has dark crown, white stripe from eye arcs back over neck-sides, prominent chestnut neck plumes with yellowish scapus, extraordinary cinnamon display "sails" that are held erect concealing wings. Female is mostly grey-brown with elegant white eyering and rear eyestripe, pale stripes and spots on the breast and flanks. Non-breeding male resembles female but has red bill. Eyes brown; bill red (male) and grey (female); tarsi yellowish.

Habitat and Behavior Inhabits forested rivers, streams, ponds and lakes. Gregarious in non-breeding season. Vigilant, immediately flies with call when alarmed. Head under wings when sleeping. Omnivorous, nests in holes in trees or banks.

Distribution NE Asia, Eastern China and Japan. Introduced elsewhere. At Maoping in Mt. Jinggang.

Status of Population Generally rare in Mt. Jinggang. China Key List: II.

3.15 斑嘴鸭 *Anas zonorhyncha* Swinhoe, 1866

识别特征 体型较大、雌雄同色的鸭类，约60 cm。大体为斑驳的褐色，头颈颜色较其余部分明显淡。头顶、过眼纹和下颊纹深灰褐色，眉纹浅色。飞行时，白色的三级飞羽和浅色翼底明显，蓝紫色翼镜具狭窄白色后缘。虹膜褐色；嘴黑色而端黄；脚珊瑚红色。

生境与习性 栖息于湖泊、稻田、溪流及城市池塘等各类湿地生境。除繁殖期外常成群活动，也和其他鸭类混群。振翅沉重而缓慢。

分布 繁殖于东北亚和东亚，越冬于东亚和东南亚。在井冈山见于厦坪和宁冈等地。

种群状况 在井冈山迁徙季常见。

Identification L ca. 60 cm. Large, mottled brown duck with yellow tip on black bill. Sexes similar; head and neck paler, contrast with dark brown upperparts and underparts. Crown, eyestripe and lower cheek bar dark ashy-brown; supercilium whitish. In flight, show white tertials, conspicuous black and white pattern on underwing, dark blue or purple speculum with narrow white borders. Eyes dark reddish-brown; bill black with yellow tip; tarsi coral-red.

Habitat and Behavior Inhabits wetlands, from lakes and ricefields, agricultural ditches and streams to smallest urban ponds. Usually in groups when non-breeding, sometimes mixed with other ducks. Flight heavy with slow beats.

Distribution Breeds in NE and E Asia. Winters in E and SE Asia. At Xiaping and Ninggang in Mt. Jinggang.

Status of Population Common on migration in Mt. Jinggang.

3.14b

3.15a

3.15b

隼形目 FALCONIFORMES
鹰科 Accipitridae

3.16 黑冠鹃隼 *Aviceda leuphotes* (Dumont, 1820)

识别特征　体型略小的黑白色鹰类，约 32 cm。头顶具长而直立的黑色羽冠。雄性的次级飞羽、次级覆羽及肩羽有较多白色，下体栗色横纹较少。雌性仅肩羽上有白色，下体具较多栗色横纹。翼宽，翼端略尖，飞行时可见黑色衬，翼灰而端黑。虹膜深褐色；嘴深灰色，蜡膜灰色；脚肉灰色。

生境与习性　栖于低地或山区的落叶阔叶林或针阔混交林。单独或成小群活动，迁徙时形成大群。振翼似乌鸦，作短距离飞行。于空中或地面捕捉大型昆虫。营巢于高大的树上。似海鸥的咪咪叫。

分布　印度、中国南部、东南亚。越冬在大巽他群岛。在井冈山广泛分布。

种群状况　在井冈山夏季常见。CITES 附录 II。国家 II 级保护野生动物。

Identification　L ca. 32 cm. Smallish mostly black and white eagle with long erectile crest. Male has many white on secondaries, secondary coverts and scapulars, and few chestnut bands below. Female has white on scapulars only, and several chestnut bands below. In flight, wings broad and paddle-shaped but rather pointed. From below note the dark-tipped pale grey primaries and black secondaries and wing-lining. Eyes dark brown; bill dark grey, cere blue-grey; tarsi pink-grey.

Habitat and Behavior　Inhabits lowland and montane deciduous broadleaf or mixed broadleaf /coniferous forests. Crow-like in flight with slow deliberate beat, circles in small groups, or large flocks on migration. Catches insects in aerial sallies. Breeds in small stick nest highs in forest tree. Shrill whinnying.

Distribution　India, Southern China, SE Asia. Winter to Greater Sunda Islands. Widespread in Mt. Jinggang.

Status of Population　Common in summer in Mt. Jinggang. CITES App. II. China Key List: II.

3.17 凤头蜂鹰 *Pernis ptilorhynchus* (Temminck, 1821)

识别特征　体型略大、色型多变的鹰，约 58 cm。头型似鸽，停栖时翅尖不及尾端，尾具不规则横纹。所有型均具对比性浅色喉块，缘以浓密的黑色纵纹，并常具黑色中线。成鸟具灰色蜡膜，头和背深色，羽冠或有或无。雄鸟眼深色，脸灰色；雌鸟眼黄色，脸棕色或有图案。幼鸟头浅色，眼和蜡膜黄色，上体具浅色杂斑。飞行时特征为头相对小而颈显长，两翼及尾均狭长。虹膜橘黄，嘴灰色，尖端深色；脚黄色。

生境与习性　栖息在岩岸、近海潟湖及河口。较隐秘的林鸟，大群集体迁徙。飞行具特色，振翼几次后便做长时间滑翔，两翼平伸翱翔高空。常在树洞和地面搜掘膜翅目昆虫幼虫，也捕食小动物，兼食部分果实。

分布　古北区东部、印度及东南亚至大巽他群岛。在井冈山见于荆竹山、八面山、湘洲等地。

3.16a

3.16b

3.17a

种群状况 在井冈山迁徙季及冬季常见。CITES 附录 II。国家 II 级保护野生动物。

Identification L ca. 58 cm. Extremely variable large raptor with an odd "pigeon-headed" look (lacks bony "brows"); wing-tips fall short of tail-tip when perches, and tail is usually irregularly banded. All forms can have a contrasting pale gular patch outlined with heavy black streaks and often a black mesial streak. Adult has grey cere, dark head and back, and nape-crest (racially variable). Male has dark eye and typically grey face; female has yellow eye and browner and/or patterned face. Juvenile has pale head, variously dark-marked, with dark eye and extensive yellow cere, and pale mottled back and upperwings. In flight, looks long-necked and small-headed, with long broad wings and tail. Eyes yellow; bill grey with blackish tip; tarsi yellow.

Habitat and Behavior Inhabits rocky coasts, large coastal lagoons and near river mouths. Concealed, large flocks on migration. Characteristic flight of a few wing beats followed by a long glide. Soars on flat wings. Diet largely hymenopteran larvae extracted from combs dug from ground and tree-holes, also other small prey and fruit, and makes aerial sallies for insects.

Distribution E Palearctic, India and SE Asia to Greater Sunda Islands. At Bamianshan, Xiangzhou and Jingzhushan in Mt. Jinggang.

3.17b

Status of Population Common in winter and on migration in Mt. Jinggang. CITES App. II. China Key List: II.

3.18 黑翅鸢 *Elamus caeruleus* (Desfontaines, 1789)

识别特征 体小、大体灰白色的鸢，约 30 cm。头宽、眼大似猫头鹰。停栖时翅尖超过尾长，黑色的肩部斑块显著。成鸟有黑色眼罩，翼上覆羽和初级飞羽的腹面黑色，与浅灰色的上体、白色的下体及其余翼下羽毛形成强烈的对比。雌雄相似。幼鸟似成鸟但沾褐色。虹膜红色；嘴黑色，蜡膜黄色；脚黄色。

生境与习性 栖息于开阔原野、农田、疏林和草原地区。常单独或集小群在晨昏活动。飞行稳定如海鸥，滑翔时双翼上举作深"V"形，常定点振翅，也是唯一一种振羽停于空中寻找猎物的白色鹰类。喜立在突出的位置，如死树或电线杆上，尾上下摆动。以田间鼠类、小鸟、爬行动物和昆虫为食。

分布 非洲、欧亚大陆南部、印度、中国南部、菲律宾及印度尼西亚及新几内亚。在井冈山见于湘洲等地。

3.18a

种群状况 在井冈山迁徙季常见。CITES 附录 II。国家 II 级保护野生动物。

Identification L ca. 30 cm. Small, mostly grey and white kite with broad owl-like head and large eyes. Wingtips extend beyond tail tip when perched. Adult has black eye patch, upperwing coverts and underside of primaries black, contrasting strongly with pale grey upperparts, clean white underparts and otherwise white underwing. Sexes similar. Juvenile appear less clean than adult, more brown. Eyes red; bill black, cere yellow; tarsi yellow.

Habitat and Behavior Inhabits woodland edges, grasslands, savanna, cultivation and scrubs. Rather crepuscular activity alone or in small groups. Glides with wings held in flat "V", also hovers like

3.18b

a kestrel. Sits on exposed perches such as dead trees or telegraph poles, where may cock tail up and down over back. Takes mostly mammals, birds, some reptiles and insects.

Distribution　Africa, S Eurasia, India, Southern China, the Philippines and Indonesia to New Guinea. At Xiangzhou in Mt. Jinggang.

Status of Population　Common on migration in Mt. Jinggang. CITES App. II. China Key List: II.

3.19a

3.19 蛇雕 *Spilornis cheela* (Latham, 1790)

识别特征　中等体型的雕，约 50 cm。蓬松的黑色冠羽末端白色。成鸟上体暗褐色或灰褐色，下体褐色，腹部、两胁及臀具白色点斑。飞行时，腹部与翼下满布白点，宽阔的双翼有明显的翼指，宽阔的白色横斑与黑色翼后缘对比显著。宽短的尾中段有白色横斑。幼鸟似成鸟但褐色较浓，体羽多白色。虹膜黄色，脸黄色；嘴灰褐色；脚黄色。

生境与习性　栖息于山地森林及其林缘的开阔地带。单独或成对活动，常于森林或人工林上空盘旋，互相召唤。多栖于森林中有阴的大树枝上监视地面，有时也停栖在显眼的树尖或电杆上。主要以各种蛇类为食，也吃蜥蜴、蛙、鼠、鸟和甲壳类动物。求偶期成对作懒散的体操表演。营巢于树林顶端的枝杈上。

分布　印度、中国南部、东南亚、巴拉望岛及大巽他群岛。在井冈山广泛分布。

种群状况　在井冈山全年常见。CITES 附录 II。国家 II 级保护野生动物。

3.19b

Identification　L ca. 50 cm. Mid-sized dark eagle with large white-tiped black crown. Adult entirely dark with heavily white spots. Sexes similar. In flight, broad wings with clear "finger", flight-feathers have broad white band across bases, black trailing edge, and broad black tail has white central band. Spotted underparts. Juvenile paler, underparts entirely cream, upperparts mottled. Eyes yellow, facial skin yellow; bill morphs; tarsi yellow.

Habitat and Behavior　Inhabits well-wooded hills, lowlands, margins of agricultural land and mangroves. Pairs often calling to each other. Courting pairs perform sluggish aerobatics. Often soars, but also perches openly on trees or poles. Feeds on snakes, lizards, frogs, birds and mice. Breeds in stick nest lined with green leaves high in forest tree.

Distribution　India, Southern China, SE Asia, Palawan and Greater Sundas Islands. Widespread in Mt. Jinggang.

Status of Population　Common all year round in Mt. Jinggang. CITES App. II. China Key List: II.

3.20 凤头鹰 *Accipiter trivirgatus* (Temminck, 1824)

识别特征　体型较大的鹰，约 42 cm。头大呈方形，翅短，站立时翼尖仅达尾基部。黑色的喉中线和髭纹明显。成鸟具灰色的头和羽冠，上体灰褐，胸部具深色纵纹，腹部及大腿白色具深褐色粗横斑。幼鸟似成鸟但下体纵纹及横斑近黑，上体褐色较淡。飞行时，浅色的翼下覆羽具棕色斑点，飞羽具 4 条以上深色横斑，灰色的尾具 3 条深色宽横斑，蓬松的尾下覆羽白色，两翼较其他同属鹰类显短圆。虹膜褐色至橘黄色；嘴灰色、蜡膜黄色；腿及脚黄色。

3.20

生境与习性 多栖于山地森林和山脚林缘地带，也见于城市公园及低地。性机警，善藏匿。常躲在树叶丛中，有时也停于空旷处孤立的树枝上。多单独活动，叫声较为沉寂。多数情况下在林中静候猎物，主要以蛙、蜥蜴、鼠类、昆虫等动物为食，也吃鸟和其他小型兽类。繁殖期常在森林上空翱翔，同时发出响亮叫声。常营巢于近水的针叶林或阔叶林中高大的树上。

分布 中国（西南、台湾）。印度、东南亚。在井冈山见于八面山、湘洲等地。

种群状况 在井冈山全年常见。CITES 附录 Ⅱ。国家 Ⅱ 级保护野生动物。

3.21a

Identification L ca. 42 cm. Chunky and rather dark accipiter with a large square head, short primary, strong black mesial throat-streak and malar streak. Wingtips reach just beyond tail base when perches. Adult has grey head and crest, brown mantle and wings; breast heavily streaked dark brown; belly and flanks boldly barred brown on cream. Juvenile as adult but streaks and bars of underside are blackish and upperparts are paler brown. In flight, underwing coverts pale with brown spots, 4+ dark bars on pale flight feathers, three broad dark barks on grey tail, vent feathers fluffy white. Wings rather short and round. Eyes orange to brown; bill grey with yellow cere; tarsi yellow.

Habitat and Behavior Inhabits forested montane areas, urban parks, gardens and lowlands. Mostly solitary and concealed. Still-hunts from perch. Feeds on frogs, lizards, mice, birds, mammals and insects. Often soars over forest canopy calling loudly when breeding. Nests high in trees near water areas.

Distribution SW China, Taiwan China. India. SE Asia. At Bamianshan and Xiangzhou in Mt. Jinggang.

Status of Population Common all year round in Mt. Jinggang. CITES App. II. China Key List: II.

3.21b

3.21 赤腹鹰 *Accipiter soloensis* (Horsfield, 1821)

识别特征 中等体型的灰色鹰，约 33 cm。成鸟上体淡灰色，翼、上背及尾颜色稍深；下体白，胸及两胁略沾粉色，两胁及腿略具横纹。飞行时除初级飞羽末端黑色外，几乎全白。幼鸟上体褐色，胸部及腿上具褐色横斑，尾具深色横斑。虹膜红或褐色；嘴深灰色具黑端，蜡膜橘黄；脚橘黄。

生境与习性 栖息于山地森林和林缘地带，也见于低山丘陵和山麓平原地带的小块丛林、农田地缘和村庄附近。常单独或成小群活动。通常从栖处捕食，捕食动作快，有时在上空盘旋。常追逐小鸟，也吃蛙、鼠、蜥蜴、昆虫等其他动物性食物。营巢于树上，有时也利用喜鹊废弃的旧巢。

分布 繁殖于东北亚及中国。冬季南迁至东南亚、菲律宾、印度尼西亚及新几内亚。在井冈山见于白银湖、黄洋界、宁冈等地。

种群状况 在井冈山夏季和迁徙季常见。CITES 附录 Ⅱ。国家 Ⅱ 级保护野生动物。

Identification L ca. 33 cm. Mid-sized grey accipiter. Adult has plain grey head and upperparts, darker wings, mantle and tail. Underparts white with pale rufous wash or bars, dim brown bars on belly and thighs. In flight, black primaries contrast with strongly pale other parts. Juvenile dark brown with blackish streaks and bars on underside. Eyes yellow to dark brown; bill grey with black tip, cere orange; tarsi yellow-orange.

Habitat and Behavior Inhabits forested montane areas, open fields near wooded hills, and wooded country. Solitary, migrates in flocks. Soars with wings in a dihedral. Usually hunts frogs and lizards from a perch in a swift dash but sometimes circles overhead, also after birds in open wooded areas. Nests in trees, sometimes uses old nests of magpies.

Distribution Breeds in NE Asia and China. Migrating in winter to SE Asia, the Philippines, Indonesia and New Guinea. At Baiyinhu, Huangyangjie and Ninggang in Mt. Jinggang.

Status of Population Common in summer and on migration in Mt. Jinggang. CITES App. II. China Key List: II.

3.22 日本松雀鹰 *Accipiter gularis* (Temminck & Schlegel, 1844)

识别特征 体型较小的雀鹰，约 27 cm。停栖时翼达尾部一半，尾上具 4 条较窄的深色横斑。成年雄鸟上体深蓝灰，下体略白，胸、胁、腹染棕且具砖红色横斑。雌鸟上体深灰褐色，下体乳白色具浓密的褐色横斑。幼鸟胸具纵纹而非横斑，多棕色。

飞行时，略带锥形的双翼具窄而圆的翼尖。虹膜黄（幼鸟）至红色（成鸟）；嘴蓝灰而端黑，蜡膜绿黄；脚绿黄色。

生境与习性　栖息于山地针叶林和混交林中，也出现在林缘和疏林地带，是典型的森林猛禽。多单独活动，常在林缘上空捕猎食物。主要以小型鸟类为食，也吃昆虫、蜥蜴等。振翼迅速，结群迁徙。营巢于针叶林或针阔叶混交林中的河谷、溪流附近的高大树上。

分布　繁殖于古北区东部。越冬于东南亚。在井冈山地区见于大井、七溪岭等地。

种群状况　在井冈山地区迁徙季常见。CITES 附录 II。国家 II 级保护野生动物。

Identification　L ca. 27 cm. Small accipiter with medium primary projection and four narrow dark bands on largely pale tail. Adult male has dark bluish-grey back, underparts whitish, breast, flanks and belly variably washed and barred brick red. Female has dark grey-brown back, heavy greyish-brown barring on creamy-white underpart. Juvenile streaked rather than barred on chest and more rufous. In flight, rather tapered wings with narrow rounded tips and uniformly patterned underwing. Eyes yellow (juv.) to red (ad.); bill blue-grey, tipped black, with green-yellow cere; feet green-yellow.

Habitat and Behavior　Typical forest sparrowhawk. Inhabits montane coniferous forest and mixed forests, and open fields near wooded hills. Highly migratory, in flocks. Wings beat fast and fluttery. Makes surprise attacks on small birds during low-level flight, and is somewhat crepuscular. Nests in high trees near streams and valleys.

Distribution　Breeds E Palearctic. Winters in SE Asia. At Dajing and Qixiling in the Jinggangshan Region.

Status of Population　Common on migration in the Jinggangshan Region. CITES App. II. China Key List: II.

3.23 苍鹰 *Accipiter gentilis* (Linnaeus, 1758)

识别特征　体大的鹰，约 56 cm。胸部丰满，头部明显突出，翼宽阔，尾长。无冠羽或喉中线，具深色的头顶和耳羽，以及显著的白色宽眉纹。成鸟上体深灰色，下体白，具深灰色细横斑，灰色的尾有 3 条深色宽横斑。飞行时，翼后缘相当弯曲，翼尖窄且具显著"手指"，白色的尾下覆羽常常很明显。雌鸟较大。幼鸟上体褐色浓重，羽缘色浅成鳞状纹，下体具偏黑色粗纵纹。成鸟虹膜橙红（雄）或橙黄（雌），幼鸟黄色；嘴角质灰色；脚黄色。

生境与习性　见于亚高山森林和种植园。性羞怯，见人很远即飞。林地鹰类，能做快速翻转扭绕。振翅强而有力。主要食物为鸽类，也捕食其他鸟类及哺乳动物，如野兔。

分布　北美洲、欧亚大陆、北非。在井冈山地区见于南风面等地。

种群状况　在井冈山地区罕见于迁徙季。CITES 附录 II。国家 II 级保护野生动物。

Identification　L ca. 56 cm. Large accipiter with long broad wings and tail. Lacks crest and mesial throat-stripe, but has dark crown and ear-covers, and very prominent white brows. Adult upperparts rather dark grey, underparts white with fine dark grey barring from breast to belly, undertail coverts long, white, often conspicuous; grey tail

3.22a

3.22b

3.23

has three broad dark bands. In flight, wings bulge at secondaries, hand somewhat narrower at tip and clearly fingered. Female larger. Juvenile brown upperparts and buff underparts heavily streaked brown. Eyes orange-red (male) or orange-yellow (female) or pale yellow (juv.); bill grey, greenish-yellow cere; tarsi yellow.

Habitat and Behavior Inhabits forests and plantations. Rather shy, views often distant. A hawk of woodlands. Flight aggressive, turns and accelerates with remarkable speed in pursuit of prey, also glides and soars over territory. Wingbeats stiff, powerful and steady. Preys largely on pigeons but also can take gamebirds and mammals as large as hares.

Distribution North America, Eurasia, N Africa. At Nanfengmian in the Jinggangshan Region.

Status of Population Rare on migration in the Jinggangshan Region. CITES App. II. China Key List: II.

3.24a

3.24 普通鵟 *Buteo japonicus* Temminck & Schlegel, 1844

识别特征 体型略大的红褐色鵟，约 55 cm。栗色的髭纹显著。成鸟羽色多变，大多次级覆羽中部有白色斑块，深色的下胸和下腹与浅色的上腹和臀部形成对比。幼鸟下体具更多纵纹。飞行时从腹面看，翼后缘、翼尖和翼角黑色，尾羽具窄的次端横斑。虹膜黄色至褐色；嘴灰色，端黑，蜡膜黄色；脚黄色。

生境与习性 栖息于山地森林，喜开阔原野且在空中热气流上高高翱翔，在裸露树枝上歇息。飞行时振翅缓慢、僵硬，常在空中定点振羽。滑翔时双翼平直或向下倾斜。捕食哺乳动物等。营巢于树上或岩石上，用树枝搭建并铺满绿叶。

分布 繁殖于古北区及喜马拉雅。北方鸟至北非、印度及东南亚越冬。在井冈山广泛分布。

种群状况 在井冈山冬季和迁徙季常见。CITES 附录 II。国家 II 级保护野生动物。

Identification L ca. 55 cm. Largish reddish-brown buzzard with prominent chestnut moustachial. Adult variable, often with a paler bar on median secondary coverts, dark brown lower chest and lower belly contrasting with pale upper belly and vent. Juvenile more heavily streaked below. In flight, underwing flight feathers pale, coverts pale brown, large carpal patch and primary tips black; undertail pale with fine grey bars. Eyes yellow to brown; bill grey, tip black, cere yellow; tarsi yellow.

Habitat and Behavior Inhabits forest, well-wooded hills, prefers open country where it circles on thermals high overhead or rests on exposed tree branches. Wing beats slow, stiff, sometimes hover. Glides on flat wings or with lowered "hands". Takes mostly mammals. Nests in trees or cliffs, building large stick nest lined with green leaves.

Distribution Breeds Palearctic and Himalaya. Some winter to N Africa, India and SE Asia. Widespread in Mt. Jinggang.

Status of Population Common in winter and on migration in Mt. Jinggang. CITES App. II. China Key List: II.

3.25 林雕 *Ictinaetus malaiensis* (Temminck, 1822)

识别特征 体型较大的黑色雕，约 70 cm。头和嘴显小，而翼

3.24b

种群状况　在井冈山迁徙季常见。

Identification　L ca. 33 cm. Breeding adult has dark chocolate-brown above, black below with very long curved tail, white head and foreneck, white wings, golden nape-patch outlined in black. Non-breeding adult is white below, lacks tail-streamers, head with dark crown and broad dark stripe through eye ending in a breast-band bordered above by yellowish supercilium and neck-stripe. Juvenile as non-breeding adult but duller, with buff-fringed upperparts, wavy lines on wing coverts, narrowly barred breast and more rufescent cap. Eyes dark (ad.) or yellowish (juv.); bill slaty-blue (breeding) or dark yellow (non-breeding); tarsi greenish-grey, toes extremely long.

Habitat and Behavior　Inhabits wet crop fields, ponds, and lakes with extensive surface vegetation. Takes insects and invertebrates from water's surface and floating plants. Forms single-species flocks, takes flight in response to predators. Polyandry, female fights for mating success. Nests a flimsy mat on floating vegetation in still water. Clutch size is 4 eggs. Incubation period is 26 days. Male incubate and take care of offspring.

Distribution　Indian subcontinent to China and SE Asia. Migrating south as far as the Philippines and Greater Sunda Islands. At Dajing and Xinchengqu in Mt. Jinggang.

Status of Population　Common on migration in Mt. Jinggang.

3.46b

鸻形目 CHARADRIIFORMES
鸻科 Charadriidae

3.47 凤头麦鸡 *Vanellus vanellus* (Linnaeus, 1758)

识别特征　体型略大的黑白色鸻类，约 30 cm。具狭长的黑色上翘羽冠。上体具绿黑色金属光泽，下体白色具黑色胸带及红棕色臀部。繁殖期雄性脸和喉黑色，繁殖期雌性喉部为斑驳的白色，非繁殖期成鸟脸淡黄色，眼下具黑色斑，喉白色。幼鸟色暗，羽冠较短。飞行时，翼黑色具白端，尾白而具宽的黑色次端带；从腹面看，翼内缘白色。虹膜黑色；嘴近黑；脚粉褐色。

生境与习性　栖息于水域附近的沼泽、草地、水田、旱田、河滩和盐碱地等。迁飞时常结成大群。飞行时机械地振翅，但能灵活地转向。以动物性食物为主，也食小麦、草茎、草籽等植物性食物。

分布　古北区。冬季南迁至印度及东南亚的北部。在井冈山见于新城区、黄坳等地。

种群状况　在井冈山迁徙季和冬季常见。

3.47a

Identification　L ca. 30 cm. Largish black and white plover with long wispy black upturned crest. Dark green upperparts, black breast-band and white underparts with rufous vent. Breeding male has black face and throat; breeding female throat is speckled white. Non-breeding adult has buffy face with dark patch below eye and white throat. Juvenile is duller and shorter-crested with buff fringes above. In flight, wings dark with white-tipped outermost primaries, and tail white with broad black terminal band; from below, wing-linings white. Eyes black; bill black; tarsi dull brownish-pink.

Habitat and Behavior　Inhabits marshland fringes or wet meadows. Usually gregarious, typically in flocks; feeds both night and day.

3.47b

Normal flight has distinctive mechanical flapping action. Aerially agile and capable of dramatic dives and switchbacks, Mostly animal-based foods, also feeds on grass seeds and blades.

Distribution Palearctic. South in winter to Indian subcontinent and northern of SE Asia. At Xinchengqu and Huang'ao in Mt. Jinggang.

Status of Population Common in winter and on migration in Mt. Jinggang.

3.48 灰头麦鸡 *Vanellus cinereus* (Blyth, 1842)

识别特征 体大的灰色麦鸡，约35 cm。成鸟头、颈及上胸灰色，胸带黑色，上体褐色，腹部白色。幼鸟似成鸟但褐色较浓而无黑色胸带。飞行时，从背面看，黑色的翼尖与白色翼中部、褐色翼上覆羽及背形成对比，白色的尾具黑色次端斑；从腹面看，翼内侧大片白色部分与黑色初级飞羽形成对比。虹膜橘红色；眼周裸出部及眼先肉垂黄色；嘴黄色，端黑；脚黄色。

生境与习性 栖息于开阔的沼泽、水田、耕地、草地、河畔或山中池塘。迁飞时常10余只结群，有时多至40-50只，也与其他水鸟混群。常涉水取食，多以昆虫、水蛭、螺类、水草及杂草籽为食。

分布 繁殖于中国（东北）及日本。冬季南迁至印度东北部及东南亚。在井冈山见于新城区、罗浮等地。

种群状况 在井冈山迁徙季和冬季常见。

Identification L ca. 35 cm. Large pale lapwing. Adult has grey head, neck and upper chest, dark breast-band, brown upperparts, white belly. Juvenile lacks breast-band, and has narrow buff fringes on upperparts. In flight, from above, separating black wing tip contrast with white secondaries, brown wing coverts and back, white rump and tail with black subterminal band. From below, wings white with black tip. Eyes reddish-orange, narrow eyering yellow; bill yellow with black tip; tarsi bright yellow.

Habitat and Behavior Inhabits wet rice fields, grasslands and marshes, in winter also riverside. Usually gregarious, occurring in flocks of up to 40-50, and mixing with other shorebirds. Often wades while feeding, feeds on insects, leeches, mollusc and grass seeds.

Distribution Breeds in NE China and Japan. Migrating in winter to NE Indian subcontinent and SE Asia. At Xinchengqu and Luofu in Mt. Jinggang.

Status of Population Common on migration and in winter in Mt. Jinggang.

3.48a

3.48b

3.49 金眶鸻 *Charadrius dubius* Scopoli, 1786

识别特征 体小、嘴短的沙褐色鸻，约16 cm。繁殖期成鸟枕部至上体褐色，飞羽深色，但较长的三级飞羽覆盖了初级飞羽。前额白色，额基具黑纹，并经眼先和眼周延至耳后形成黑色贯眼纹，其上缘白色，下缘亦白并形成颈环。下体白色，具黑色或褐色的全胸带。成鸟黑色部分在幼鸟或非繁殖期为褐色，金黄色的眼圈不显著。飞行时翼上无白斑，尾羽两侧浅色、尾端深色。虹膜暗褐色，眼周金黄色；嘴黑色，下嘴基橘红色；腿黄色。

生境与习性 栖息于近水的草地、盐碱滩、多砾石的河滩、沼泽、水田及沿海海滨等地。常单独或成对活动，性活泼，边走

3.49a

边觅食。常急速奔跑一段距离后，稍事停息，然后再向前走。飞行迅速而灵活。主要以昆虫、甲壳类、软体动物等小型无脊椎动物为食，亦吃少量草籽等植物性食物。营巢于河心沙洲或近水滩地地面凹陷处，非常简陋。窝卵数常 4 枚。孵化期 18-22 天，雏鸟早成性。

分布　北非、古北区、东南亚至新几内亚。北方的鸟南迁越冬。在井冈山见于厦坪等地。

种群状况　在井冈山夏季和迁徙季常见。

Identification　L ca. 16 cm. Small, sandy brown, short-billed plover. Breeding adult has upperparts including rear crown mid brown, flight feathers darker, but very long tertials cover primaries. Face, lores and forecrown black with white fringe above, forehead white, chin, throat white extending to hindneck forms white collar. Underparts white with prominent black breast band. Juvenile and non-breeding adult brown where adult black, has less prominent eyering, but also has face patch extending in point below ear coverts. In flight, lacks wing bar, tail plain, but paler at sides and darker at tip. Eyes dark brown, with bright yellow eyering; bill black with small orange base to lower mandible; tarsi yellow.

Habitat and Behavior　Inhabits coastal, riverine and inland wetlands, often where substrate is muddy, sandy or shingle. Mostly solitary or in pairs, not usually mixing with other waders. Vocal, very active. Flight fast and agile. When feeding, often vibrates one foot on wet mud. Nest is shallow unlined scrape on gravel bed. Clutch size is 4 eggs. Incubation period is 18-22 days. Chicks are precocial.

Distribution　N Africa, Palearctic, SE Asia to New Guinea. Northern populations winter south. At Xiaping in Mt. Jinggang.

Status of Population　Common in summer and on migration in Mt. Jinggang.

3.50 长嘴剑鸻 *Charadrius placidus* Gray, JE & Gray, GR, 1863

识别特征　体型略大的鸻，约 22 cm。略长的嘴全黑，黄色的脚较长，翼折合后翼尖离尾末端甚远。只有隐约的淡色翼带。繁殖期成鸟体羽特征为具黑色的前额横纹和全胸带，但贯眼纹灰褐而非黑。非繁殖期成鸟头部较平淡，褐色胸带镶黑色上缘。幼鸟似剑鸻及金眶鸻。虹膜褐色；嘴黑色；腿及脚暗黄。

生境与习性　喜活动于内陆水域附近的沼泽、河滩、田埂上，也见于沿海滩涂的多砾石地带。多单只或 3-5 只结群活动。以昆虫、蜘蛛、甲壳类等动物性食物为食，冬季食大量植物性食物。

分布　繁殖于东北亚，中国（华东及华中）。冬季至东南亚。在井冈山见于新城区等地。

种群状况　在井冈山迁徙季常见。

Identification　L ca. 22 cm. Largish plover with long bill and black breast band. Breeding adult as black breast-band and broad black crown-patch, and thin dull eyering. Non-breeding adult has less contrast on face and only upper margin of collar black, the remainder brown. Juvenile as female but often even duller. Eyes dark brown, with narrow pale yellow eyering; bill black; tarsi dull ochre to yellow.

Habitat and Behavior　Inhabits predominantly rivers with

gravel or rocky bars and banks, lakeside habitat, wetlands and wet fields. Mostly solitary, sometimes in loose groups. Feeds on insects, spiders, crustaceans, also plants in winter.

Distribution Breeds in NE Asia, E and C China. South in winter to SE Asia. At Xinchengqu in Mt. Jinggang.

Status of Population Common on migration in Mt. Jinggang.

3.51a

鸻形目 CHARADRIIFORMES
鹬科 Scolopacidae

3.51 丘鹬 *Scolopax rusticola* Linnaeus, 1758

识别特征 体大而矮胖的鹬，约35 cm。头略成三角形，嘴长且直。头顶及颈背具深色宽横斑。上体以淡黄褐色为主，在上背形成宽阔的栗黑色纵带，下体色浅密布暗色横斑。双翼圆阔。飞行时，翼上具宽阔的栗色横斑。虹膜深褐色；嘴黄褐色，端黑；脚粉灰色。

生境与习性 主要栖息于丘陵、山区潮湿的针叶林、混交林和阔叶林中。多单只活动。白天隐蔽，伏于地面，晨昏或夜晚飞至开阔地觅食。性隐秘，靠近时僵直不动，不得已时起飞，振翅嗖嗖作响。飞行径直而缓慢，双翼甚弯曲，头高抬。取食蚯蚓、昆虫幼虫、蛙类等，也吃部分绿色植物及其种子。

分布 古北区。于东南亚为候鸟。在井冈山见于大坝里、荆竹山等地。

种群状况 在井冈山迁徙季常见。

Identification L ca. 35 cm. Large, round snipe with triangular head, thick, long bill and short leg. Forehead grey, broad black transverse bars on rear crown and hind neck. Upperparts pale yellowish-brown with broad cinnamon-black streaks on mantle; underparts narrowly barred dark brown. In flight, show broad rufous-banded wings. Eyes dark brown; bill yellow brown with dark tip; tarsi pinkish-grey.

3.51b

Habitat and Behavior Inhabits deciduous and mixed forest from lowlands to hills. Feeds mostly at night by probing in soft mud, woodland soil and leaf-litter. Extraordinarily cryptic, freezes when approached, and flushes with noisy wings at close range. Flight slow, straight, on deeply bowed wings, head held high. Feeds on earthworms, insect larvae and frogs, sometimes plants.

Distribution Palearctic. Migrant in SE Asia. At Dabali and Jingzhushan in Mt. Jinggang.

Status of Population Common on migration in Mt. Jinggang.

3.52a

3.52 扇尾沙锥 *Gallinago gallinago* (Linnaeus, 1758)

识别特征 中等体型、嘴长的沙锥，约26 cm。脸皮黄色，具深色的侧冠纹、贯眼纹和颊纹；上体深褐，具白及黑色的细纹及蠹斑；胸皮黄色具褐色纵纹，腹部白色，胁部具深褐色横斑；外侧尾羽几乎与中央尾羽等宽。飞行时，次级飞羽具白色宽后缘，翼下具白色宽横纹。站立时，背部浅色纵纹明显。虹膜褐色；嘴褐色，端黑；脚橄榄色。

生境与习性 栖于沼泽地带及稻田，通常隐蔽在高大的芦苇草丛中。被驱赶时跳出并作曲折飞行，同时发出警叫声。飞行较迅速、较高、较不稳健。空中炫耀为向上攀升并俯冲，外侧尾

3.52b

羽伸出，颤动有声。啄食泥土中的软体动物、昆虫幼虫、蠕虫、植物种子等。

分布　繁殖于古北区。南迁越冬至非洲、印度及东南亚。在井冈山地区见于黄洋界、南风面等地。

种群状况　在井冈山地区迁徙季常见。

Identification　L ca. 26 cm. Mid-sized snipe with extra-long bill and point wings. Head buff with dark lateral crown-stripe, eyestripe and moustachial stripes; upperparts deep brown with white and black streaks and bars, forms two pale buff lines on mantle when standing. Underparts dark brown, heavily barred on flanks, but belly white. Outer tail feathers almost as broad as central rectrices. In flight, mostly dark underwing has white bars along covert tips. Secondaries have broad white tips. Eyes black; bill brown with dark tip; tarsi greenish-yellow.

Habitat and Behavior　Inhabits damp, swampy areas from tundra to steppe. Usually skulking, sometimes coming into open. On approach typically freezes, then explodes into flight at short range, zigzag flight for a considerable distance. May forage nocturnally; feeds with rapid jerky vertical probing. Feeds on mollusks, worms, larvae and seeds.

Distribution　Breeds Palearctic. Migrates south in winter to Africa, India and SE Asia. At Huangyangjie and Nanfengmian in the Jinggangshan Region.

Status of Population　Common on migration in the Jinggangshan Region.

3.53a

3.53 青脚鹬 *Tringa nebularia* (Gunnerus, 1767)

识别特征　中等体型的鹬，约32cm。嘴长而粗且略向上翘，形长的腿近绿。上体灰褐具白色羽缘，翼尖及尾部横斑近黑；下体白色，喉、胸及两胁具褐色纵纹。飞行时，背部白色长斑明显，翼下具深色细纹，脚超出尾羽(小青脚鹬翼下白色，脚不超过尾)。幼鸟上体较褐。虹膜暗褐色；嘴基部绿色，端黑；脚黄绿色。

生境与习性　喜沿海和内陆的沼泽地带及大河流的泥滩。通常单独或两三成群。以水生昆虫、甲壳类、小鱼及水生植物为食。进食时嘴在水里左右甩动寻找食物。头紧张地上下点动，受惊扰即向远方低飞而去，飞时常发出口哨一般独特的鸣声，飞出一段距离后落下继续觅食。

分布　繁殖于古北区，从英国至西伯利亚。越冬在非洲南部、印度次大陆、东南亚至澳大利亚。在井冈山见于茨坪、新城区、荆竹山等地。

种群状况　在井冈山迁徙季常见。

Identification　L ca. 32 cm. Mid-sized sandpiper with long stout, slightly upturned bill. Upperparts pale grey and underparts white. Breeding adult has fine streaks on dark grey crown, face and neck, feathers on grey-brown upperparts with blackish centers and black marginal triangles. Non-breeding paler with less streaking on underparts. In flight, white wedge from rump to up back, fine dark bars on white tail, tarsi protrude beyond tail. Eyes dark brown; bill greyish with dark tips; tarsi dull green.

Habitat and Behavior　Inhabits marsh and flats on coasts and

3.53b

inland. Mostly feeds singly or in small groups, flocking at roosts; feeds by wading and picking, sometimes running after prey. Alarm call when flushed to flight, and lands at a short distance. Feeds on crustaceans, fish, aquatic insects and plants.

Distribution　Breeds in Palearctic from Britain to Siberia. Wintering to S Africa, Indian subcontinent, SE Asia, Malaysia to Australia. At Ciping, Xinchengqu and Jingzhushan in Mt. Jinggang.

Status of Population　Common on migration in Mt. Jinggang.

3.54 林鹬 *Tringa glareola* Linnaeus, 1758

识别特征　体型略小而纤细的鹬，约20 cm。白色眉纹长、背褐灰色具显眼的白色斑点，胸具黑褐色纵纹，腹部及臀偏白，尾白而具褐色横斑。飞行时尾部的横斑，白色的腰部及翼下，以及翼上无横纹为其特征。脚远伸于尾后。虹膜褐色；嘴黑色；脚淡黄至橄榄绿色。

生境与习性 主要栖于各种淡水和盐水湖泊、水塘、水库、沼泽和水田地带。多单只活动，亦结成松散小群。惊起时向上高飞，也能潜泳以逃避敌害。觅食水生昆虫、蠕虫、虾等，也吃部分植物。

分布 繁殖于欧亚大陆北部。冬季南迁至非洲、印度次大陆、东南亚及澳大利亚。在井冈山见于新城区等地。

种群状况 在井冈山迁徙季和冬季常见。

Identification L ca. 20 cm. Mid-sized sandpiper. Adult brown-grey with long whitish supercilium. Prominent white spots on brownish-grey upperparts. Drak brown streaked breast contrasts with pale belly and vent. In flight, underwing pale grey, rump white, several narrow brown bars on tail, toes protrude well beyond tail. Eyes brown; bill black; tarsi yellow to olive-green.

Habitat and Behavior Inhabits inland freshwater and coastal brackish wetlands, wet fields, rivers and mudflats. Usually solitary or in loose small groups. Flies when threatened, sometimes dives to escape. Feeds by picking from mud. Feeds on aquatic insects, worms and shrimp, sometimes plants.

Distribution Breeds in N Palearctic. Migrates south in winter to Africa, Indian subcontinent, SE Asia and as far as Australia. At Xinchengqu in Mt. Jinggang.

Status of Population Common on migration and in winter in Mt. Jinggang.

3.55 白腰草鹬 *Tringa ochropus* Linnaeus, 1758

识别特征 中等体型、矮壮的鹬，约 23 cm。繁殖期成鸟上体黑褐色具细小的白色斑点，白色眼圈显著；下体白色，胸具黑褐色纵纹。飞行时，翼下近黑、腰白色、尾部具深色宽横斑，脚伸至尾后。冬羽颜色较灰，纵纹不明显。虹膜暗褐色；嘴基部暗橄榄色，端黑；脚橄榄绿色。

生境与习性 广泛栖息于各类海岸和内陆湿地。多单只或结小群。行动时尾巴上下摆动，受惊扰时频频点头，身体也会摆动。惊飞时发出响亮的尖声，先锯齿形飞行，然后绕着圆圈高飞，最后急速下降，在远处着陆。主要以各种昆虫和小型水生无脊椎动物为食。

分布 繁殖于欧亚大陆北部。冬季南迁远及非洲、印度次大陆、东南亚及加里曼丹岛。在井冈山见于新城区、宁冈等地。

种群状况 在井冈山迁徙季和冬季常见。

Identification L ca. 23 cm. Mid-sized and stocky sandpiper. Breeding adult has black-brown upperparts with fine white spectacles, and narrow white eyering; underparts white with dense black streaks on breast. In flight, wings blackish above and below, rump contrastingly white, tail white with broad black bars, toes protrude just beyond tail. Non-breeding more grey, less streaks. Eyes black-brown; bill black, dark olive-green at base; tarsi dark greenish.

Habitat and Behavior Inhabits forests with pools, migration or winter on streams, rivers, freshwater wetlands. Solitary or in

3.54a

3.54b

3.55

small groups. Walks by flicking tail. Nodding head frequently when disturbed. Usually steep climbing zigzag snipe-like flight with flight call. Feeds on insects and aquatic invertebrates.

Distribution　Breeds in N Palearctic. Migrates south in winter as far as Africa, Indian subcontinent, SE Asia, and kalimantan Island. At Xinchengqu and Ninggang in Mt. Jinggang.

Status of Population　Common on migration and in winter in Mt. Jinggang.

3.56a

3.56 矶鹬 *Actitis hypoleucos* (Linnaeus, 1758)

识别特征　矮小而活跃的鹬，约 20 cm。嘴短，翼不及尾。繁殖期成鸟上体深褐色，具狭窄白眼圈，翼角有"几"字形白斑；下体白，胸侧具褐灰色斑块。飞行时，翼上具宽阔白色翼带，翼下具黑色及白色横纹，腰无白色。非繁殖期成鸟较平淡，胸部斑块不明显。秋季幼鸟翼上覆羽具微带褐色波浪形横斑。虹膜暗褐色；嘴黑褐色；脚暗橄榄绿。

生境与习性　栖于从沿海滩涂和沙洲至海拔 1500 m 的山地稻田及溪流、河流两岸等不同生境。常单独活动，性活跃，走动时头部和尾部不停地上下摆动。贴近水面，快速地扇动翅膀飞行。以昆虫、蠕虫、小鱼、水藻等为食。

分布　繁殖于古北区及喜马拉雅。冬季至非洲、印度次大陆、东南亚并远至澳大利亚。在井冈山见于新城区、宁冈等地。

种群状况　在井冈山迁徙季和冬季常见。

3.56b

Identification　L ca. 20 cm. Smallish sandpiper. Breeding adult upperparts dark brown with narrow white eyering, and white wedged patch before carpal; underparts white with brownish-grey breast patches. Non-breeder plainer, breast patches less distinct. Juvenile has scaled upperparts, scapulars and wing coverts have pale buff fringes. In flight, upperwings has broad white band, underwings has white and black bars, lacks white rump patch. Eyes black brown; bill dark grey; tarsi dull olive-green.

Habitat and Behavior　Inhabits freshwater, brackish and coastal wetlands, often along lakeshores, streams and rivers. Solitary mostly, vivacious, walks with nobbing head and flicking tail. Flies over surface and wings beats fast. Feeds on insects, worms, fish and algae.

Distribution　Breeds across Palearctic and in Himalaya. Migrates south in winter to Africa, Indian subcontinent, SE Asia and as far as Australia. At Xinchengqu and Ninggang in Mt. Jinggang.

Status of Population　Common on migration and in winter in Mt. Jinggang.

鸻形目 CHARADRIIFORMES
彩鹬科 Rostratulidae

3.57 彩鹬 *Rostratula benghalensis* (Linnaeus, 1758)

识别特征　体型略小、形似沙锥，约 25 cm。雌鸟色彩鲜艳，头及胸深栗色，顶冠纹黄色，白色眼圈延至枕侧；背及两翼偏绿色，白色条带从肩部绕至下背并渐变为褐黄色；下体白色。雄鸟体型较小，图案似雌鸟但较暗淡，且多具杂斑，翼覆羽具金色点斑。虹膜红褐色；嘴橙色至黄褐色；脚黄绿色。

3.57a

生境与习性　栖息于水草茂密的沼泽、池塘、稻田、河滩草丛和灌丛。性隐秘，喜晨昏活动。行走时尾上下摇动，飞行时双腿下悬如秧鸡。以昆虫、软体动物、蚯蚓和植物等为食。

分布　非洲、印度至中国及日本、东南亚、巽他群岛及澳大利亚。在井冈山见于新城区。

种群状况　在井冈山迁徙季可见。

Identification　L ca. 25 cm. Smallish, snipe-like shorebird. Female colorful, head to breast deep chestnut with yellow crown-stripe, and white eyering extending to nape-side; upperparts greenish-brown with white stripe curving up breast-side extending into brownish-yellow on lower back; underparts white. Female smaller, pattern similar to male but duller, and more irregular markings, gold spots on wing coverts. Eyes red-brown; bill orange to yellow-brown; tarsi greenish-grey.

Habitat and Behavior　Inhabits swamps, reed beds, rice fields, ponds, damp grassland, and cover along streams and rivers. Usually solitary or in pairs. Very secretive, crepuscular or nocturnal. Flight like rails with legs hanging down. Omnivorous, feeds on invertebrates, seeds of grasses and cultivated grain. Usually uses a scything action of the bill to sift food in soft mud or shallow water.

Distribution　Africa, India to China and Japan, SE Asia, Sunda Islands and Australia. At Xinchengqu in Mt. Jinggang.

Status of Population　Visible on migration in Mt. Jinggang.

鸻形目 CHARADRIIFORMES
燕鸥科 Sternidae

3.58 白翅浮鸥 *Chlidonias leucopterus* (Temminck, 1815)

识别特征　体小、腰部白色的燕鸥，约 23 cm。浅灰色的尾浅分叉、近方形。繁殖期成鸟的头、背及下体黑色，与白色翼上覆羽、腰、臀及浅灰色尾成明显反差。非繁殖期成鸟上体浅灰，耳羽处有一黑色斑点，以狭窄的黑带与灰色的头顶相接，白色的枕部具狭窄黑带，颈侧无黑斑，下体白。幼鸟似非繁殖期成鸟，但上背和翼上覆羽具深褐色扇形斑纹。初级飞羽超出尾长。虹膜深褐；嘴繁殖期深血红色，非繁殖期黑色；脚繁殖期橙红色，非繁殖期暗红色。

生境与习性　喜沿海地区、港湾及河口，也至内陆稻田及沼泽觅食。以小群活动，常与其他浮鸥混群；常栖于杆状物或石块上。飞行灵活轻快，取食时低低掠过水面，顺风而飞捕捉昆虫，或把嘴伸入水中。以小鱼、虾、昆虫及其幼虫和其他水生动物为食。

分布　繁殖于南欧及波斯湾，横跨亚洲至俄罗斯中部及中国。冬季南迁至非洲南部，并经印度尼西亚至澳大利亚，偶至新西兰。在井冈山见于新城区等地。

种群状况　在井冈山迁徙季常见。

Identification　L ca. 23 cm. Small tern with white rump and nearly square, pale grey tail. Breeding adult has jet-black head, mantle and underparts, contrasting strongly with white shoulder and wing coverts, white rump and vent, and grey tail. Non-breeding adult has pale grey upperparts with dark spot on ear coverts, at most narrowly connected to

3.57b

3.57c

3.58

grey crown by black band, narrow dark band on white nape, lacks dark patches on grey breast-sides just forward of wings, underparts white. Juvenile like winter adult, but mantle and wing coverts have dark brown scalloping and lack conspicuous pale tertial tips. Long primaries extending well beyond tail. Eyes dark brown; bill dark blood-red (breeding) or black (non-breeding); tarsi orange (breeding) or dull orange/red (non-breeding).

Habitat and Behavior Inland freshwater habitats, such as taiga swamps, river slacks and wet/flooded agricultural land. Migration or winter on inland lakes, coasts, coastal wetlands and mangroves. Often in flocks, regularly mixes with other marsh terns. Flight buoyant, frequently banking and changing direction, dipping to water to pick prey on or near surface. Feeds on fish, shrimp and aquatic insects.

Distribution Breeds in S Europe and Persian Gulf across Asia to C Russia and China. Migrates in winter to S Africa and through Indonesia to Australia and occasionally New Zealand. At Xinchengqu in Mt. Jinggang.

Status of Population Common on migration in Mt. Jinggang.

3.59a

3.59 须浮鸥 *Chlidonias hybrida* (Pallas, 1811)

识别特征 体型略小、尾浅开叉、双翼显圆的燕鸥，约 25 cm。繁殖期成鸟前额至枕部黑色，上体灰色，仅颊部白色，下体深灰黑色，尾下覆羽白色。非繁殖期成鸟上体灰，下体白，头顶具细纹、前额和眼先白色，眼后至枕部有一黑斑。幼鸟似非繁殖期成鸟但上体具褐色杂斑，腰及尾浅灰色。虹膜深褐；嘴血红色（繁殖期）或黑色；脚红色。

生境与习性 栖息于沼泽地、湖泊、水塘等地，常在内陆漫水地和稻田觅食。一般成群，频繁地在水面上空飞舞，行动敏捷轻快。取食时低掠水面，或俯冲入水捕捉食物，但又不会全身浸于水中。以小鱼、虾、水生昆虫和其他水生动物为食，也吃部分水生植物。

分布 繁殖在非洲南部、西古北区、南亚及澳大利亚。在井冈山见于新城区等地。

种群状况 在井冈山迁徙季及冬季常见。

3.59b

Identification L ca. 25 cm. Small, grey tern with shallow-forked grey tail and rounded wings. Breeding adult has full black crown from forehead to nape, grey upperparts with only cheeks and chin white, dark grey underparts with white undertail coverts. Non-breeding adult dark with extensive white forehead and white lores, narrow black streak from eye to nape; and all-white underparts. Juvenile resembles winter adult, but mantle and wing coverts have dark brown scalloping contrasting with paler grey rump and tail; flight feathers blacker. Eyes deep brown; bill deep blood-red (breeding) or black (non-breeding); tarsi blood-red.

Habitat and Behavior Inhabits swamps and lagoons, lakes and pools, or flooded fields. Usually in groups. Flight very light, changes direction frequently. Typically dips and hawks for food from surface or above. Insects important in diet, also eat fish, shrimp and some aquatic plants.

Distribution Breeds in S Africa, W Palearctic, S Asia and Australia. At Xinchengqu in Mt. Jinggang.

Status of Population Common on migration and in winter in Mt. Jinggang.

3.59c

鸻形目 CHARADRIIFORMES
鸥科 Laridae

3.60 红嘴鸥 Chroicocephalus ridibundus (Linnaeus, 1766)

识别特征 体型略小的灰白色鸥类，约 40 cm。繁殖期成鸟具深褐色头罩，并具不完整的狭窄白眼圈，上背和翼灰色，具黑色翼尖，其余体羽白色。非繁殖期成鸟无深色头罩，但颊部有一黑斑，眼上方或有灰色污迹。亚成鸟似非繁殖期成鸟，但三级飞羽及翼上覆羽褐色，翼尖及尾羽末端黑色。飞行时，浅灰色的翼窄而尖，从背面看外侧初级飞羽白色，初级飞羽后缘黑色。虹膜褐色；嘴红色（繁殖期）或橘红色具黑端；脚红色（幼鸟色较淡）。

生境与习性 栖于平原和低山丘陵地带的湖泊、河流、水库、河口及海滨等。常浮于水上或立于漂浮物及岸边岩石、沙滩上。上下翻飞，俯冲取食。主要以小鱼、虾、水生昆虫、甲壳类、软体动物等水生无脊椎动物为食，也吃鼠、蜥蜴等小型陆生脊椎动物及小型动物尸体。

分布 繁殖于古北区。南迁至印度及东南亚越冬。在井冈山见于新城区等地。

种群状况 在井冈山冬季常见。

3.60a

Identification L ca. 40 cm. Smallish grey and white gull. Breeding adult has dark brown hood with narrow broken white eyering, mantle and wings grey with black wing-tips, the rest white. Non-breeding adult lacks hood but has blackish spot on rear cheek and grey smudge above eye. Juvenile resembles winter adult, but have brown tertials and wing coverts, black wingtips and narrow black terminal tail-band. In flight, wings narrow, with pointed tip; mostly pale grey but outer primaries white, forming broad white leading edge, though all outer primaries have small black tip. Eyes brown; bill dark blood-red (breed), orange-red with black tip; tarsi dull red, paler in winter, orange in juvenile.

Habitat and Behavior Inhabits wetlands from coastal saltmarshes to inland lakes, rivers and swamps. May also frequent parks. Often floats on the surface, perch on floats, or reef. Aerially agile, hovering frequently and with frequent changes of direction. Feeds on fish and invertebrates, sometimes corpses of animals.

Distribution Breeds Palearctic. Migrant to India and SE Asia. At Xinchengqu in Mt. Jinggang.

Status of Population Common in winter in Mt. Jinggang.

3.60b

鸽形目 COLUMBIFORMES
鸠鸽科 Columbidae

3.61 山斑鸠 Streptopelia orientalis (Latham, 1790)

识别特征 中等体型、粗壮的斑鸠，约 32 cm。颈侧有具黑白色条纹的块状斑。上体的深色鳞片状体羽缘棕色，腰蓝灰，尾羽近黑，尾梢浅灰。下体多偏粉色。虹膜橙黄色；嘴铅灰色；脚粉红色。

生境与习性 栖息于多树地区，常在丘陵、山脚及平原。常结

3.61a

群活动，亦见与珠颈斑鸠混群。食物主要为植物性，包括植物种子、幼芽、嫩叶、果实及农作物等，也食蜗牛、昆虫等动物性食物。起飞时猛然向上，振翅有声，尾呈扇形，然后滑翔。营巢于树上或灌木丛间。窝卵数常 2 枚。孵化期约 18 天。双亲均参加孵卵和育雏。

分布　喜马拉雅、印度、东北亚、日本、中国。北方鸟南下越冬。在井冈山见于八面山、湘洲、厦坪等地。

种群状况　在井冈山全年常见。

Identification　L ca. 32 cm. Mid-sized, rather dumpy dove with white and black neck bars. Head pinkish-brown, upperwing coverts dark with scaly rufous fringes, rump blue-grey, tail-tips pale grey. Underparts pink. Eyes orange-yellow; bill dark grey; tarsi dull pink.

Habitat and Behavior　Inhabits forest, woodland edge, farmland, parkland and urban gardens. Usually in groups, sometimes mixed with spotted dove. Mostly phytophagous, sometimes snails and insects. In aerial display swoops steeply upward with loud wing clapping and fanned tail, then gliding. Builds nests at mid-height in tree, extralimitally sometimes on the ground. Clutch size is 2 eggs. Incubation period is 18 days. Both parents incubate and breed.

Distribution　Himalaya, India, NE Asia, Japan, China. Northern birds migrate to south of range. At Bamianshan, Xiaping and Xiangzhou in Mt. Jinggang.

Status of Population　Common all year round in Mt. Jinggang.

3.62 珠颈斑鸠 *Spilopelia chinensis* (Scopoli, 1786)

识别特征　中等体型、略纤瘦的斑鸠，约 30 cm。颈侧具缀满白点的黑色块斑。上体灰褐、较山斑鸠单调，下体粉红。飞行时，翼内缘青灰色，深褐色的尾略显长，外侧尾羽末端白色明显。虹膜橙色；嘴深灰色；脚暗粉红色。

生境与习性　栖息于有疏树的草地、丘陵、郊野农田或住家附近，也见于潮湿的阔叶林。常结成小群，有时和山斑鸠等其他鸠类混群。在树上停歇或在地面觅食，受惊时飞到附近的树上，拍翼咔嗒有声。食物以植物种子，特别是农作物种子为主。巢通常位于树上或在矮树丛和灌木丛间，也见于山边岩石的裂缝中。窝卵数常 2 枚。孵化期约 18 天。雌雄均参加孵卵和育雏。

分布　常见并广布于东南亚。经小巽他群岛引种其他各地远及澳大利亚。广泛分布于井冈山。

种群状况　在井冈山全年常见。

Identification　L ca. 30 cm. Mid-sized, rather slender dove with prominent white-spotted black patches on neck-sides. Head grey, upperparts plain grey-brown, but somewhat scalloped; underparts dark pinkish-grey. In flight, long graduated tail mainly blackish-brown with white tips to outer feathers, wing-lining slaty. Eyes orange; bill dark grey; tarsi dull dark pink.

Habitat and Behavior　Inhabits lowlands, farmland, villages, urban and suburban areas and parks, also moist deciduous forest. Usually in small groups, may mix with Oriental. Largely terrestrial, also perch on trees. Flight typically rather slow and low with wing-hit noise at the beginning. Feeds on seeds, especially grains. Nests in low trees or bushes, sometimes in crevices of mountains. Clutch size is 2 eggs.

3.61b

3.62a

3.62b

Incubation period is 18 days. Both parents brood and raise offsprings.

Distribution　Widely distributed and common in SE Asia to Lesser Sunda Islands and introduced elsewhere as far as Australis. Widespread in Mt. Jinggang.

Status of Population　Common all year round in Mt. Jinggang.

3.63 火斑鸠 *Streptopelia tranquebarica* (Hermann, 1804)

3.63a

识别特征　体型较小、体色较深的斑鸠，约 23 cm。后颈具黑色半领环。雄鸟头部蓝灰，上体酒红，初级飞羽近黑，腰和尾深石板色，外侧尾端白色，下体浅棕红色。雌鸟色较暗淡，头暗棕色，体羽红色较少。虹膜暗褐色；嘴灰色；脚褐红色。

生境与习性　常结群活动于有稀树的开阔田野、村庄附近，有时与山斑鸠、珠颈斑鸠混群。在地面急切地边走边找食物。主要啄食植物性食物。营巢于丛林里的乔木上或稀疏的树林中，一般置于树冠外围较茂密绿叶中，隐蔽很好。窝卵数常 2 枚。

分布　喜马拉雅、印度、中国至东南亚。在井冈山见于八面山等地。

种群状况　在井冈山全年可见。

Identification　L ca. 23 cm. Small dove with rather dark plumage and black hind-collar. Male has pale blue-grey head, deep brownish-pink upperparts and pale grey rump; flight feathers and tail blackish-grey with white tips to black outer tail feathers. Underparts pale brownish-pink. Female generally duller, less pinkish-brown. Eyes brown; bill grey; tarsi dull grey to dark purplish-red.

3.63b

Habitat and Behavior　Inhabits scrub, at woodland edge in open country and farmland with trees, hedges and isolated stands of trees. Usually forage on ground. Phytophagous. Nests placed fairly high in trees or bushes, often covered by dense leaves. Clutch size is 2 eggs.

Distribution　Himalaya, India, China to E Asia. At Bamianshan in Mt. Jinggang.

Status of Population　Visible all year round in Mt. Jinggang.

3.64 斑尾鹃鸠 *Macropygia unchall* (Wagler, 1827)

识别特征　体大、尾长的深褐色鸠，约 38 cm。雄鸟头灰，颈背呈亮绿色，背及尾褐色，满布黑色或深褐色横斑。颈至胸侧具狭窄波状纹，胸腹偏粉，至臀部渐变为白色。雌鸟无亮绿色。虹膜黄色或浅褐色；嘴黑色；脚紫褐色。

生境与习性　栖于丘陵的森林中。成对或结小群活动。疾速穿越树冠层，落地时尾上举。常倒挂着觅食。以野果等植物性食物为主。营巢于树上或灌木丛间。窝卵数仅 1 枚，也有 2 枚的记录。

分布　喜马拉雅至东南亚、爪哇（印尼）及巴厘岛。在井冈山见于八面山、湘洲等地。

种群状况　在井冈山罕见。国家 II 级保护野生动物。

Identification　L ca. 38 cm. Large, long-tailed, dark brown dove with pale vent. Male head pinkish-grey, with iridescent green on hindneck; upperparts and tail blackish-brown with heavy rufous-brown bars, narrow black scalloping from neck to breast-sides; breast pinkish to pale vent. Female lacks iridescent green. Eyes yellow or pale brown; bill black; tarsi dull purplish-brown.

3.64

Habitat and Behavior　Inhabits dense subtropical forest on montane slopes. In pairs or small flocks. Flies fast through canopy; on ground may raise long tail. Acrobatic while feeding, even hanging upside-down. Largely on small fruit. Builds nests on trees or bushes. Clutch size is 1 or 2.

Distribution　Himalaya to SE Asia, Java and Bali Island. At Bamianshan and Xiangzhou in Mt. Jinggang.

Status of Population　Rare in Mt. Jinggang. China Key List: II.

鹃形目 CUCULIFORMES
杜鹃科 Cuculidae

3.65 褐翅鸦鹃 *Centropus sinensis* (Stephens, 1815)

识别特征　体大而尾长的褐、黑色杜鹃，约 52 cm。成鸟上背、翼为纯栗红色，余部黑色而带有光泽。亚成体具多少不一的横纹。虹膜红色（成鸟）或灰蓝至暗褐色（幼鸟）；嘴黑色；脚黑色。

生境与习性　喜林缘地带、次生灌木丛、多芦苇河岸及红树林。单只或成对活动。在矮丛顶鸣叫，常下至地面活动或在浓密灌丛中攀爬。晨昏常见在芦苇顶上晒太阳。善走而拙于飞行。动物性食物为主，包括昆虫、蚯蚓、软体动物、蜥蜴、蛇、田鼠、鸟卵、雏等。巢成粗糙球状，出入口开于侧方。窝卵数约 5 枚。

分布　印度、中国、东南亚及大巽他群岛。在井冈山见于宁冈等地。

种群状况　在井冈山全年可见。国家 II 级保护野生动物。

Identification　L ca. 52 cm. Large cuckoo with long graduated tail. Adult mantle and wings chestnut, the rest glossy black. Juvenile has fine dark bars on wings. Eyes red (ad.) or greyish-blue to brown (juv.); bill black; tarsi black.

Habitat and Behavior　Inhabits forest edges, dense scrub and reedy areas, mangrove. Usually solitary or in pairs. Often skulking, clambering awkwardly in cover, or may sun itself, flopping on top of bush with song; partly predatory, taking small prey such as snail, some carrion, robbing nests. Nest bulky, usually domed with a entrance on the side. Clutch size is 5 eggs.

Distribution　India, China, SE Asia, Greater Sunda Islands. At Ninggang in Mt. Jinggang.

Status of Population　Visible all year round in Mt. Jinggang. China Key List: II.

3.66 小鸦鹃 *Centropus bengalensis* (Gmelin, JF, 1788)

识别特征　似褐翅鸦鹃但体型较小，约 42 cm，嘴和尾亦显短。繁殖期成鸟头、下体及尾乌黑，上背暗栗色，两翼及翼下栗色，肩和翼上覆羽具浅色矛状纹。非繁殖期成鸟上体褐色，具密集的矛状纹，翼红棕色，下体色浅，胸胁具细横纹。幼鸟似非繁殖期成鸟，但双翼和尾多褐色横纹，黄褐色的头、颈部具浅色纵纹。虹膜红褐色；嘴黑色（成鸟）或角质色（幼鸟）；脚铅黑色。

生境与习性　喜山边灌木丛、沼泽地带及开阔的草地。常栖地面，有时作短距离的飞行，由植被上掠过。性机警而隐蔽，稍受惊就奔入密丛深处。食物主要为昆虫和其他小型动物。营巢于茂密的矮植物丛中，巢圆球形。常 3 枚卵，雌雄均参与育雏。

分布　印度、中国及东南亚。在井冈山见于新城区、黄坳等地。

种群状况　在井冈山偶见。国家Ⅱ级保护野生动物。

Identification　L ca. 42 cm. Very similar to Greater Coucal, but smaller with shorter bill and tail. Breeding adult dull or dirty black, dull chestnut mantle, warm rufous-brown wings with pale shaft-streaks on scapulars and wing coverts, and underwing coverts chestnut. Non-breeding adult upperparts brown with heavy white shaft-streaks, wings rufous, underparts otherwise rufous-white barred dusky. Juvenile resembles non-breeding adult but has dark barring on brown wings and tail, tawny head and neck with dark streaks, and tawny underparts. Eyes reddish-brown; bill black (ad.) or greyish-horn (juv.); tarsi blackish.

Habitat and Behavior　Inhabits scrub, farmland, grassland and marsh edges. Alert, usually stays hidden in grass, sometimes perching above it. Builds domed nest of grass, incorporating standing grasses. Both sexes participate in care of young. Clutch size is 3 eggs.

Distribution　India, China and SE Asia. At Xinchengqu and Huang'ao in Mt. Jinggang.

Status of Population　Rare in Mt. Jinggang. China Key List: II.

3.66b

3.67 大杜鹃 *Cuculus canorus* Linnaeus, 1758

识别特征　中等体型的杜鹃，约32 cm。上体灰色，尾偏黑色，腹部近白而具黑色横斑，横纹较细。棕红色变异型雌鸟为棕色，背部具黑色横斑。幼鸟上体灰褐色，有白色细横纹和淡色枕斑。虹膜及眼圈黄色；嘴黑褐色，基部及嘴角处近黄色；脚黄色。

生境与习性　喜开阔的有林地带、农田、草地及大片芦苇地。多单独活动，有时成对，难见集群。有时停在电线上鸣叫。飞行快速而有力，循直线前进，嗜吃各种毛虫，特别是繁殖期间几乎纯以毛虫为食，并吃柔软昆虫。巢寄生，产一枚卵寄孵于大苇莺、麻雀、灰喜鹊、棕头鸦雀、棕扇尾莺、北红尾鸲、红尾伯劳、灰头鹀、白鹡鸰等鸟巢中。卵先产于地面，后用嘴衔着放入小鸟巢中。叫声响亮清澈，二声一度。

分布　繁殖于欧亚大陆。迁徙至非洲及东南亚。在井冈山见于小溪洞等地。

种群状况　在井冈山罕见。

3.67a

Identification　L ca. 32 cm. Mid-sized cuckoo. Adult upperparts grey and tail blackish. Underparts white, greyish-black bars on breast and flanks narrower than Oriental, vent and undertail coverts whiter (not buff). Juvenile finely barred brown and grey across upperparts, rump and tail, with white nape patch. Eyering orange (ad.) or dark (juv.); bill grey above and yellowish below and at gape; tarsi yellow.

Habitat and Behavior　Inhabits open grassland, farmland, reed beds, marshes, also parkland, low woodland and taiga forest to limit of tundra. Each female parasitizes a single species and lays eggs to match that species. Parasitic on a variety of small passerines, mostly chats, also pipits and buntings. Two-noted phrase.

Distribution　Breeds in Eurasia. Migrating to Africa and SE Asia. At Xiaoxidong in Mt. Jinggang.

Status of Population　Rare in Mt. Jinggang.

3.67b

3.68 中杜鹃 *Cuculus saturatus* Blyth, 1843

识别特征　体型略小的杜鹃，约 26 cm。似大杜鹃，但臀部微沾褐色，腹部的横纹色深且较宽，翼下有一条显眼的浅色带。赤色型雌鸟常见。虹膜红褐色，眼圈黄色；上嘴暗角绿色，下嘴尖端角绿色，其余大部角黄色；脚橘黄色。

生境与习性　在高地林区繁殖，栖息过境时广泛分布于各类生境。常隐于林冠，除春季繁殖期叫声非常频繁外很难见到。嗜吃柔软的昆虫和毛虫。产卵于灰背燕尾、冠纹柳莺、缝叶莺、灰头鹪莺等巢中。卵的颜色随寄主的不同而变异，只有形状和大小可区别。音为四声，第一声高调，随着 3 个同音调的音节。

分布　繁殖于欧亚大陆北部及喜马拉雅。冬季至东南亚及大巽他群岛。在井冈山地区见于黄洋界、大院农场等地。

种群状况　在井冈山地区春季常见。

Identification　L ca. 26 cm. Smallish cuckoo with brown wash on vent. Similar to Eurasian Cuckoo, but darker and broader bars on underparts, pale bared underwing. "Rufous" type female more common. Eyes orange to brown (ad.), or dark (juv.), yellow eyering; bill yellow with dark green tip; tarsi yellow.

Habitat and Behavior　Inhabits mixed forests and orchards in hilly country. Favors canopy. Secretive but highly vocal. Feeds on caterpillars and other large insects within foliage, and in sally-flight. Brood parasite of forktails and warblers. Four-noted phrases.

Distribution　Breeds in N Eurasia and Himalaya. Migrating in winter to SE Asia and Greater Sunda Islands. Resident races occur on the Greater Sundas. At Huangyangjie and Dayuannongchang in the Jinggangshan Region.

Status of Population　Common in spring in the Jinggangshan Region.

3.68

3.69

3.69 小杜鹃 *Cuculus poliocephalus* Latham, 1790

识别特征　体型较小的杜鹃，约 26 cm。上体灰色，头、颈及上胸浅灰色。白色腹部具间隔较宽的深色横纹，臀部及尾下覆羽皮黄色，无横纹。尾灰，无横斑，具狭窄白端，尾下色深具白点。棕色型雌鸟全身具黑色条纹。虹膜褐色，眼圈黄色；上嘴深灰，下嘴及嘴角黄色，端黑；脚黄色。

生境与习性　栖于低地和山地的落叶林或常绿阔叶林。飞行敏捷，常隐匿于稠密的树叶间。昼夜皆鸣叫，鸣声独特易辨。食物以昆虫为主。产卵于鹪鹩、白腹蓝鹟、莺亚科及画眉亚科小鸟等巢中。卵颜色多与寄主相似。

分布　喜马拉雅至印度、中国中部及日本。越冬在非洲、印度南部及缅甸。在井冈山地区见于荆竹山、牛石坪等地。

种群状况　在井冈山地区春季常见。

Identification　L ca. 26 cm. Smallish cuckoo with yellow eyering. Hood and upper breast pale grey, upperparts grey. Blackish barring on white underparts wide spaced, enhanced by unbarred vent and buffy-grey undertail coverts. Undertail blackish with white spots, uppertail grey with narrow white tip. Hepatic female bright rufous with indistinct barring entirely. Eyes brown; bill grey above, yellowish below and at gape; tarsi yellow.

Habitat and Behavior　Inhabits deciduous and evergreen broadleaf forest in lowlands and hills. Flight quick over canopy, often perches in dense trees. Elusive, but highly vocal day and night. Feeds on insects. Brood parasite of laughingthrush and cettia warblers.

Distribution　Himalaya to India, Central China and Japan. Winters to Africa, S India and Myanmar. At Jingzhushan and Niushiping in the Jinggangshan Region.

Status of Population　Common in spring in the Jinggangshan Region.

3.70 四声杜鹃 *Cuculus micropterus* Gould, 1838

识别特征　中等体型的杜鹃，约 30 cm。似大杜鹃，但成年雄鸟头颈灰色，与略染棕的深灰色上体型成对比；下体白，具间距较宽的黑横斑；尾灰并具宽阔的黑色次端斑及狭窄白端，尾下白点形成横纹。雌鸟头、背多褐色。幼鸟头及上背具偏白的皮黄色鳞状斑纹，有宽阔尾带。虹膜暗褐色；嘴深灰色，基部黄色；脚黄色。

生境与习性　栖息于平原至高山森林中，有时也出现在农田边缘。甚隐蔽，常只闻其声不见其影。食物以昆虫为主，特别是毛虫。卵寄孵于雀形目鸟类巢中，其寄主国内记录有大苇莺、灰喜鹊、灰卷尾等。鸣声洪亮，四声一度，每度反复间隔 2-3 秒，常从早到晚经久不息，又以天亮时为甚。叫声似"gue-gue-gue-guo"。

分布　东亚、东南亚、菲律宾、加里曼丹岛、苏门答腊并附近岛屿及爪哇（印尼）西部。在井冈山见于白银湖、茅坪、湘洲等地。

种群状况　在井冈山春季常见。

3.70

Identification　L ca. 30 cm. Mid-sized cuckoo with yellow eyering. Similar to Himalayan Cuckoo. Adult male hood to lower neck grey, contrast with brownish-dark grey upperpart. Underparts white, with widely spaced black bands from the breast to the vent. Tail long, dark grey above with broad blackish subterminal band and white tip, below pale to dark grey with bands of white spots. Female has brownish-grey hood and dark brown upperparts. Juvenile has whitish or buff scaling on head and back, and dark bands on tail. Eyes brown; bill dark grey, and yellow at gape; tarsi yellow.

Habitat and Behavior　Inhabits lowland to hilly deciduous and evergreen forest and forest edges, sometimes occurs at farmland edge. Elusive, but highly vocal. Giving a repetitive, loud, four-syllable call. Feeds on insects, especially caterpillars. Brood parasite of Oriental Reed Warbler, Magpie and Ashy Drongo.

Distribution　S and SE Asia, the Philippines, Kalimanta Island, Sumatra with offshore islands and W Java. At Baiyinhu, Maoping and Xiangzhou in Mt. Jinggang.

Status of Population　Common in spring in Mt. Jinggang.

3.71 噪鹃 *Eudynamys scolopaceus* (Linnaeus, 1758)

识别特征　体型较大、尾长的杜鹃，约 42 cm。雄鸟全身黑色带钢蓝色光泽。雌鸟深灰色染褐，并具大量白斑，在腹部形成横纹。虹膜深红色；嘴暗绿色；脚蓝灰色。

生境与习性　栖于稠密或开阔的森林，也常出现在果园、灌丛或园林。常隐蔽于大树顶层密集的叶簇中，若不鸣叫则很难发现，受惊时立即飞离远去。飞行快速而无声。食物比一般杜鹃杂，野果、种子、其他植物性食物及昆虫都吃。卵寄孵在黑领椋鸟、喜鹊、红嘴蓝鹊等巢中。迁来时可闻其嘹亮"koe-wow"声，重音在第二音节，重复 5 至 10 次，音速音高渐增。也有更尖声刺耳、速度更快的"快,快,快,快"声。

分布　印度、中国、东南亚及印度尼西亚。在井冈山见于长古岭等地。

种群状况　在井冈山夏季可见。

3.71a

Identification　L ca. 42 cm. Large cuckoo with long, broad tail. Male glossy bluish-black. Female brownish-grey with numerous white spots on upperparts and neck, barring across entire underparts. Eyes red; bill dull-green; tarsi blue-grey.

Habitat and Behavior　Inhabits dense forest, open forest, plantations, orchards, scrub and gardens. Skulking in leaves, flies away immediately and quietly when disturbed. Feeds on insects and fruit. Brood parasite of Large-billed Crow and Common Magpie. Giving shrill, fast, oft-repeated slurred whistles ko-el rising in pitch and frequency, a loud repetitive koe-wow, by day or at night.

Distribution　India, China, SE Asia and Indonesia. At Changguling

3.71b

in Mt. Jinggang.

Status of Population　Visible in summer in Mt. Jinggang.

鸮形目 STRIGIFORMES
草鸮科 Tytonidae

3.72 草鸮 *Tyto longimembris* (Jerdon, 1839)

识别特征　中等体型的猫头鹰，约 35 cm。面庞心形，脚长。似仓鸮，但脸及胸部的褐色浓重，上体深褐。全身多具点斑、杂斑或蠕虫状细纹。虹膜暗褐色；嘴偏白色；脚褐色。

生境与习性　栖息于山坡草地或开阔草原。夜行性，多在黄昏及夜间活动，白天隐藏在茂密的草灌丛中。受干扰时，只能作短距离飞行，飞时左右摇摆，很不规则。主食鼠类、麻雀等小动物，进食 2-3 h 后吐出残留食块，内含不能消化的毛和骨。

分布　非洲、新几内亚、日本、澳大利亚、印度次大陆至中国（西南及华南）、东南亚。在井冈山见于大井等地。

种群状况　在井冈山可见。CITES 附录 II。国家 II 级保护野生动物。

Identification　L ca. 35 cm. Mid-sized owl with heart-shaped buff facial disc. Similar to Barn Owl, but much darker brown facial and breast. Upperparts dark brown and tawny, with some fine silvery spots and dark smudges on crown, mantle and wings. Eyes dark brown; bill pale; tarsi pale brown.

Habitat and Behavior　Inhabits open areas with tall grassland. Crepuscular, roosts in grasses by day. Various shrieking and screaming sounds in flight. Feeds on rats and birds. Disgorges pellets of fur, bones or feathers after 2-3 hours.

Distribution　Africa, New Guinea, Japan, Australia, Indian subcontinent to SW and S China and SE Asia. At Dajing in Mt. Jinggang.

Status of Population　Visible in Mt. Jinggang. CITES App. II. China Key List: II.

鸮形目 STRIGIFORMES
鸱鸮科 Strigidae

3.73 领角鸮 *Otus lettia* (Hodgson, 1836)

识别特征　体型略大的角鸮，约 24 cm。具明显耳羽簇及特征性的浅沙色颈圈。上体偏灰或沙褐，具黑色及皮黄色蠹纹；下体浅褐色，具黑色纵纹。虹膜暗褐色；嘴角质色；脚污黄。

生境与习性　栖于森林、林缘、农田、公园及有树的街道。夜行性鸟类，白天大都隐藏在具浓密枝叶的树冠上，或其他阴暗的地方，一动不动，黄昏至黎明前较活跃。常鸣叫。主要以鼠类、小鸟、昆虫为食。常产卵于天然树洞、啄木鸟的旧洞或喜鹊的旧巢中。窝卵数 2-5 枚，以 3 枚居多。发出轻柔的 "wuh" 声，约 4-10 s 重复一次。

分布　印度次大陆、东亚及东南亚。在井冈山见于茨坪、茅坪、湘洲等地。

种群状况　在井冈山全年常见。CITES 附录 II。国家 II 级保护野生动物。

Identification　L ca. 24 cm. Rather large scops owl with particularly

large head and conspicuous spotted ear-tufts. Upperparts generally greyish-brown or dull brown mottled black and buff, with pale brown nuchal collar; underparts pale brown with black streaks. Eyes yellowish-brown; bill grey horn; tarsi feathered, toes pale brown.

Habitat and Behavior Inhabits farmland, tree-lined streets, parks, woodland edge and forest. Crepuscular and nocturnal. Feeds on large insects and small vertebrates. Breeds in hole in tree trunk, or scrap nests of Magpie. Clutch size is 2-5 eggs, mostly 3. Gives a mellow, resonant, down slurred wuh! Repeated every 4-10 seconds.

Distribution Indian subcontinent, E Asia and SE Asia. At Ciping, Maoping and Xiangzhou in Mt. Jinggang.

Status of Population Common all year round in Mt. Jinggang. CITES App. II. China Key List: II.

3.74a

3.74 红角鸮 *Otus sunia* (Hodgson, 1836)

识别特征 体小的角鸮，约 19 cm。分灰色型及棕色型。肩羽大都有一条显眼白色带，下体满布黑色条纹。与领角鸮区别在于体小，眼为黄色且无浅色颈圈；与黄嘴角鸮的区别在于胸具黑色条纹，下体灰色重。虹膜黄色至浅橙色；嘴角质灰色；脚灰。

生境与习性 夜行性，于林缘、林中空地及次生植被的小矮树上捕食甲虫及其他昆虫。飞行迅速有力。受惊时竖起角羽。窝卵数 3-5 枚。孵卵期 22 天以上。育雏期约 25 天，25 日龄的雏鸟已能飞翔。

分布 繁殖于喜马拉雅、印度次大陆、东亚及东南亚。有些至南方越冬。在井冈山地区见于茨坪、荆竹山、八面山、大院农场等地。

种群状况 在井冈山地区全年常见。CITES 附录Ⅱ。国家Ⅱ级保护野生动物。

3.74b

Identification L ca. 19 cm. Small scops owl with prominent ear-tufts. Two color forms, grey morph and rufous morph. Marked white band from shoulder to lower back, grey-brown or rufous underparts with strong black streaks. Smaller than Collared Scops Owl, eye yellow, and lack pale nuchal collar; greyer than Mountain Scops Owl, and more dark streaks on chest. Eyes yellow to pale orange; bill pale grey-horn; toes grey to greyish-flesh colored.

Habitat and Behavior Inhabits montane forest, woodland edge, parks, large gardens and coniferous taiga forest. Crepuscular and nocturnal, roosts by day in thick vegetation. Erects ear-tufts when alarmed. Feeds mainly on beetles and other insects. Clutch size is 3-5 eggs. Incubation period over 22 days. Brooding period is 25 days.

Distribution Breeds Himalayas, Indian subcontinent, E and SE Asia. Some winter south. At Ciping, Jingzhushan, Bamianshan and Dayuannongchang in the Jinggangshan Region.

Status of Population Common all year round in the Jinggangshan Region. CITES App. II. China Key List: II.

3.75 黄嘴角鸮 *Otus spilocephalus* (Blyth, 1846)

识别特征 体型较小的褐色角鸮，约 18 cm。耳羽簇较短，肩部有一排白斑。亚种 *latouchi* 上体红棕色，具细小的深色杂斑，肩部白斑较模糊，浅褐色的脸盘具深棕色边缘，下体黄褐色具

3.75

银色斑点和深褐色斑纹。亚种 *hambroecki* 上体深褐色，肩部白斑明显。眼黄色；嘴角质色至乳白色；趾灰色。

生境与习性　栖息于山地常绿阔叶林或混交林中。夜行性，白天常栖居在树洞中。在森林中下层觅食大型昆虫和小型哺乳类。全年均有轻柔、悠远的双音节金属哨声"plew-plew"，每隔 5-10s/ 次。

分布　喜马拉雅、印度次大陆的东北部、中国南部、东南亚、苏门答腊及加里曼丹岛北部。在井冈山见于茨坪、茅坪、双溪口等地。

种群状况　在井冈山全年常见。CITES 附录 Ⅱ。国家 Ⅱ级保护野生动物。

Identification　L ca. 18 cm. Small brown scops owl with short ear-tufts, white bar across scapulars. *O. s. latouchi* upperparts rufous-brown with fine dark mottling, a faint hind-collar, facial disc pale brown rimmed blackish-brown, and underparts buff-brown with silver spots and dark brown fleck. *O. s. hambroecki* upperparts dark brown with prominent pale hind-collar. Eyes yellow; bill pale horn to creamy white; toes grey.

Habitat and Behavior　Inhabits montane broadleaf and mixed evergreen forest. Strictly nocturnal, roosting in tree-hollow by day. Forages below canopy and understory, hunting for large insects and small mammals. Calls frequently, a level fluty two-note whistle, "plew-plew", repeated monotonously at 5-10 second intervals for several minutes.

Distribution　Himalaya, NE Indian subcontinent, southern China, SE Asia, Sumatra and N Kalimanta Island. At Ciping, Maoping and Shuangxikou in Mt. Jinggang.

Status of Population　Common all year round in Mt. Jinggang. CITES App. II. China Key List: II.

3.76 领鸺鹠 *Glaucidium brodiei* (Burton, 1836)

识别特征　体型纤小、浑圆的猫头鹰，约 16 cm。头大、无耳羽簇，尾显长、具狭窄横斑。上体红褐色或灰褐色，头顶具浅色细斑，面盘具白色短眉纹和显著领圈，头后具一对假眼；下体白色，胸、胁具宽阔的褐色纵纹。虹膜黄色；嘴浅黄绿色至角质色；脚黄绿色。

生境与习性　栖息于山脚至高山的森林，偏好高大树木。多在日间及晨昏活动。性勇猛，常攻击猎取几乎和它等大的猎物，主要以鼠类、小鸟及昆虫为食。营巢于天然洞穴，或强占拟啄木鸟或啄木鸟的巢繁殖。窝卵数常 4 枚，有时 3 枚或 5 枚。昼夜发出圆润的单一哨音"pho, pho-pho, pho"。仿其叫声可非常容易地招引此鸟，也会引来那些围攻领鸺鹠的小型鸣禽。

分布　喜马拉雅至中国南部、东南亚、苏门答腊及加里曼丹岛。在井冈山见于八面山、五指峰、湘洲等地。

种群状况　在井冈山全年常见。CITES 附录 Ⅱ。国家 Ⅱ级保护野生动物。

Identification　L ca. 16 cm. Tiny plump owl with large head (lacks ear-tufts), and rather long tail. Upperparts reddish- or greyish-brown, tail narrowly banded. Fine pale speckle on crown. Facial disc with short white eyebrows and white collar. Broad pale hindcollar with large black "eyes", creating prominent false face. Underparts white with broad brown streaks on chest, belly and flanks. Eyes yellow;

3.76

bill horn to pale yellow-green; toes yellow.

Habitat and Behavior　Inhabits forest from foothills to high mountains, preferring taller trees. Diurnal and crepuscular, often hunts birds, murines and insects. Frequently mobbed by small birds. Flight fast. Nests in tree cavities, which it may usurp from barbet or woodpecker. Clutch size is 3-5 eggs, mostly 4. Soft series of clear, level three- or four-note hoots, "pho, pho-pho, pho", repeated for several minutes day or night. Attracted by mimic easily.

Distribution　Himalaya to Southern China, SE Asia, Sumatra and Kalimanta Island. At Bamianshan, Wuzhifeng and Xiangzhou in Mt. Jinggang.

Status of Population　Common all year round in Mt. Jinggang. CITES App. II. China Key List: II.

3.77 斑头鸺鹠 *Glaucidium cuculoides* (Vigors, 1830)

识别特征　体较小的鸺鹠，约 24 cm。无耳羽簇，白色的颏纹明显；上体褐色且具浅黄色横斑，沿肩部有一道白色线；下体几全褐，具深褐色横斑；臀部白，两胁栗色；近黑色的尾具间距较宽的白色细横斑。虹膜橙黄色；嘴偏绿而端黄；脚绿黄色。

生境与习性　栖息的生境较广泛，森林、农田、村庄、公园都可见。昼夜都活动。站姿比其他鸺鹠接近水平方向。低空起伏飞行。能像鹰那样在空中捕捉小鸟和大型昆虫，也吃蛙、鼠等。营巢于天然洞穴，有时也抢占其他鸟类的洞巢。晨昏时发出快速的颤音，调降而音量增。另发出一种似犬叫的双哨音。

分布　喜马拉雅、印度东北部至中国南部及东南亚。在井冈山见于湘洲、宁冈等地。

种群状况　在井冈山全年常见。CITES 附录 Ⅱ。国家 Ⅱ级保护野生动物。

Identification　L ca. 24 cm. Small, plump brown owl with heavy buff bars on head and upperparts. Narrow white eyebrows joined white whisker lines; broad collar white; narrow pale bar on scapulars. Underparts barred rufous-brown and pale rufous on breast and flanks, vent pale. Widely spaced white bars on blackish tail. Eyes orange-yellow; bill horn; toes yellow.

Habitat and Behavior Inhabits gardens, parks, secondary and primary forest, from lowlands to mountains. Diurnal and nocturnal; slightly level stance, wags tail when alarmed. Flight undulating. Feeds on birds and insects, frogs and rats. Nests in natural hollow, sometimes appropriated from another hole-nesting bird. Quavering trill increasing steadily in volume. "Wu u u u u u u u u", also various squawks, churrs and barks.

Distribution Himalaya, NE India to Southern China and SE Asia. At Xiangzhou and Ninggang in Mt. Jinggang.

Status of Population Common all year round in Mt. Jinggang. CITES App. II. China Key List: II.

3.77a

咬鹃目 TROGONIFORMES
咬鹃科 Torgonidae

3.78 红头咬鹃 *Harpactes erythrocephalus* (Gould, 1834)

识别特征 体大、尾长的棕褐色咬鹃,约33 cm。雄鸟头颈深红色,眼周裸皮蓝灰色,白色胸斑将鲜红的下体与头颈分离开来;上体棕褐色,翼覆羽有白色波浪纹;中央尾羽红棕色具黑端,外侧尾羽黑色,尾下黑白相间。雌鸟头黄褐色,下体浅红色。虹膜深红色;嘴蓝灰色;脚灰色。

生境与习性 栖息于热带、亚热带的密林。常单个或成对活动。树栖性,或静立于树枝上伺机突袭飞过的昆虫,或攀于小乔木的顶枝间啄食野果。飞行力较差,虽快而不远,略呈波浪状。以天然树洞或啄木鸟废弃的巢洞为巢。窝卵数3-4枚。雌雄均参与孵卵、育雏。

分布 喜马拉雅至中国南部、东南亚及苏门答腊。在井冈山见于水口、西坪等地。

种群状况 在井冈山春夏季常见。

3.77b

Identification L ca. 33 cm. Large rufous trogon with long tail. Male has deep red hood with blue-grey eyering skin, divided from red underparts by narrow white breast band. Upperparts and uppertail rufous-brown, wings black with narrow white barring on wing coverts. Outer rectrices black; undertail graduated, appears white and black. Female has yellowish-brown hood and chest, rest of underparts pale red. Eyes dark red; bill blue-grey; tarsi grey.

Habitat and Behavior Inhabits dense tropical and subtropical forest. Solitary or in pairs. Perches still and very upright in lower and middle canopy, sallying out for insects. Sometimes snatch small fruit from tree. Flight fast and undulating. Nests in excavated cavities near the top of rotten tree-stumps. Clutch size is 3-4 eggs. Both parents incubate and raise chicks.

Distribution Himalaya to Southern China, SE Asia and Sumatra. At Shuikou and Xiping in Mt. Jinggang.

Status of Population Common in spring and summer in Mt. Jinggang.

3.78

夜鹰目 CAPRIMULGIFORMES
夜鹰科 Caprimulgidae

3.79 普通夜鹰 *Caprimulgus jotaka* Temminck & Schlegel, 1845

识别特征　中等体型、深灰褐色的夜鹰，约 28 cm。全身满布深色蠹状纹，尾上具宽阔黑横斑。头顶平而色深，嘴小而口裂大，深色的脸上具白色短髭纹，具一白色喉斑（有时从中断开）。飞行时，雄鸟尾部近末端有一条中间断开的白色横带，初级飞羽有小白斑；雌鸟无。虹膜黑色；嘴乌黑色；脚深灰色。

生境与习性　栖息于山区森林及林缘。白天伏贴于多树山坡的地面或树枝，黄昏时开始活跃。夜间常蹲在路上，两眼展开甚大，暗中闪亮。飞行飘忽无声，两翼鼓动缓慢。在森林上空飞捕昆虫为食。不营巢，卵产在地面或岩石上。窝卵数多 2 枚，雌雄共同孵卵。

分布　印度次大陆、中国及东南亚。南迁至印度尼西亚及新几内亚。在井冈山见于八面山、荆竹山等地。

种群状况　在井冈山夏季常见。

Identification　L ca. 28 cm. Mid-sized, dark grey-brown nightjar with black heavily vermiculated entirely, and broad black bands on uppertail. Crown flat and dark, face dark with white malar curves, large extensible gapes but tiny bills, greyish-black chin and neck-sides with broad white throat patch (sometimes broken in center). In flight, male has narrow white slash on wings and white subterminal tail-band, female has not. Eyes black; bill black; tarsi dark grey.

Habitat and Behavior　Inhabits broadleaf evergreen forest and forest edges. Roosts on the ground or lengthwise on tree branches by day, becoming active at dusk. Flight erratic and noiseless, when flushed flies rather lazily. Often on the road at night, show large and blink eyes. Feeds on insects over trees. Nests on bare ground. Clutch size is 2 eggs. Both parents incubate.

Distribution　Indian subcontinent, China and SE Asia. Migrates to Indonesia and New Guinea. At Bamianshan and Jingzhushan in Mt. Jinggang.

Status of Population　Common in summer in Mt. Jinggang.

雨燕目 APODIFORMES
雨燕科 Apodidae

3.80 白腰雨燕 *Apus pacificus* (Latham, 1801)

识别特征　体型略大、纤长的雨燕，约 18 cm。双翼狭长成镰刀状，尾叉较深。通体黑褐色，腰部白斑较窄，喉部色略浅；下体具皮黄色鳞状纹。虹膜深褐色；嘴黑色；脚黑色。

生境与习性　出现在热带至北极的低地至高山。大多集飞在近山地带，也常与其他雨燕混合，飞行时常发出高频的颤音。进食时做不规则的振翅和转弯。食物几纯为昆虫。结群营巢在山洞、悬崖峭壁或人不易接近的建筑上。窝卵数 2-3 枚。

分布　繁殖于西伯利亚及东亚。迁移经东南亚至印度尼西亚、新几内亚及澳大利亚越冬。在井冈山见于井冈冲水库等地。

种群状况　在井冈山夏季常见。

Identification L ca. 18 cm. Largish, slender swift with long, narrow sickle-shaped wings and deeply forked tail. Overall blackish-brown with narrow white rump patch and whitish throat, buff scaling on flanks, belly and vent. Eyes brown; bill black; tarsi black.

Habitat and Behavior Occurs tropics to Arctic, lowlands to high mountains. Often forages over diverse habitats in mixed flocks and usually highly vocal. Flight erratic, frequently turning. Feeds on insects. Nests colonially in cliff crevices and man-made structures. Clutch size is 2-3 eggs.

Distribution Breeds in Siberia and E Asia. Migrates in winter through SE Asia to Indonesia, New Guinea and Australia. At Jinggangchong Reservoir in Mt. Jinggang.

Status of Population Common in summer in Mt. Jinggang.

3.81a

佛法僧目 CORACIIFORMES
翠鸟科 Alcedinidae

3.81 普通翠鸟 *Alcedo atthis* (Linnaeus, 1758)

识别特征 体小、鲜艳的翠鸟,约15 cm。成鸟头、翼金属蓝绿色,头顶具亮蓝色鳞纹,橘黄色条带横贯眼部及耳羽,颈侧及颏白;背、腰及尾亮蓝色,翼上覆羽具亮蓝色斑点;下体及翼下橘红色。幼鸟色黯淡。虹膜褐色;嘴黑色(雌鸟下嘴橘黄色);脚红色。

生境与习性 单独或成对栖息于池塘、水库、湖泊、小溪等临近水的岩石或探出的树枝上。一见有饵,迅速直扑入水中叼取。有时悬停于空中,俯头注视水中,然后猛冲捕食。飞行疾速而径直,常低掠水面而过,并发出尖锐的叫声。主要捕食小鱼,兼食一些甲壳类和水生昆虫。营巢于田野堤基的沙土中,掘作隧道,通常距水较远。隧道深60cm左右。窝卵数6-7枚。

分布 广泛分布于欧亚大陆、东南亚、印度尼西亚至新几内亚。在井冈山见于茨坪、新城区、荆竹山、宁冈等地。

种群状况 在井冈山全年常见。

3.81b

Identification L ca. 15 cm. Small kingfisher with rufous-orange ear coverts. Adult has metallic greenish-blue head and wings, with bright blue scales on crown and spotted upperwing coverts; back, rump and tail shining blue. Orange spot on lores and band across ear coverts, white neck-sides and chin; underparts and underwing coverts bright orange. Juvenile duller. Eyes brown; bill black (orange-based in female); tarsi red.

Habitat and Behavior Inhabits wetlands, lakes and ponds, wooded streams and rivers. Solitary or in pairs, sits in exposed vantage points overlooking water, plunge-dives for aquatic prey. Sometimes hovering flight. Flight very fast, direct and often very low over water with high-pitched vocal. Feeds on small fish, crustaceans and aquatic insects. Nests in excavated earthen tunnels (about 60 cm long). Clutch size is 6-7 eggs.

Distribution Widespread in Eurasia, SE Asia, Indonesia to New Guinea. At Ciping, Xinchengqu, Ninggang and Jingzhushan in Mt. Jinggang.

Status of Population Common all year round in Mt. Jinggang.

3.82a

3.82 白胸翡翠 *Halcyon smyrnensis* (Linnaeus, 1758)

识别特征 体略大的翠鸟，约 27 cm。上体和尾鲜蓝略带绿色、颏、喉及胸中部白色，头、颈、肩及下体余部赭褐色。飞行时，现出白色翼斑和黑色翼端，翼上覆羽深褐色。虹膜深褐色；嘴珊瑚红色；脚红色。

生境与习性 栖息于各种淡水或海岸湿地，如沼泽、湖泊、池塘、河流、水田等。常见停息在电线上。觅食并不仅在水中，还常在地面上。主食昆虫、螃蟹、蛙、蜥蜴、蠕虫等。巢营于较大的溪流或河流堤岸，或在距水甚远的山坡崩塌处或山丘坟墓的隧道中。窝卵数 5-7 枚。

分布 中东、印度、中国及东南亚。在井冈山见于茅坪、宁冈等地。

种群状况 在井冈山全年常见。

Identification L ca. 27 cm. Large kingfisher. Head, shoulder and underparts deep chocolate-brown, contrast with white chin and center of breast. Back to tail bright blue. In flight, show dark brown wing coverts, blue secondaries, and white primaries with black tips. Eyes brown; bill red; tarsi red.

Habitat and Behavior Inhabits freshwater and coastal wetlands: marshes, lakes, ponds, rivers and wet rice fields. Often perches on wires. Forages over water and on the ground. Feeds on fish, crabs, frogs, lizards and worms. Nests in tunnel riverside, or far away from water. Clutch size is 5-7 eggs.

Distribution Middle East, India, China and SE Asia. At Maoping and Ninggang in Mt. Jinggang.

Status of Population Common all year round in Mt. Jinggang.

3.82b

3.83 蓝翡翠 *Halcyon pileata* (Boddaert, 1783)

识别特征 体大的翠鸟，约 30 cm。头黑，上体及尾深蓝色，颏、颈圈、胸部白色，腹部淡红棕色。飞行时白色翼斑显见。亚成鸟色暗淡。虹膜深褐色；嘴红色；脚红色。

生境与习性 喜大河流两岸、河口及红树林。常停息在电线或电杆上。主要吃鱼，也吃蛙、蟹、昆虫等。巢营于水平的隧道洞穴中。窝卵数 3-5 枚，通常 5 枚。雌雄共同孵卵。

分布 繁殖于中国及朝鲜半岛。南迁越冬远至印度尼西亚。在井冈山见于笔架山、新城区等地。

种群状况 在井冈山夏季及迁徙季常见。

Identification L ca. 30 cm. Large kingfisher. Head black, collar, throat and breast white; upperparts and tail deep blue, underparts rufous-orange. Juvenile duller. In flight, show white wing-bar. Eyes brown; bill red; tarsi red.

Habitat and Behavior Inhabits coastal mangroves, wetlands and woods to inland forest clearings and cultivated areas with water. Often perches on wires or poles. Mostly feeds on fish, also frogs, crabs and insects. Nests in horizontal tunnels. Clutch size is 3-5 eggs, mostly 5. Both parents incubate.

Distribution Breeds in China and Korean Peninsula. Migrating south in winter as far as Indonesia. At Xinchengqu and Bijiashan in Mt. Jinggang.

Status of Population Common in summer and on migration in

3.83a

3.83b

Mt. Jinggang.

3.84 斑鱼狗 *Ceryle rudis* (Linnaeus, 1758)

识别特征 中等体型的黑白色翠鸟，约 27 cm。头黑色，较小的冠羽偏至后枕，白色眉纹显著。上体白而多具黑点；下体白，具两道黑色胸斑（雌鸟有一道断开的胸斑）。飞行时，可见初级飞羽的大白斑，及白色尾羽上的宽阔黑色次端斑。虹膜淡褐色；嘴黑色；脚黑色。

生境与习性 喜鱼塘、湖泊、沼泽地及红树林。性嘈杂。多在水域附近飞行，速度较缓慢。常在空中定点振翅和俯冲潜入水中捕食。捕捉小鱼、虾、蟹及水生昆虫和蝌蚪等为食。以堤岸或断崖的土洞为巢。窝卵数多 4-5 枚。

分布 印度东北部、斯里兰卡、缅甸、中国、中南半岛及菲律宾。在井冈山见于宁冈、新城区等地。

种群状况 在井冈山全年常见。

Identification L ca. 27 cm. Mid-sized, black and white kingfisher with loose crest restricted to hindcrown. Head black with broad white brows, upperparts white with heavy black spots, underparts white with two black breast-bands (single broken on female). In flight, large white patches on primaries and forewing, white tail with broad black subterminal band. Eyes brown; bill black; tarsi black.

Habitat and Behavior Inhabits rivers, ponds, lakes, marshes and mangroves. Noisy. Fly slowly. Often hovers conspicuously high above waterbodies and plunge-dive for prey. Feeds on fish, shrimp, crabs and aquatic insects. Nests in hollows. Clutch size is 4-5 eggs.

Distribution NE India, Sri Lanka, Myanmar, China, Indo-China Peninsula and the Philippines. At Ninggang and Xinchengqu in Mt. Jinggang.

Status of Population Common all year round in Mt. Jinggang.

3.85 冠鱼狗 *Megaceryle lugubris* (Temminck, 1834)

识别特征 体型硕大的黑白色翠鸟，约 41 cm。冠羽长而蓬松，头黑而下颊白；上体黑色并多具白色横斑和点斑；下体白色，黑色髭纹沿至胸斑。飞行时，可见雌鸟翼下覆羽红棕色。虹膜褐色；嘴黑色；脚灰色。

生境与习性 常光顾流速快、多砾石的清澈河流及溪流。单独或成对，栖于大块岩石或低枝。常沿着溪河中央直飞，边飞边叫，飞行慢而有力且不盘飞。不定点振翅。食物为鱼、虾等。常营巢于山区溪流、湖泊等的陡岸和悬崖上，有时也在堤坝和田坎上挖洞为巢。窝卵数 3-7 枚，以 5-6 枚较为常见。

分布 喜马拉雅及印度北部山麓地带、中南半岛北部、中国南部及东部。在井冈山见于茅坪、湘洲等地。

种群状况 在井冈山全年常见。

Identification L ca. 41 cm. Large black and white kingfisher with erectile crest. Large head black with white under-cheek, upperparts black with heavy white spots and bars, underparts white with grey streaks from malar region merging with breast-band. In flight, female shows rufous underwing coverts. Eyes brown; bill black; tarsi grey.

Habitat and Behavior Prefers cold, fast-flowing rivers and

3.84a

3.84b

3.85a

3.85b

streams in forested mountains. Solitary or in pairs. Plunge-dives from a rock or low branch. Often bobs head and cocks tail when perched. Fly directly and slowly along rivers and giving sharp calls. No hovering. Feeds on fish, shrimp. Nests in hollow of steep bank, cliff, or dam. Clutch size is 3-7 eggs, mostly 5-6.

Distribution Himalaya and foothills in N India, N Indo-China Peninsula, Southern and Eastern China. At Maoping and Xiangzhou in Mt. Jinggang.

Status of Population Common all year round in Mt. Jinggang.

佛法僧目 CORACIIFORMES
蜂虎科 Meropidae

3.86 蓝喉蜂虎 *Merops viridis* Linnaeus, 1758

识别特征 中等体型的蜂虎，约 28 cm（含延长的中央尾羽）。成鸟头顶及上背赭褐色，喉蓝色，过眼线黑色，翼蓝绿色，腰及长尾浅蓝，下体浅绿。幼鸟尾羽无延长，头及上背绿色。虹膜红色或褐色；嘴黑色；脚灰色或褐色。

生境与习性 栖息于常绿阔叶林林缘，也见于有林的开阔村庄和沿海。喜群居，常在林缘空旷处或村庄附近上空飞捕昆虫，并不时发出上下喙碰击声。除觅食蜂类、白蚁类外，也吃很多其他的昆虫。繁殖期群鸟聚于多沙地带，筑巢在沙质峭壁或地面洞穴。窝卵数多 4 枚。

分布 中国南部及东南亚。在井冈山见于五指峰、荆竹山等地。

种群状况 在井冈山迁徙季常见。

Identification L ca. 28 cm. Mid-sized bee-eater with blue under-cheek and throat, and greatly elongated central tail-feathers. Adult crown to mantle chocolate-brown with black eyestripe, upperparts blue-green, and underparts pale green. Rump and tail blue. Juveniles crown to mantle green and lack tail streamers. Eyes reddish-brown; bill black; tarsi grey.

Habitat and Behavior Inhabits edges of broadleaved evergreen forest, also open country with woods and along coasts. Sociable. Hunts for large flying insects, caught on the wing. Feeds on termites, bees and other insects. Excavate burrows in sandy banks to nest, often colonially. Clutch size is 4 eggs.

Distribution Southern China and SE Asia. At Wuzhifeng and Jingzhushan in Mt. Jinggang.

Status of Population Common on migration in Mt. Jinggang.

佛法僧目 CORACIIFORMES
佛法僧科 Coraciidae

3.87 三宝鸟 *Eurystomus orientalis* (Linnaeus, 1766)

识别特征 中等体型、头大的鸟，约 30 cm。整体暗蓝绿色，头和尾近黑，喉钴蓝色。飞行时，蓝紫色的飞羽上有浅蓝色翼斑。虹膜褐色具狭窄红色眼圈；嘴珊瑚红色（幼鸟嘴黑色）；脚红色。

生境与习性 栖息于阔叶林或混交林林缘，尤其是农田或村庄附近。常单只笔直站立于显眼位置，如树顶、电线等，偶尔起飞追捕过往昆虫。飞行速度不甚快，路线颠簸不定，还不时翻

3.86a

3.86b

3.86c

动着身体前进。不自营巢，卵常产于天然的、啄木鸟废弃的树洞中或鹊巢中。窝卵数 3-4 枚，通常 4 枚。雌雄均参与孵卵、育雏等活动。

分布 广泛分布于东亚、东南亚及新几内亚和澳大利亚。在井冈山地区见于黄洋界、大船、黄坳、七溪岭等地。

种群状况 在井冈山地区夏季常见。

Identification L ca. 30 cm. Mid-sized large-headed roller. Overall dark greenish-blue with cobalt-blue throat, darkest on head and tail. In flight, purple-blue flight-feathers with whitish-blue patches. Eyes brown with narrow red eyering; bill red (black in young); tarsi red.

Habitat and Behavior Occurs at edges of broadleaved evergreen and mixed forests, especially adjacent farmland and open country. Sits upright singly on prominent perch, frequently on treetops or wires, sallies for large insects. Flight loose, with somewhat floppy beats. Dependant on natural tree cavities for nesting. Clutch size is 3-4 eggs, mostly 4. Both parents incubate and raise chicks.

Distribution Widespread from E Asia, SE Asia, to New Guinea and Australia. At Huangyangjie, Dachuan, Huang'ao and Qixiling in the Jinggangshan Region.

Status of Population Common in summer in the Jinggangshan Region.

戴胜目 UPUPIFORMES
戴胜科 Upupidae

3.88 戴胜 *Upupa epops* Linnaeus, 1758

识别特征 中等体型、特征鲜明的鸟类，约 30 cm。具长而端黑的粉棕色丝状冠羽，嘴长且下弯。头、上背、肩及下体粉棕色，两翼具黑白相间的条纹，尾黑色具白色横斑。虹膜褐色；嘴黑色；脚铅黑色。

生境与习性 栖于山区或平原的开阔林地、林缘、河谷、耕地、果园等。平时羽冠低伏，惊恐或飞行降落时羽冠竖直。于地面用长嘴翻动觅食昆虫。营巢于树洞或岩壁、堤岸、墙垣的洞、缝中，也见营巢于建筑物的缝隙中。窝卵数 5-9 枚。孵化期 15-17 天。

分布 非洲、欧亚大陆、中南半岛。在井冈山见于宁冈等地。

种群状况 在井冈山迁徙季可见。

Identification L ca. 30 cm. Unmistakable mid-sized bird with distinct erectile, black-tipped crest and long decurved bill. Overall pinkish-brown, wings barred black and white, tail black with white band. Eyes brown; bill black; tarsi black.

Habitat and Behavior Inhabits open woodland, forest edge, groves and thickets, especially in river valleys, and in parks and gardens. Usually holds crest flat, erects fan-like crest on alighting and in aggression. Largely dependent on natural nest cavities, but sometimes utilizes holes in buildings. Clutch size is 5-9. Incubation period is 15-17 days.

Distribution Africa, Eurasia, Indo-China Peninsula. At Ninggang in Mt. Jinggang.

Status of Population Visible on migration in Mt. Jinggang.

3.87a

3.87b

3.88a

3.88b

鴷形目 PICIFORMES
拟鴷科 Megalaimidae

3.89 大拟啄木鸟 *Psilopogon virens* (Boddaert, 1783)

识别特征　体型较大的棕、绿色拟啄木鸟，约 30 cm。具有显著增大的头和嘴。头钢蓝色，前额、眼先及颏色深，上背棕色，双翼、腰及尾绿色。下体黄色而带深绿色纵纹，尾下覆羽亮红色。雌雄相似。亚成体颜色较暗。虹膜棕褐色；嘴浅黄色或褐色而端黑；脚灰色。

生境与习性　栖于落叶或常绿林中，多停息在山顶的阔叶树上。一般多成对或 5、6 只一起活动。有时数鸟集于一棵树顶鸣叫。具绿色保护色，在林间活动难于发现。飞行如啄木鸟，升降幅度大。食昆虫、植物果实等。巢营于树洞中。窝卵数 3-5 枚。发出不断重复的悠长 "peeao-peeao" 声，也发出其他叫声。

分布　喜马拉雅至中国南部及中南半岛北部。广泛分布于井冈山林区。

种群状况　在井冈山全年常见。

Identification　L ca. 30 cm. Large green and brown barbet with distinctive big head and large pale bill. Head deep steel-blue with darker forehead, lores and chin, mantle brown, wings, rump and tail green. Underparts yellow with deep green streaks, vent red. Juvenile duller. Eyes dark brown; bill pale horn with dark tip to upper mandible; tarsi grey.

Habitat and Behavior　Inhabits deciduous and evergreen forests, usually singing in canopy and difficult to see. In pairs or small groups, forages for fruit high up. Flight deeply undulating like large woodpecker. Nests in tree holes. Clutch size is 3-5 eggs. Various calls include "peeao-peeao", and loud territorial kay-oh.

Distribution　Himalaya to Southern China and N Indo-China Peninsula. Widespread in Mt. Jinggang.

Status of Population　Common all year round in Mt. Jinggang.

3.90 黑眉拟啄木鸟 *Psilopogon faber* (Swinhoe, 1870)

识别特征　体型略小、大体绿色的拟啄木鸟，约 20 cm。头部色彩鲜艳，其余大部为绿色。前额具非常狭窄的黑色，头顶前部浅黄色、后部深蓝色，眉黑，颊部的天蓝色延至颈侧，喉黄，眼先、枕部、颈侧具红点。幼鸟色彩较黯淡。虹膜褐色；嘴黑色，下嘴基灰白，嘴须明显；脚深灰色。

生境与习性　丛林鸟类，典型的冠栖型。单独或成群在树上活动，在远处亦可听见其连续而洪亮的 "咯咯咯" 的鸣叫声。只作短距离飞行，不能持久。食物主要为野果，也吃少量昆虫。营巢于树洞内。窝卵数约 3 枚。

分布　中国南部至东南亚。在井冈山见于八面山、双溪口、湘洲、茅坪等地。

种群状况　在井冈山全年常见。

Identification　L ca. 20 cm. Smallish, mainly green barbet with colorful head. Deep blue crown with very narrow black forehead and pale yellow fore-crown, black brow from eye to nape sides, cheeks

3.89a

3.89b

3.90a

3.90b

to neck-sides blue, throat yellow, small red patch on loral, mantle and bib. Remaining parts largely green. Juvenile duller. Eyes brown; bill black with prominent rictal bristles; tarsi dark grey.

Habitat and Behavior Inhabits all forest types. Usually solitary or forms flocks. Often perches motionless in foliage, but sound is obvious and unique. Flight rapid, audibly whirring and woodpecker-like. Forages on fruit, also some insects. Nests in tree holes. Clutch size is 3 eggs.

Distribution Southern China to SE Asia. At Bamianshan, Shuangxikou, Xiangzhou and Maoping in Mt. Jinggang.

Status of Population Common all year round in Mt. Jinggang.

鴷形目 PICIFORMES
啄木鸟科 Picidae

3.91 蚁鴷 *Jynx torquilla* Linnaeus, 1758

识别特征 体小的啄木鸟, 约17 cm。体羽为斑驳杂乱的黑褐色, 后枕至上背有一暗黑色菱形斑块。下体浅色具深色小横斑。嘴相对短, 呈圆锥形。较其他啄木鸟尾长, 具不明显的横斑。虹膜淡褐色;嘴角质色;脚灰褐色。

生境与习性 栖息于低山丘陵和山脚平原的阔叶林或混交林的树上, 有时也在河滩。性孤独, 多单个活动。常在地面觅食, 行走跳跃式, 像麻雀一样, 但尾上翘。也站在树枝或电线上, 不錾木。嗜食蚁类。舌长, 先端具钩并有黏液, 能伸入树洞或蚁巢中取食。

分布 非洲、欧亚大陆、印度、东南亚、中国。在井冈山见于新城区等地。

种群状况 在井冈山迁徙季和冬季常见。

Identification L ca. 17 cm. Small woodpecker with rather short bill and rather long tail. Overall grey-brown, upperparts greyer with broad blackish-brown patch from crown to back, also from lores to ear coverts. Underparts warmer buffy-grey with narrow dark brown bars on chin and flanks. Tail grey and rounded at tip, with 3-4 narrow blackish-brown bars. Eyes brown; bill horn; tarsi grey-brown.

Habitat and Behavior Inhabits open deciduous woodland, riparian thickets, forest edges and reed beds. Solitary, usually feeds on ground. Jumps like sparrow with cocking tail. Dose not drum. Licks ants using long tongue.

Distribution Africa, Eurasia, India, China and SE Asia. At Xinchengqu in Mt. Jinggang.

Status of Population Common on migration and in winter in Mt. Jinggang.

3.92 斑姬啄木鸟 *Picumnus innominatus* Burton, 1836

识别特征 体型纤小、尾短的啄木鸟, 约10 cm。头顶深栗红色, 与白色长眉纹、深色贯眼纹、白色颊纹及深色髭纹形成对比, 颏喉白色。上体橄榄绿色, 下体灰白色具显著鳞状斑, 尾黑白相间。雄鸟前额橘黄色。虹膜红褐色;嘴灰黑;脚铅灰色。

生境与习性 栖息于山地灌丛、竹林或混交林间。形小而敏捷,

3.91a

3.91b

3.92a

常单个或成对与鹟莺、山雀等小鸟混群。多攀缘于低矮的小树和灌丛的枝条上觅食。食物主要是蚁类及蚁卵等。巢营于树洞中。

分布 喜马拉雅至中国南部、东南亚、加里曼丹岛及苏门答腊。在井冈山见于水口、八面山等地。

种群状况 在井冈山全年常见。

Identification L ca. 10 cm. Tiny woodpecker with short tail. Crown dark chestnut contrasting with long white supercilium, black eyestripe and submalar stripe, chin white. Upperparts olive-green, underparts pale grey with prominent black scaling from breast to flanks; tail black and white. Male has orange forehead. Eyes brown; bill grey; tarsi blue-grey.

Habitat and Behavior Inhabits thickets, bamboo or mixed deciduous forest. Active, often joins in mixed-species flocks. Forages on bark of branches and twigs. Feeds on ants and ant eggs. Nests in hollows of trees.

Distribution Himalaya to Southern China, SE Asia, Kalimanta Island and Sumatra. At Shuikou and Bamianshan in Mt. Jinggang.

Status of Population Common all year round in Mt. Jinggang.

3.93 大斑啄木鸟 *Dendrocopos major* (Linnaeus, 1758)

识别特征 体型中等的黑白色啄木鸟,约 24 cm。前额、颊部和下体白色,肩部的大块白斑独特易认。雄鸟头顶黑,枕部具红色斑块而雌鸟无。雌雄臀部均为红色。幼鸟头顶红色。虹膜近红;嘴灰色;脚灰色。

生境与习性 栖于山地和平原的园圃、村寨、树丛及森林间。在树干上一边攀登,一边以嘴叩树。常见于裸露的树顶竖木或鸣叫。飞行略呈波状,两翼一张一合。吃昆虫及树皮下的蛴螬,冬季也食植物性食物。营巢于树洞里,一般啄凿已腐败的树干为巢洞。窝卵数 4-5 枚。孵卵由雌雄亲鸟共同承担。孵化期 10-12 天。育雏期 23-30 天。

分布 欧亚大陆的温带林区、印度东北部、缅甸西部、北部及东部、中南半岛北部。在井冈山见于龙市、茅坪等地。

种群状况 在井冈山全年常见。

Identification L ca. 24 cm. Mid-sized black and white woodpecker with large white shoulder patch. Forehead, cheeks and underparts white. Tail black with large white spots at sides. Male crown black with bright red nape patch, female lacks red on nape. Juvenile has reddish crown. Eyes red-brown; bill dark grey; tarsi blue-grey.

Habitat and Behavior Inhabits several woodland types, including gardens and riparian scrub. Agile, forages on trunks, limbs, even amongst fruit on outer twigs, occasionally on ground. Drums and calls from tops of bare trees. Undulating flight with fold and unfold wings. Excavate holes in live or dead trees for nesting and roosting. Clutch size is 4-5. Incubation period is 10-12 days, chicks leave nest after 23-30 days. Both parents incubate.

Distribution Eurasian temperate forest zone, NE India, W, N and E Myanmar and N Indo-China Peninsula. At Longshi and Maoping in Mt. Jinggang.

Status of Population Common all year round in Mt. Jinggang.

3.94 白背啄木鸟 *Dendrocopos leucotos* (Bechstein, 1802)

识别特征　中等体型的黑白型啄木鸟，约25 cm。雄鸟顶冠绯红，雌鸟顶冠黑，额白。宽阔的黑色髭纹与颈侧的黑带相连，但不与枕部黑斑相接。上体黑色，翼上具白色横斑，飞行时白色的下背明显。下体白而具黑色纵纹，臀部浅绯红。虹膜褐色；嘴深灰色；脚灰色。

生境与习性　喜栖于老朽树木。性活跃，不怯生。常从下到上沿着树干呈螺旋式攀缘。食物为昆虫和野果。营巢在木质较软的树干上，尤以腐朽和枯死树上较多。常自己凿洞。窝卵数3-5枚。

分布　东欧至日本及中国。在井冈山见于湘洲等地。

种群状况　在井冈山全年可见。

Identification　L ca. 25 cm. Mid-sized black and white woodpecker. Head white with red crown on male (black on female), forehead white, bold black malar stripe connects to neck bar, but not to nape. Upperparts black with broad white transverse bars on wing. Lower back and rump white. Underparts white with black streaks. Vent pale red. Eyes brown; bill dark grey; tarsi grey.

Habitat and Behavior　Inhabits deciduous woodland, often in damp areas, near lakes, with many dead and dying trees. Active and bold, often corkscrew from bottom to top of trees. Feeds on insects and fruit. Nests on dead trees. Clutch size is 3-5 eggs.

Distribution　E Europe to Japan and China. At Xiangzhou in Mt. Jinggang.

Status of Population　Visible all year round in Mt. Jinggang.

3.95 星头啄木鸟 *Dendrocopos canicapillus* (Blyth, 1845)

识别特征　体略小的黑白型啄木鸟，约15 cm。头顶灰色，颊部的褐色斑块与黑色髭纹相接，形成白色的颊纹。上体棕黑色，背及翼上具白色斑点；下体皮黄色具暗色纵纹，臀部无红色。雄鸟眼后上方具红色条纹。虹膜红褐色；嘴深灰色；脚灰色。

生境与习性　栖息地与大斑啄木鸟相似，见于各类型的林地。飞行迅速，也呈波状，常单只或成对地在树林间觅食。主食昆虫，冬季也吃植物性食物。营巢于树洞里。窝卵数约4枚。

分布　巴基斯坦、中国、东南亚、加里曼丹岛及苏门答腊。在井冈山见于茅坪、黄坳、荆竹山等地。

种群状况　在井冈山全年常见。

Identification　L ca. 15 cm. Smallish black and white woodpecker. Crown to nape dark grey, brownish facial patch merges with black malar, giving a white moustached appearance. Upperparts mostly brownish-black with white back and wing spotting, contrasts with pale, finely streaked buff underparts. Lacks red on vent. Male has red streaks on sides of rear crown. Eyes reddish-brown; bill dark grey; tarsi grey.

Habitat and Behavior　Common resident of broadleaf evergreen forest, forest edges from lowlands to mountains, sometimes in woodlands. Flight undulated and fast, forage in the tree solitarily or in pairs. Feeds on insects, also plants in winter. Nests in holes of

3.94a

3.94b

3.95a

trees. Clutch size mostly 4.

Distribution　Pakistan, China, SE Asia, Kalimanta Island and Sumatra. At Maoping, Huang'ao and Jingzhushan in Mt. Jinggang.

Status of Population　Common all year round in Mt. Jinggang.

3.95b

3.96 灰头绿啄木鸟 *Picus canus* Gmelin, JF, 1788

识别特征　体型较大的绿色啄木鸟，约 27 cm。上体暗灰绿色，下体暗灰染些许绿色。头灰色，枕部、眼先及细髭纹黑色。初级飞羽黑色具白斑。腰亮绿色。尾暗绿色且无横斑。雄鸟前顶冠猩红，而雌鸟为灰色。虹膜红褐；嘴近灰，下嘴基黄色；脚蓝灰。

生境与习性　栖于低地至丘陵的开阔林地及林缘。攀树索虫为食，亦在地面觅食蚂蚁等，秋冬时则兼吃植物性食物。

分布　欧亚大陆、印度、中国、东南亚及苏门答腊。在井冈山见于锡坪等地。

种群状况　在井冈山全年可见。

Identification　L ca. 27 cm. Large green woodpecker. Upperparts dull greyish-green with grey head and black nape. Loral stripe and narrow malar black, male has bright red forehead patch, female lacks. Primaries black with white spots. Rump bright green. Tail dull dark green and unbarred. Underparts dull grey with green wash. Eyes brown; bill dark grey, yellow at base below; tarsi grey.

3.96a

Habitat and Behavior　Inhabits open deciduous and mixed broadleaf forest, hill forest and swamp forest, from lowlands to foothills. Often feeds on trunk, or ground for ants and termites, also fruit and nectar.

Distribution　Eurasia, India, China, SE Asia and Sumatra. At Xiping in Mt. Jinggang.

Status of Population　Visible all year round in Mt. Jinggang.

3.97 竹啄木鸟 *Gecinulus grantia* (Horsfield, 1840)

识别特征　体型中等的红棕色啄木鸟，约 25 cm。头淡皮黄色。上体红褐色，较黄嘴栗啄木鸟少黑色横纹；下体浅灰褐色。飞羽及尾具深色横斑。雄鸟前顶红色。虹膜红褐色；嘴黄褐色，基部灰色；脚灰色。

生境与习性　喜单个栖于竹林及山地杂有竹林的次生林。甚喧闹。食物为甲壳昆虫及蚂蚁。

分布　尼泊尔至中国南部、缅甸北部及中南半岛。在井冈山见于八面山、草坪等地。

种群状况　在井冈山偶见。

3.96b

Identification　L ca. 25 cm. Mid-sized, rufous-brown woodpecker. Head pale buff with red crown patch in males. Upperparts rufous, and less barred than Bay Woodpecker. Underparts grey-brown. Wing and tail dark barred. Eyes dark reddish-brown; bill yellowish-horn with grey base; tarsi grey.

Habitat and Behavior　Inhabits broadleaved evergreen and mixed forests, usually in bamboo thickets. Often in pairs, foraging low in dense bamboo clumps. Noisy. Feeds on beetles and ants.

Distribution　Nepal to Southern China, N Myanmar and Indo-China Peninsula. At Bamianshan and Caoping in Mt. Jinggang.

Status of Population　Rare in Mt. Jinggang.

3.97

3.98 黄嘴栗啄木鸟 *Blythipicus pyrrhotis* (Hodgson, 1837)

识别特征 体型略大的红棕色啄木鸟，约30 cm。头浅褐色，形长的嘴黄色。体羽红棕色具浓重的暗色横斑。雄鸟颈侧及枕具绯红色块斑，而雌鸟无。虹膜红褐色；嘴淡绿黄色；脚深灰色。

生境与习性 栖息于阔叶乔木林中，常单个或成对活动。叫声频繁而嘈杂，易与其他啄木鸟区别。不凿击树木。食物为蠕虫。营巢于离地面较高的树洞中。窝卵数5-7枚，通常6枚。

分布 尼泊尔至中国（华南）、东南亚。在井冈山见于茨坪、湘洲、五指峰、小溪洞等地。

种群状况 在井冈山全年常见。

Identification L ca. 30 cm. Large, rufous-brown woodpecker with long bright yellow bill. Head pale brown, rest of upperpart, breast and belly rufous-brown with black barring. Males have bright red nape patch. Eyes reddish-brown; bill pale greenish-yellow; tarsi dark grey.

Habitat and Behavior Inhabits broadleaved evergreen forest from lowlands to low mountains. Shy but highly vocal, often seen in pairs or single bird. Generally does not excavate trees for food. Feeds on worms. Nests in tree hollows. Clutch size is 5-7 eggs, mostly 6.

Distribution Nepal to S China and SE Asia. At Ciping, Xiangzhou, Wuzhifeng and Xiaoxidong in Mt. Jinggang.

Status of Population Common all year round in Mt. Jinggang.

3.98

雀形目 PASSERIFORMES
八色鸫科 Pittidae

3.99 仙八色鸫 *Pitta nympha* Temminck & Schlegel, 1850

识别特征 中等体型、色彩艳丽的八色鸫，约20 cm。头部色彩对比显著，有宽阔的黑色贯眼纹、皮黄色细眉纹、褐色侧冠纹和黑色顶冠纹。上体蓝绿色，翼及腰部斑块天蓝色；喉白色；下体淡灰棕色，臀部鲜红。飞行时翼上白斑很小。虹膜褐色；嘴偏黑；脚粉褐色。

生境与习性 喜低地灌木丛及次生林。性机警而胆怯、善跳跃。常见单个在林下地面落叶层觅食，也飞落在乔木树上停歇。飞行直而低，速度较慢。主要以昆虫为食，也吃蚯蚓等其他无脊椎动物，有迁徙性。

分布 繁殖于日本、朝鲜半岛、中国（华东及华南）。越冬在加里曼丹岛。在井冈山见于黄洋界、茅坪等地。

种群状况 在井冈山迁徙季常见。IUCN 红色名录（2014）：易危。CITES 附录Ⅱ。国家Ⅱ级保护野生动物。

Identification L ca. 20 cm. Brightly colored pitta. Crown rich chestnut with black median crown-stripe, supercilium buff and black patch across eye to nape. Upperparts green-blue, with bright blue scapulars and rump. Underparts pale buff, lower belly and vent red. Eyes brown; bill dark; tarsi fresh-brown.

Habitat and Behavior Inhabits broadleaved evergreen and

3.99

mixed forests from lowlands to hills. Forages singly on ground for invertebrates such as insects and earthworms. Flight low and straight, sometimes perches on trees.

Distribution Breeds in Japan, Korean Peninsula and E/S China. Winter to Kalimantan Peninsula. At Huangyangjie and Maoping in Mt. Jinggang.

Status of Population Common on migration in Mt. Jinggang. IUCN Red List (2014): VU. CITES App. II. China Key List: II.

雀形目 PASSERIFORMES
百灵科 Alaudidae

3.100 小云雀 *Alauda gulgula* Franklin, 1831

识别特征 体小、全身斑驳褐色的百灵类，约15 cm。头具浅色眉纹及短羽冠，嘴较鹨厚重。似云雀但更小、耳羽暖褐色更重，初级飞羽和次级飞羽更多红棕色，三级飞羽几乎达初级飞羽端部。飞行时翼后缘偏褐色而非白色。虹膜深褐色；嘴角质色；脚肉色。

生境与习性 栖于长有短草的开阔地区。从不停栖树上。非繁殖季节集群。杂食性，以杂草种子、稗子、谷物等植物性食物和昆虫、蜘蛛、虫卵等动物性食物为食。巢置于地面稍凹处，有时有杂草掩盖。窝卵数 3-4 枚。于地面及向上炫耀飞行时发出高音的甜美鸣声。

分布 繁殖于古北区。冬季南迁。在井冈山见于黄坳、宁冈等地。

种群状况 在井冈山迁徙季及冬季常见。

Identification L ca. 15 cm. Mid-sized heavily streaked, brown lark with short crest. Heavier bill than pipit. Similar to Eurasian Skylark, but smaller, warmer brown ear coverts, richer rufous in primaries and secondaries, shorter primary projection (tertial tip reach to primary projection almost). In flight, indistinct trailing edge to wings brownish, instead of white edge in Eurasian Skylark. Eyes dark brown; bill horn; tarsi dull grey-pink.

Habitat and Behaviour Inhabits dry agricultural land and steppe. Never perches on trees. Gregarious in winter. Omnivorous, feeds on seeds and insects. Nests in cup on ground, usually under clump or clod. Clutch size is 3-4 eggs. Song-flight high, soaring and circling with rapid fluttering wingbeats.

Distribution Breeds in E Palearctic. Migrates south in winter. At Huang'ao and Ninggang in Mt. Jinggang.

Status of Population Common on migration and in winter in Mt. Jinggang.

3.100a

3.100b

雀形目 PASSERIFORMES
燕科 Hirundinidae

3.101 家燕 *Hirundo rustica* Linnaeus, 1758

识别特征 中等体型、尾深分叉的燕，约 20 cm（含尾后饰羽）。成鸟上体辉钢蓝色，下体大部白色。前额、颏及喉红棕色，镶蓝黑色胸带。飞行时从腹面看，近尾端具白色带。幼鸟体羽色暗，尾无延长，暗色胸带不甚清晰。虹膜黑色；嘴及脚黑色。

生境与习性 见于从城市到农村的各种人居生境。飞行敏捷，全天结群在田间、水面及空旷地上空急速地飞行，捕食飞虫，刮风下雨也不例外。常见成群地停栖在电线、电杆上。巢多置于屋檐下或梁上，呈浅皿状或半碗状。窝卵数 4-5 枚。孵化期14-15 天。雏鸟由雌雄亲鸟共同喂养，约 20 天后出飞。

分布 几乎遍及全世界。在井冈山广泛分布。

种群状况 在井冈山夏季及迁徙季常见。

Identification L ca. 20 cm (including elongated tail feathers). Mid-sized swallow with deeply forked tail. Adult has glossy steely-blue upperparts and largely white underparts. Forehead, chin and throat brick-red, breast-band blue-black, tail with white spots near tips of feathers. Juvenile is much duller overall with shorter tail-streamers, ill-defined dusky breast-band. Eyes black; bill black; tarsi black.

3.101a

3.101b

Habitat and Behavior Generally in lowland habitat from urban to rural areas. Spends much of the day in steady relentless cruising flight low over ground or water in pursuit of insects, but often rising to mid-heights to join other hirundines. Congregates in roosting flocks, especially poles and telegraph wires. Nest an untidy halt-cup of mud and plant stems adhering inside under eaves of buildings. Clutch size is 4-5 eggs. Incubation period is 14-15 days. Both parents brood. Brooding period is about 20 days.

Distribution Nearly worldwide. Widespread in Mt. Jinggang.

Status of Population Common in summer and on migration in Mt. Jinggang.

3.102a

3.102 金腰燕 *Cecropis daurica* (Laxmann, 1769)

识别特征 体略小、腰浅栗色、尾深分叉的燕，约 18 cm。成鸟上体深钢蓝色，下体偏白而多具黑色细纹，耳羽、枕侧、腰及臀部浅棕色。幼鸟上体较暗淡，翼覆羽及三级飞羽具浅色羽端，下体纵纹较弱，尾羽无延长。虹膜黑色；嘴及脚黑色。

生境与习性 常栖于山间村镇附近的树枝或电线上。全天大部分时间在原野飞行，张口捕食飞虫。有时与家燕混群飞行。性喜结群，平日结小群，秋末南迁时常结成数百只的大群。每年常繁殖两次。营巢于住户横梁上、屋檐下、天花板上，巢呈半葫芦状。窝卵数 4-6 枚，通常 5 枚。孵卵和育雏由雌雄亲鸟共同承担。雏鸟 19 天出壳，22 天后出飞。

分布 繁殖于欧亚大陆及印度的部分地区。冬季迁至非洲、印度南部及东南亚。在井冈山广泛分布。

种群状况 在井冈山夏季及迁徙季常见。

Identification L ca. 18 cm. Smallish swallow with black deeply forked tail and pale rufous rump. Adult has glossy blue-black upperparts and fine dark streaked whitish underparts with rufous ear coverts, nape-sides, rump and vent. Juvenile duller above, with buff tips to wing coverts and tertials, weaker streaking below and shorter tail-streamers. Eyes black; bill black; tarsi black.

3.102b

Habitat and Behavior Occurs in lowland habitats typically close to water, prefers wires and trees near rural areas to perch. Flight slower with more prolonged glides in pursuit of insects. Congregates on or prior to migration. Breeds usually solitarily or in small loose groups. Nests are mud bottles under overhang, in crevice or on building, attempting nest construction inside occupied house. Clutch size is 4-6 eggs, usually 5. Incubation period is 19 days. Both parents incubate and brood chicks. Brooding period is 22 days.

Distribution Breeds in Eurasia and parts of India. Migrating south in winter to Africa, S India and SE Asia. Widespread in Mt. Jinggang.

Status of Population Common in summer and on migration in Mt. Jinggang.

3.103 烟腹毛脚燕 *Delichon dasypus* (Bonaparte, 1850)

识别特征 体小而矮壮的深色燕，约 13 cm。腰白，尾叉较浅。上体暗钢蓝色，下体烟灰白色，雄性下体较雌性白。飞行时，翼下覆羽黑色。虹膜褐色；嘴黑色；脚粉红，被白色羽至趾。

3.103a

生境与习性　栖息于多悬崖的山区或海岸。常集大群，也与其他燕或雨燕混群。比其他燕更喜留在空中，多见其于高空翱翔。飞行时发出兴奋的嘶嘶叫声。营巢于悬崖上、桥梁下或隧道中，巢成半球状，入口狭窄。

分布　繁殖于喜马拉雅至日本。越冬南迁至东南亚。在井冈山繁殖于井冈冲水库大坝。

种群状况　在井冈山夏季及迁徙季常见。

Identification　L ca. 13 cm. Small, compact, black martin with moderately forked tail and white rump. Upperparts dull steel blue-black, underparts dusky grey-white. Male is whiter below than female. In flight, underwing coverts black. Eyes brown; bill black; tarsi pink with white feathering as far as the toes.

Habitat and Behavior　Inhabits montane or coastal regions with cliffs. Often in large flocks, mixes with other swallows or swiftlets. More aerial than other swallows, tends to stay high in sky. Excited whickering calls when flying. Builds hemispherical mud-pellet nest with narrow round entrance tube on crags, under bridges or in tunnels.

Distribution　Breeds Himalaya to Japan. Winters south to SE Asia. Breeds under the dam at Jinggangchong Reservoir in Mt. Jinggang.

Status of Population　Common in summer and on migration in Mt. Jinggang.

3.103b

3.103c

雀形目 PASSERIFORMES
鹡鸰科 Motacillidae

3.104 山鹡鸰 *Dendronanthus indicus* (Gmelin, JF, 1789)

识别特征　中等体型的鹡鸰，约 17 cm。上体橄榄褐色，具醒目的白色眉纹；翼黑色具两条黄白色翼斑；尾羽褐色，外侧尾羽白色。下体偏白，具两条黑色胸带，下方的胸带有时不完整。幼鸟更偏褐色，胸带更不完整。虹膜深褐色；上嘴灰色，下嘴肉红色；脚偏粉色。

生境与习性　常栖息于林间空地、林缘、果园及村落附近。单独或成对在地面行走，或在较粗的树枝上驰走，尾不断左右摆动。飞行呈波浪式曲线，一高一低，常伴随着鸣叫。被驱赶时迅速落到地面或树上，较好地隐蔽在落叶层里。主要以昆虫为食，也食小的蜗牛、蛞蝓等。常在葡萄架或大树的水平枝上筑巢，巢呈杯状。窝卵数 4-5 枚。

分布　繁殖在亚洲东部。冬季南迁至印度、中国（东南）、东南亚。在井冈山见于茨坪、罗浮等地。

种群状况　在井冈山夏季常见。

Identification　L ca. 17 cm. Mid-sized wagtail. Upperparts olive-brown with prominent white supercilium, black wing coverts with broad yellowish-white double wing bar; long tail brown with white outer rectrices. Underparts whitish with black double breast bands, sometimes the lower one broken. Juvenile browner, with less complete breast bands than adult. Eyes dark brown; bill grey above, pinkish below; tarsi pink.

Habitat and Behavior　Inhabits shaded open forest floor, woodland, orchards and plantations. Singly or in pairs on ground

3.104a

3.104b

commonly, sometimes walk along thick branches, sways rear and tail from side to side. Wave-like in flight. Well-camouflaged against leaf litter; when flushed may alight quickly on ground or in trees. Feeds on insects and mollusks. Builds cup nest on horizontal branch of large trees or grape trellis near streams or rivers. Clutch size is 4-5 eggs.

Distribution　Breeds in E Asia. Migrates south in winter to India, SE China and SE Asia. At Ciping, Luofu in Mt. Jinggang.

Status of Population　Common in summer in Mt. Jinggang.

3.105a

3.105 白鹡鸰 *Motacilla alba* Linnaeus, 1758

识别特征　中等体型的黑白灰色鹡鸰，约 20cm。不同亚种羽色不一，但均无全黑色的耳羽。多数繁殖期成体具黑色喉部，所有非繁殖成体喉部白色。在井冈山分布的亚种 *leucopsis*，雄鸟头顶中部至腰黑色，前额、脸及颏部白色，黑色胸斑不与黑色颈背相连；下体白色。雌鸟上体较灰，黑色胸斑较小。第一年冬羽前额至腰灰色或石板色，深色胸斑新月形。虹膜黑色；嘴及脚黑色。

生境与习性　出现在河岸、农田至海岸的各种生境。多单独或 3-5 只结群活动，在地面或水边奔驰觅食，尾上下摆动不已，有时在空中捕食昆虫。飞行呈波浪式。受惊扰时飞行骤降并发出尖锐示警叫声。几乎纯食昆虫。筑巢在洞穴、石缝、河边土穴及灌丛中，有时筑巢在居民点屋顶、墙洞等处。窝卵数 4-5 枚。

分布　非洲、欧洲及亚洲。繁殖于东亚的鸟南迁至东南亚越冬。在井冈山广泛分布。

种群状况　在井冈山全年常见。

Identification　L ca. 20 cm. Mid-sized white, grey and black wagtail. Various races that never has all-black ear coverts. Most breeding adults have black throat, all non-breeding adults have white throat. *M. a. leucopsis* (which race in Mt. Jinggang) is black from mid-crown to rump, with white forehead, face and chin, and black chest patch (especially in breeding season) unconnected to black of neck; underparts white. Female is grey on crown and upperparts with small black chest patch. First-winter has grey to slaty upperparts including forehead, and blackish chest-crescent. Eyes black; bill black; tarsi black.

Habitat and Behavior　Inhabits wide range of habitats from riversides and cultivated land to coasts. Often solitary, sometimes in small groups. Walks on the ground or along river edges for foods with vertical tail-wagging, sometimes fly to catch insects. Flies with low dipping flight giving alarm call when disturbed. Nests on ground under rock, bush or other shelter, or in hole of wall or bank. Clutch size is 4-5 eggs.

Distribution　Africa, Europe and Asia. Birds breeding in E Asia winter south to SE Asia. Widespread in Mt. Jinggang.

Status of Population　Common all year round in Mt. Jinggang.

3.105b

3.106 黄鹡鸰 *Motacilla tschutschensis* Gmelin, JF, 1789

识别特征　中等体型的鹡鸰，约 18 cm。羽色多变，繁殖期成鸟均有橄榄绿色的上体，双翼黑褐色具两道白色或黄白色翼斑，尾黑褐色具白色外侧尾羽。非繁殖期成鸟相似，均有偏褐色的上体，胁部白色至浅黄色，部分具浅黄色的臀部及尾下覆羽，白色的眉纹不环绕耳羽。雌鸟及幼鸟无黄色的臀部。虹膜深褐色；嘴黑色

3.106

（繁殖期）或上嘴深灰色、下嘴角质色（非繁殖期）；脚黑色。

生境与习性　栖息于草场、稻田、原野及沼泽边缘。常成对或 3-5 只小群活动。飞行似鹨，不如白鹡鸰般起伏，栖止时尾轻微上下摆动。在地面奔走觅食，有时在空中捕食。食物几纯为动物性。

分布　繁殖于欧洲至西伯利亚及阿拉斯加。南迁至印度、中国、东南亚、新几内亚及澳大利亚。在井冈山见于厦坪等地。

种群状况　在井冈山迁徙季和冬季常见。

Identification L ca. 18 cm. Variable, but all breeding adults have olive-green upperparts, blackish wings with white fringes to coverts and short, slender, blackish tail with white outer feathers. All non-breeding adults similar, has brown upperparts, off-white to buff flanks, some with pale yellow lower vent and undertail coverts; white supercilium does not wrap around ear coverts. Female and juvenile lack yellow vent. Eyes dark brown; bill black (breeding) or dark grey above, horn below (non-breeding); tarsi black.

Habitat and Behavior Inhabits wet meadows, rice fields, grassy swamps and wetland margins. Usually in pairs or small groups. Pumps tail less strongly, and flight less undulating and more pipit-like than in White Wagtail. Forage by walking, sometimes flies to catch insects.

Distribution Breeds Europe to Siberia and Alaska. Migrates south to India, China, SE Asia to New Guinea and Australia. At Xiaping in Mt. Jinggang.

Status of Population Common in winter and on migration in Mt. Jinggang.

3.107a

3.107b

3.107 灰鹡鸰 *Motacilla cinerea* Tunstall, 1771

识别特征　中等体型、长尾的鹡鸰，约 19 cm。具狭长的白色眉纹、深灰色的上体、黑色的双翼及黄色的腰。繁殖期雄鸟喉部黑色，与白色髭纹相接，胸至臀部明黄色。繁殖期雌鸟颏及喉白色，杂有些许黑色，下体黄色较浅。非繁殖期成鸟喉部白色，雌鸟较雄鸟色浅，且胸部更多皮黄色。幼鸟似雌鸟但上体偏橄榄色，腰部黄色较浅。虹膜深褐色；嘴黑色（雄）或深灰色（雌）；脚肉色。

生境与习性　栖息于溪流、河谷、湖泊、水塘、沼泽等水域岸边或水域附近的草地、农田及林区居民点。常单独或成对活动。多在水边行走或跑步捕食，尾羽上下弹动，有时也在空中捕食。主要以昆虫为食，也吃蜘蛛等其他小型无脊椎动物。营巢于河流两岸的各式生境，如河边土坑、石头缝隙、河岸倒木树洞、房屋墙壁缝隙等。隐蔽较好。窝卵数 4-6 枚，多 5 枚。孵化期 12 天。雏鸟晚成。

分布　繁殖于欧洲至西伯利亚及阿拉斯加。南迁至非洲、印度、东南亚至新几内亚及澳大利亚。广泛分布于井冈山。

种群状况　在井冈山全年常见。

Identification L ca. 19 cm. Long-tailed wagtail with narrow white supercilium, plain dark grey upperparts, black wings, and yellow rump. Breeding male has diagnostic black throat bordered by broad white malar, and bright yellow underparts from chest to vent. Breeding female has white chin and throat mottled black, and generally paler underparts. Non-breeding adult has white throat, female paler with more buff on breast than male. Juvenile as female but more olive upperparts and less yellow rump. Eyes dark brown;

3.107c

bill black (male) or dark grey (female); tarsi flesh-pink.

Habitat and Behavior Inhabits streams, rivers, and lakes edges, also wetlands and coasts. Usually solitary or in pairs. Forages preferentially on rocks in water, forward-bent posture with slightly cocked tail. Feeds on insects and other invertebrates. Nest placed in bank, amid stones or roots, next to stream. Clutch size is 4-6 eggs, usually 5. Incubation period is 12 days. Chicks are altricial.

Distribution Breeds Europe to Siberia and Alaska. Migrates south to Africa, India, SE Asia to New Guinea and Australia. Widespread in Mt. Jinggang.

Status of Population Common all year round in Mt. Jinggang.

3.108a

3.108 树鹨 *Anthus hodgsoni* Richmond, 1907

识别特征 中等体型、背橄榄绿色的鹨，约15 cm。具显著的浅色眉纹（于眼先为皮黄色，眼后为乳白色，且上缘镶一条黑色细纹），颊深色，颊后缘上方有一白点。上体无纵纹或仅有少量纵纹。颏及喉皮黄色，黑色细髭纹汇入胸部浓密的深色纵纹，两胁亦具较多纵纹。幼鸟眉纹较暗淡，上体多纵纹而胁部少纵纹。虹膜深褐色；上嘴深灰，下嘴角质色；脚粉红色。

生境与习性 多见于杂木林、针叶林、阔叶林、灌木丛及其附近的草地，也见于居民点、田野等地。多在地上奔驰、觅食。繁殖期常成对活动，迁徙期间多结小群。受惊立刻飞到附近树上，停栖时尾常上下摆动。主要以昆虫及其幼虫为食，冬季兼食一些植物性食物。巢营在林间空地或林缘。窝卵数4-6枚，多为5枚。孵化期13-15天。

分布 繁殖于喜马拉雅及东亚。冬季迁至印度及东南亚。在井冈山广泛分布。

种群状况 在井冈山全年常见。

3.108b

Identification L ca. 15 cm. Mid-sized olive-backed pipit with prominent supercilium (buffish before eye, pale cream behind eye, bordered above by narrow blackish line) and pale spot above dark rear edge of cheek. Upperparts lack streaks or has light streaks. Chin and throat whitish, split by narrow black malar stripe, merging into heavy black breast-streaks, and bold flank-streaks. Juvenile supercilium duller and less distinct, upperparts more distinctly streaked, and flanks less distinctly streaked. Eyes dark brown; bill dark grey above, dark horn below; tarsi pink.

Habitat and Behavior Inhabits woodland, parks, farmlands and grasslands near forest, not shy. Forages by walking. Mostly in pairs in breeding season, and gregarious on migration. Commonly lands in trees or wires when flushed, with tail wagging vertically. Feeds on insects and larvae, also seeds in winter. Clutch size is 4-6 eggs, usually 5. Incubation period is 13-15 days.

Distribution Breeds Himalaya and E Asia. Migrates in winter to Indian subcontinent and SE Asia. Widespread in Mt. Jinggang.

Status of Population Common all year round in Mt. Jinggang.

3.109 理氏鹨 *Anthus richardi* Vieillot, 1818

识别特征 体大、腿长、站姿直挺的鹨，约18 cm。成鸟上体褐色具深色宽纵纹，眉纹、眼先、颏、喉浅皮黄色，双翼深色

3.109a

具皮黄色羽缘；下体偏白，胁部和具深色细纵纹的上胸染黄褐色。第一年冬羽更干净，大覆羽、中覆羽、三级飞羽羽缘白色。虹膜深褐色；上嘴灰色，下嘴粉色；脚粉色。

生境与习性　喜开阔沿海或山区草甸、火烧过的草地及干稻田。单独或成小群活动。站于地面时姿势甚直。飞行呈波状，每次跌飞均发出叫声。几纯以昆虫为食，偶尔也吃植物性食物。营巢于河川、沼泽地附近的草地上凹坑内或草丛旁。窝卵数 4-6 枚。

分布　中亚、印度、中国、蒙古及西伯利亚和东南亚。在井冈山见于大井、黄坳等地。

种群状况　在井冈山迁徙季常见。

Identification　L ca. 18 cm. Large sandy-brown pipit with long leg and upright stance. Adult upperparts mid-brown with broad dark streaks, supercilium, lores, chin and throat pale buff, wings dark with buff fringes; underparts off-white with buffy-brown streaked upper chest and flanks. 1st winter cleaner, with narrow white fringes to median and greater coverts and tertials. Eyes dark brown; bill grey above, pink below; tarsi pink.

Habitat and Behavior　Inhabits steppe, semi-desert, grassland and cultivated areas. Usually solitary or in small groups. Deep undulations in flight, calling with each dip. Feeds on insects, occasionally vegetable foods. Nests in depressions on the ground or next to grasses near water. Clutch size is 4-6 eggs.

Distribution　C Asia, India, China, Mongolia and Siberia through SE Asia. At Dajing and Huang'ao in Mt. Jinggang.

Status of Population　Common on migration in Mt. Jinggang.

3.110 黄腹鹨 *Anthus rubescens* (Tunstall, 1771)

识别特征　略单调的灰褐色鹨，约 15 cm。上体暗褐色仅具模糊纵纹，下体皮黄色，胸及两胁具或多或少的纵纹，颈侧具近三角形的黑色斑块。繁殖期成鸟眉纹、眼先、下体略带橘黄色，胸部纵纹较浅。非繁殖期成鸟下体偏白，胸胁纵纹浓密。其余多数鹨上体和下体都具显著纵纹。虹膜深褐色；嘴深褐色，下嘴基黄褐色；脚粉色至褐色。

生境与习性　夏季栖于多岩石的高山或亚高山苔草地带，迁徙时或冬季栖息于沿溪流的湿润多草地区、稻田、湿地和林缘。多成松散小群在地面活动，行走轻快。性羞怯不易靠近。

分布　繁殖于古北区西部、东北亚及北美洲。越冬南迁。在井冈山见于大井、黄坳等地。

种群状况　在井冈山迁徙季及冬季常见。

Identification　L ca. 15 cm. Rather drab, grey-brown pipit. Upperparts dull with indistinct streaks, underparts buff with variable streaking from breast to flanks, triangular blackish patch on side of neck. Breeding adult has warm orange-buff supercilium, lores and underparts, and light streaks or spots on breast. Non-breeding adult underparts whiter with heavily streaks; other species heavily streaked below and also heavily streaked above, except some Olive-backed. Eyes dark brown; bill blackish-brown, yellow-brown base; tarsi pink to brown.

Habitat and Behavior　In summer, rocky alpine and subalpine tundra; on migration or in winter damp grasslands, fields,

3.109b

3.110a

3.110b

wetlands and woodland edges. Terrestrial, gait brisker. Shy and unapproachable, and often loosely gregarious. Flight call is sharp jeet-eet, less shrill than Water Pipit. Song is rapid series of chee or cheedle notes.

Distribution Breeds W Palaearctic, NE Asia and North America. Wintering southward. At Dajing and Huang'ao in Mt. Jinggang.

Status of Population Common on migration and in winter in Mt. Jinggang.

3.111 粉红胸鹨 *Anthus roseatus* Blyth, 1847

识别特征 体型中等的鹨，约15 cm。上背具纵纹，宽阔的浅色眉纹上缘镶黑色，眼先及耳羽深色。繁殖期成鸟上体深橄榄褐色，灰色的头上具粉色眉纹、黑色过眼纹及髭纹；下体粉红，仅胁部具少量深色纵纹。非繁殖期成鸟上体偏灰具黑色粗纵纹，眉纹粉皮黄色；下体几无粉红，胸及两胁具浓密的黑色点斑或纵纹。在手中时，柠檬黄色的小翼羽为本种特征。虹膜深褐色；上嘴深灰，下嘴角质色或粉色；脚偏粉色。

生境与习性 繁殖于高山地带，冬季多栖息于低海拔的草地、农田，藏隐于近溪流处。站姿较多数鹨为平。常成对或结十几只小群活动，性活跃，不停在地上或灌木丛上觅食。繁殖期主要以昆虫为食。

分布 喜马拉雅、中国。越冬至印度北部的平原地带。在井冈山地区见于南风面等地。

种群状况 在井冈山地区迁徙季常见。

Identification L ca. 15 cm. Mid-sized pipit with strongly streaked mantle and bold supercilium (pale and broad with black brow above); lores and ear coverts dark. Breeding adult has dark olive-brown upperparts, with pink supercilium, black eyestripe and malar on grey head; underparts rather plain with pale vinaceous-pink flush, lightly dark streaks on flanks. Non-breeding has greyer streaked upperparts with buffy-pink supercilium; underparts lack pink with narrow black malar, heavily spotted on breast to flanks with buff wash. In hand, lemon-yellow axillaries diagnostic. Eyes dark brown; bill dark grey above, horn/pink below; tarsi pale or pinkish.

Habitat and Behavior Breeds in montane alpine zone, but winters at lower altitudes, frequenting grasslands and rice fields, where typically skulks near streams. More horizontal posture than most pipits. Usually in pairs or in small groups. Active, forages on the ground or bushes. Feeds on insects in breeding.

Distribution Himalaya and China. Wintering to N Indian plains. At Nanfengmian in the Jinggangshan Region.

Status of Population Common on migration in the Jinggangshan Region.

3.112 山鹨 *Anthus sylvanus* (Hodgson, 1845)

识别特征 体型较大、上背纵纹浓密的暖褐色鹨，约17 cm。眉纹白，嘴显粗短。上体褐色较浓，小翼羽浅黄色；下体纵纹范围较大（胁部亦有纵纹）；尾羽窄而尖。虹膜褐色；上嘴灰色，下嘴粉色；脚偏粉色。

生境与习性 栖息于山区具岩石、灌木、树丛的开阔草地，也

3.111

3.112a

3.112b

见于山区废弃梯田斜坡。单独或成对活动。常站在岩石顶，尾有力弹动而非摆动。惊起时在低空直线急飞。主要以鳞翅目、鞘翅目等昆虫为食。营巢于地上草丛中或有草丛隐蔽的石隙间。窝卵数 3-5 枚。

分布　喜马拉雅至中国(西南、东南)。在井冈山地区见于南风面、武功山等地。

种群状况　在井冈山地区全年常见。

Identification　L ca. 17 cm. Large, heavily streaked warm brown pipit with dirty white supercilium and short, thick bill. Crown to most of upperparts pale brown with bolder streaks on mantle, scapulars and wing coverts. Underparts warm brown, with thick streaks on breast sides, but thinner towards flanks. Eyes dark brown; bill grey above, pink below; tarsi brownish-pink.

Habitat and Behavior　Inhabits mountains on open grassy ground with scattered boulders, scrub and trees. Also on abandoned terrace cultivations mountain slopes. Solitary or in pairs. When alerted it often adopts an upright posture, frequently on top of a rock, and twitches its tail with small movements. When flushed it shoots off in a typical low, darting flight. Nests in the grasses or covert gaps of boulders. Clutch size is 3-5 eggs.

Distribution　Discontinuously Himalaya to SW, SE China. At Nanfengmian and Wugongshan in the Jinggangshan Region.

Status of Population　Common all year round in the Jinggangshan Region.

雀形目 PASSERIFORMES
山椒鸟科 Campephagidae

3.113 暗灰鹃鵙 *Coracina melaschistos* (Hodgson, 1836)

识别特征　体型中等的深色鹃鵙，约 23 cm。雄鸟通体瓦灰色，双翼及尾羽黑色，尾羽腹面末端白色。雌鸟似雄鸟但色浅，白色眼圈通常不完整，颊部具白色细纹，下体浅灰色具黑色横纹。飞行时现出浅色的腰和翼下的白斑。虹膜红褐色；嘴黑色，嘴尖具小钩；脚深灰色。

生境与习性　栖于甚开阔的山地森林、林缘及竹林。性活跃，单独或成对活动，也常出现在鸟浪中。时常逗留在树冠层，在树叶上搜寻昆虫。营巢于高大乔木树冠层的水平枝上，较隐蔽。窝卵数 2-4 枚。雌雄鸟轮流孵卵，雏鸟晚成。

分布　喜马拉雅、中国、东南亚。在井冈山见于双溪口等地。

种群状况　在井冈山夏季及迁徙季常见。

Identification　L ca. 23 cm. Mid-sized dark cuckooshrike with long graduated tail. Male almost entirely slate-grey with black wings, black tail with large white tips forming spots on undertail. Female like male but paler, often has broken white eyering and fine whitish streaks on cheek, underparts pale grey with dark bars, including vent. In flight, shows pale rump and white patch at base of primaries from below. Eyes reddish-brown; bill black with slight hook; tarsi dark grey.

Habitat and Behavior　Inhabits open montane forest, forest

3.113a

3.113b

edges, woodlands and bamboo groves. Singly or in pairs, fairly active and conspicuous; often in bird waves, deliberately gleaning insects from leaves. Nests high in tree. Clutch size is 2-4 eggs. Parents incubate alternately. Chicks are altricial.

Distribution　Himalaya, China and SE Asia. At Shuangxikou in Mt. Jinggang.

Status of Population　Common in summer and on migration in Mt. Jinggang.

3.114 小灰山椒鸟 *Pericrocotus cantonensis* Swinhoe, 1861

识别特征　体型较小的灰、黑白色山椒鸟，约 18 cm。雄鸟前额白色(嘴上方羽毛多为白色)，白斑延至眼后；黑色过眼纹汇入褐灰的头顶后部及枕部，白色的脸颊后缘不规则；两翼黑色，大覆羽及三级飞羽有浅色边缘；腰浅棕色，胸、胁暗棕色，黑色的尾具污白色的外侧尾羽。雌鸟较雄鸟更显褐色，前额白斑和浅黄色翼斑有时不显。虹膜黑色；嘴黑色；脚黑色。

生境与习性　多栖息于常绿、落叶阔叶林和针叶林。冬季形成较大群。飞行呈波状前进，常边飞边叫。觅食于乔木的中上层。性杂食，以昆虫为主要食物。常营巢于松树或其他高大乔木上。巢呈杯状。窝卵数 3-4 枚。

分布　繁殖于中国（华中、华南及华东）。于东南亚越冬。在井冈山见于罗浮等

种群状况　在井冈山夏季可见。

Identification　L ca. 18 cm. Small grey, black and white minivet. Male has white on forehead (feathers above bill mostly white), white patch extending behind eye; dark eyestripe merging charcoal-grey hindcrown and nape, and pale cheek has irregular dark rear margin; wings black with buff fringes to greater coverts and tertial; pale brownish rump, dingy brownish breast and flanks, black tail with sullied whitish outer rectrices. Female generally browner than male on underparts, variable forehead patch and pale yellowish wing-bar. Eyes black; bill black; tarsi black.

Habitat and Behavior　Inhabits deciduous, broadleaf evergreen and pine forests. Gregarious in winter. Calling in undulated flight. Forages in middle-upper layer. Omnivorous, feeds on insects and fruit. Builds cup nests on pines or other large trees. Clutch size is 3-4 eggs.

Distribution　Breeds C, S and SE China. Wintering to SE China. At Luofu in Mt. Jinggang.

Status of Population　Visible in summer in Mt. Jinggang.

3.115 灰山椒鸟 *Pericrocotus divaricatus* (Raffles, 1822)

识别特征　灰白色的山椒鸟，约 19 cm，次级飞羽基部通常有白斑。雄鸟白色的前额斑块及眼，黑色贯眼纹与墨黑色的头顶及后枕相接。上体及腰冷灰色，黑色的尾具白色外侧尾羽。下体白色，胸较干净仅两侧染灰。雌鸟前额白斑较狭窄，头顶、耳羽及枕部灰色，上体较雄性色浅。第一年冬羽似雌鸟但头部图案较弱，脸为杂灰色，三级飞羽具较宽的白色边缘。虹膜、嘴及脚黑色。

生境与习性　繁殖季节栖于茂密的原始落叶阔叶林和红松阔叶混交林中，非繁殖季见于多种生境。迁徙时结群。喜在树端站立或在森林上空飞翔，飞行呈波浪状，边飞边鸣叫。多在树上搜寻昆虫为食，能捕食飞虫。

分布　东北亚及中国东部。冬季往南至东南亚。在井冈山见于罗浮等地。

种群状况　在井冈山迁徙季可见。

Identification　L ca. 19 cm. Grey and white minivet with whitish patch at base of secondaries. Male has white forehead reaching to eye, black eyestripe merging jet-black hindcrown and nape; cold grey upperparts including rump; bright white outer rectrices on black tail. Underparts white, clean white breast with some cool-grey

3.114

3.115a

3.115b

on sides. Female has white forehead and grey crown, ear coverts and nape; upperparts slightly paler grey than male and underparts clean white. First-winter as female but with weaker head pattern, and face mostly mottled cold-grey; tertials broadly and contrastingly fringed white. Eyes black; bill black; tarsi black.

Habitat and Behavior　Inhabits mixed broadleaf evergreen or broadleaf deciduous forest in breeding season. Occur in almost any kind of woodland in non-breeding season. Forms small to fairly large, very active flocks, often in high bare branches. Calling in undulated flight. Hunts insects in tree canopy.

Distribution　NE Asia and Eastern China. Winters south to SE Asia. At Luofu in Mt. Jinggang.

Status of Population　Visible on migration in Mt. Jinggang.

3.116 灰喉山椒鸟 *Pericrocotus solaris* Blyth, 1846

识别特征　体型较小的鲜艳山椒鸟，约 17 cm。雌雄均有深灰色的头和上背，浅灰色的额和喉，灰黑色的翼及尾。雄鸟下背至腰、外侧尾羽及下体亮橘红色，翼黑色具"フ"字形红色翼斑。雌鸟似雄鸟但红色部位为黄色，上背至腰多橄榄灰色。虹膜深褐色，嘴及脚黑色。

生境与习性　栖息于平原和山区杂木林、阔叶林、针叶林以至茶园间。一般结小群活动，有时也集大群。繁殖季节成对。飞行时躯体与双翅相衬如十字，边飞边叫。食物几纯为昆虫。常营巢于常绿阔叶林、栎林，巢多置于树侧枝上或枝丫间，呈浅杯状。窝卵数 3-4 枚。

分布　喜马拉雅、中国南部、东南亚。在井冈山广泛分布。

种群状况　在井冈山全年常见。

Identification　L ca. 17 cm. Small colorful minivet. Both sexes have dark grey head and mantle, pale grey chin and throat, greyish-black wings and tail. Male has bright orange underparts, lower back, rump and outer tail-feathers, "フ"-like wing patch. Female is lemon-yellow where male is orange, more olive-grey on mantle to rump. Eyes dark brown; bill black; tarsi black.

Habitat and Behavior　Inhabits broadleaf evergreen and deciduous forest, sometimes pine forest. Usually in small noisy parties, sometimes in large flocks, joins bird waves, but in pairs in breeding season. Cross-shaped in flight with soft rasping. Feeds on insects. Nests in small cup towards end of high branch. Clutch size is 3-4 eggs.

Distribution　Himalaya, Southern China, SE Asia. Widespread in Mt. Jinggang.

Status of Population　Common all year round in Mt. Jinggang.

3.117 赤红山椒鸟 *Pericrocotus speciosus* (Latham, 1790)

识别特征　体型略大的鲜艳山椒鸟，约 19 cm。嘴较灰喉山椒鸟厚重。雄鸟胸腹、腰、外侧尾羽及翼上的"刁"字形斑纹猩红色，余部蓝黑色。雌鸟背部多灰色，前额黄色，眼先黑色，灰色的耳羽与黄色的喉部形成对比，余部黄色替代雄鸟的红色。虹膜

黑褐色；嘴及脚黑色。

生境与习性　栖于中低海拔的山地和平原的阔叶林，也见于松林、稀树草地或开垦的耕地。结群活动或与其他鸟混群，繁殖季节大都成对活动。性活泼，喜结集于乔木冠部觅食。主食昆虫。常营巢于森林中乔木的水平枝杈上。窝卵数 2-4 枚。雏鸟晚成，雌雄共同筑巢和育雏。

分布　印度、中国南部及东南亚。在井冈山见于大峡谷、荆竹山等地。

种群状况　在井冈山全年常见。

Identification　L ca. 19 cm. Largish colorful minivet with heavy bill. Male has red breast, belly, lower back, rump, outer tail feathers, and "刁"-like wing patch, other parts are blue-black. Female has grey above with broad yellow forehead, bold black stripe on lores, darker grey ear coverts contrasting strongly with yellow throat, bright yellow where male is red. Eyes dark brown; bill black; tarsi black.

Habitat and Behavior　Prefers primary forest where it flits between the tree tops of finer-leaved trees in pairs or small groups. Joins bird waves. Mostly feeds on insects. Nest is a disguised shallow cup in fork towards end of thin branch. Clutch size is 2-4 eggs. Both parents nest and brood. Chicks are altricial.

Distribution　India, Southern China and SE Asia. At Daxiagu and Jingzhushan in Mt. Jinggang.

Status of Population　Common all year round in Mt. Jinggang.

雀形目 PASSERIFORMES
鹎科 Pycnonotidae

3.118 领雀嘴鹎 *Spizixos semitorques* Swinhoe, 1861

识别特征　体型略大的橄榄绿色鹎，约23 cm。头黑色具短羽冠，厚重的嘴象牙色。脸颊具白色细纹，白色半颈环将灰黑色的头部与橄榄绿的下体分开，尾绿而端黑。虹膜褐色；嘴浅黄；脚偏粉色至褐色。

生境与习性　栖居于平原和山地的灌丛、竹林、次生林、林缘及果园等生境。性喜结群，也见单独或成对活动，常停栖于电线或竹林。杂食性，以植物性食物为主，也在飞行中捕捉昆虫。常营巢于溪边或路边小树侧枝梢处，也营在灌木丛中。窝卵数3-4 枚。

分布　中国南部及中南半岛北部。在井冈山广泛分布。

种群状况　在井冈山全年常见。

Identification　L ca. 23 cm. Large olive-green bulbul with unique blunt ivory bill. Black head with short crest and white-streaked ear coverts; white half-collar separates grey hood from olive-green underparts. Rather broad-tipped green tail with black tip. Eyes brown; bill pale yellow; tarsi pinkish-brown.

Habitat and Behavior　Inhabits scrub, bamboo, woodland edges, secondary growth and orchards. Usually gregarious, sometimes solitary or in pairs. Perches on telegraph wires. Omnivorous, prefers wild fruit; catches insects in flight. Nests on the branches of small

3.117b

3.118a

3.118b

trees next to streams or paths, also in bushes. Clutch size is 3-4 eggs.

Distribution　Southern China and N Indo-China Peninsula. Widespread in Mt. Jinggang.

Status of Population　Common all year round in Mt. Jinggang.

3.119 白头鹎 *Pycnonotus sinensis* (Gmelin, JF, 1789)

识别特征　体型中等的橄榄绿色鹎类，约 19 cm。成鸟头黑色，略具羽冠，眼后一白色宽纹延至枕部，眼先有一小白点，耳羽白色。上体灰绿色，飞羽外缘橄榄色，尾深灰色，外侧尾羽具黄绿色羽缘。下体污白，颏、喉及尾下覆羽白色。幼鸟较暗淡，头灰色。虹膜深褐色；嘴及脚黑色。

生境与习性　栖息于林地、灌丛、农田、市区公园等多种生境。喜结群，常数十只为群集聚枝头叶丛边鸣叫边觅食，冬季可见较大群。性活泼善鸣，不甚畏人。杂食性，常见在枝顶翻飞捕食昆虫。营巢于灌木、竹林或乔木上。窝卵数 3-5 枚，常 4 枚。

分布　中国南部、越南北部及琉球群岛。在井冈山广泛分布。

种群状况　在井冈山全年常见。

Identification　L ca. 19 cm. Mid-sized olive-green bulbul. Adult has black head with slight crest and broad white patch from eye to nape, tiny white spot on lores and larger spot on rear ear coverts, chin and throat white. Upperparts greyish-green with olive wings, tail dark grey with yellowish-green outer fringes to feathers. Underparts dusty-white with white vent. Juvenile duller with grey head. Eyes dark brown; bill black; tarsi black.

Habitat and Behavior　Inhabits woodland, scrub, cultivated area and parks. Sociable, active and noisy, often perching conspicuously on treetops, poles or wires. In large groups in winter. Omnivorous, sometimes fly catches from perch. Builds nests in bushes, bamboo or trees. Clutch size is 3-5 eggs, usually 4.

Distribution　Southern China, N Vietnam and Ryukyu Islands. Widespread in Mt. Jinggang.

Status of Population　Common all year round in Mt. Jinggang.

3.120 黄臀鹎 *Pycnonotus xanthorrhous* Anderson, 1869

识别特征　体型中等的灰褐色鹎类，约 20 cm。头黑色略具羽冠，耳羽褐色。上体及尾灰褐色。喉白，胸带灰褐，下体污白，尾下覆羽黄褐色至鲜黄色。虹膜褐色；嘴黑色；脚黑色。

生境与习性　栖于中低山的各种林地、农田、灌丛中。非繁殖季节均成群活动，有时与其他鹎混群。杂食性。常营巢于灌木或竹丛间，也见于林下小树。窝卵数 2-5 枚。

分布　中国南部、缅甸及中南半岛北部。在井冈山见于宁冈等地。

种群状况　在井冈山全年可见。

Identification　L ca. 20 cm. Mid-sized greyish-brown bulbul with slight crest. Upperparts and tail dull greyish-brown, black head with grey-brown ear coverts. Chin and throat white, underparts dusty-white with dusky grey-brown wash on breast, vent buff-ochre to yellow. Eyes brown; bill black; tarsi black.

Habitat and Behavior　Inhabits scrub, farmland and secondary

3.119a

3.119b

3.120

growth on hillsides. Gregarious in non-breeding season, sometimes mixed with other bulbuls. Noisy and conspicuous. Builds nest in bushes, bamboo and small trees. Clutch size is 2-5 eggs.

Distribution Southern China, Myanmar and N Indo-China Peninsula. At Ninggang in Mt. Jinggang.

Status of Population Visible all year round in Mt. Jinggang.

3.121a

3.121 栗背短脚鹎 *Hemixos castanonotus* Swinhoe, 1870

识别特征 体型略大的栗色鹎类，约 21 cm。上体栗褐色，头顶黑色而略具羽冠，前额、眼先、颊部及枕部栗红色。翼深褐色，覆羽、二级及三级飞羽具浅色羽缘。尾深灰褐色，具略方的黑色端。颏、喉白，胸及两胁浅灰，腹部至尾下覆羽白。虹膜红棕色；嘴黑色；脚深褐至黑色。

生境与习性 栖于中低山的常绿阔叶林、次生林及林缘，也见于山村附近路边树丛中。多成小群在高大的树上觅食，或活动于矮灌丛间。时常发出响亮的银铃般叫声，远处可闻。杂食性，以植物为主，亦吃昆虫等动物性食物。营杯状巢于小树或灌木枝杈上。窝卵数 3-5 枚。

分布 中国南部及越南西北部。在井冈山广泛分布。

种群状况 在井冈山全年常见。

3.121b

Identification L ca. 21 cm. Largish brown bulbul with slight crest. Upperparts warm chestnut-brown with black crown, and chestnut forehead, lores, cheek and nape. Wings black-brown with pale fringes on coverts, secondaries and tertials. Tail dark grey-brown with broad square black tip. Chin and throat white, breast and flanks grey, belly and vent white. Eyes reddish-brown; bill black; tarsi dark brown to black.

Habitat and Behavior Inhabits lowland evergreen broadleaf forests, secondary growth and woodlands. Lives in small active parties. Keeps to rather dense thickets. Loud sharp ringing call, can be identified from far away. Omnivorous. Builds cup nest in trees or bushes. Clutch size is 3-5 eggs.

Distribution Southern China and N Myanmar. Widespread in Mt. Jinggang.

Status of Population Common all year round in Mt. Jinggang.

3.122a

3.122 绿翅短脚鹎 *Ixos mcclellandii* (Horsfield, 1840)

识别特征 体型较大的橄榄色鹎，约 24 cm。成鸟头红褐色具蓬松、夹白色细纹的羽冠，喉灰白具褐色纵纹，亦蓬松；背部灰色，翼及尾橄榄绿色；上胸及颈背棕褐色，腹胁染灰色，臀部浅黄。幼鸟羽冠较短，头顶及胸部的纵纹较弱。虹膜红褐色；嘴近黑；脚粉褐色。

生境与习性 栖于中低山的次生阔叶林、混交林或针叶林，也见于溪流河畔或村寨附近的竹林、杂木林。成对或集小群活动，有时形成较大的群。常立于树梢，飞捕昆虫。性嘈杂，大胆围攻猛禽及杜鹃类。杂食性，食物以植物性为主，多为果实，兼食昆虫。营巢于乔木侧枝上或林下灌木和小树上。巢呈杯状、

3.122b

甚小，与体型颇不相称。窝卵数 2-4 枚。

分布　喜马拉雅至中国南部、缅甸、中南半岛及东南亚。在井冈山广泛分布。

种群状况　在井冈山全年常见。

Identification　L ca. 24 cm. Large olive-green bulbul with striated shaggy dark brown crest. Adult has rufous head, mantle to rump grey, wings and tail olive-green. Chin and throat grey-white with brown stripes, often fluffed out. Upper breast rufous, yellowish-white on belly and flanks, vent pale yellow. Juvenile has shorter crest, weaker streaking on crown and breast. Eyes reddish-brown; bill black; tarsi pinkish-brown.

Habitat and Behavior　Inhabits scrub, forest edge and forests. In pairs and small groups, sometimes forms large flocks. Found more in treetops, and may feed on the wing. Noisy, various mewing calls. Aggressively mobs raptors and cuckoos. Feeds mainly on berries, also insects. Builds a small cup nest on tree branchees, or bushes in forest. Clutch size is 2-4 eggs.

Distribution　Himalaya to Southern China, Myanmar, Indo-China Peninsula and SE Asia. Widespread in Mt. Jinggang.

Status of Population　Common all year round in Mt. Jinggang.

3.123 黑短脚鹎 *Hypsipetes leucocephalus* (Gmelin, JF, 1789)

识别特征　中等体型的黑色鹎，约 20 cm。略具松散的羽冠。有两种典型色型，通体黑色，或仅头颈部白色、余部黑色，也有两种色型的中间过渡型。幼鸟偏灰，羽冠较平。虹膜褐色；嘴及脚红色。

生境与习性　栖于中低山的常绿阔叶林、落叶阔叶林，也见于平原、河谷地带或公园等。随着季节变化而发生垂直迁移现象。冬季常集聚为上百只的大群，散落在树冠上。飞行径直而快速。杂食性，冬季以植物性食物为主，夏季多食花蜜和昆虫。营巢于山地森林中乔木的水平横枝或树杈处。窝卵数 2-4 枚。

分布　印度、中国南部（含台湾）、缅甸及中南半岛。在井冈山广泛分布。

种群状况　在井冈山全年常见。

Identification　L ca. 20 cm. Mid-sized black bulbul with rather loose crest. Two typical types, all black, or black except white head and neck, some transitional forms between the two. Juvenile more grey with flat crest. Eyes brown; bill red; tarsi red.

Habitat and Behavior　Inhabits broadleaf evergreen and mixed deciduous forests, also plains, valleys and parks. Shows some seasonal movements vertically. Gregarious in scattered treetop flocks up to 100 when not breeding. Flight rapid and direct. Food in winter mainly fruit but in summer visits tree flowers for nectar and insects. Nests in a neat large shallow cup on boughs and in forks in trees, including conifers. Clutch size is 2-4 eggs.

Distribution　India, Southern China (include Taiwan), Myanmar and Indo-China Peninsula. Widespread in Mt. Jinggang.

Status of Population　Common all year round in Mt. Jinggang.

3.123a

3.123b

3.123c

雀形目 PASSERIFORMES
叶鹎科 Chloropseidae

3.124 橙腹叶鹎 *Chloropsis hardwickii* Jardine & Selby, 1830

识别特征 体型略大的鲜艳叶鹎，约 20 cm。雄鸟上体绿色，下体浓橘黄色；赭黄色的前额与头顶染蓝色，下脸颊、喉及上胸黑色，髭纹亮蓝色，两翼及尾深蓝色。雌鸟大体绿色，髭纹蓝色，腹中央至臀部具一道狭窄的橘黄色条带。幼鸟通体暗绿色。虹膜褐色；嘴黑色；脚铅灰色。

生境与习性 栖息于亚热带落叶、常绿阔叶林以至公路边的林间。常成对或结 3-5 只的小群，多活动于乔木冠层，偶尔也到林下灌丛和地上活动、觅食。性活泼，常不停在树枝间跳动或在树间飞来飞去，并发出叫声。主要以昆虫为食，也吃部分植物花蜜、果实及种子。常营巢于森林中树上，窝卵数 3 枚。常模仿其他鸟的叫声。

分布 喜马拉雅、中国南部及东南亚。在井冈山广泛分布。

种群状况 在井冈山全年常见。

Identification L ca. 20 cm. Mid-sized colorful forest bird. Male has green upperparts and orange-ochre underparts; blue wash on ochre crown and forehead, extensive black face, throat and upper breast with broad purple-blue moustache, and deep blue flight feathers and tail. Female largely green with blue moustachial and dull yellow-orange band on central belly to vent. Juvenile entirely dull green. Eyes brown; bill black; tarsi grey.

Habitat and Behavior Inhabits deciduous to broadleaf evergreen forest, sometimes in roadside parks and woodlands. Usually in pairs or small groups, actively visits canopy layers. Often gathers with other birds at flowering trees for insects and nectar. Nests in trees in forests. Clutch size is 3 eggs. Sometimes mimic other species's voice.

Distribution Himalaya, Southern China and SE Asia. Widespread in Mt. Jinggang.

Status of Population Common all year round in Mt. Jinggang.

雀形目 PASSERIFORMES
伯劳科 Laniidae

3.125 红尾伯劳 *Lanius cristatus* Linnaeus, 1758

识别特征 体型中等的褐色伯劳，约 20 cm。上体棕褐色或灰褐色，下体皮黄；具黑色眼罩和细白色眉纹，颏、喉白色；两翼黑褐色，具浅色羽缘。成鸟前额灰，雌鸟脸部图案较雄鸟暗淡，下体具鳞状细纹。幼鸟似雌鸟但背及体侧具更多深褐色鳞状细纹。虹膜暗褐色；嘴黑色；脚铅灰色。

生境与习性 栖息于平原、丘陵和低山区的灌丛、林缘、公园、农田等地，喜开阔地带。多单独或成对活动，常在固定的栖点（树枝、电线）停栖鸣叫。主要以昆虫为食，也食蜥蜴等。常将猎物穿挂于树上的尖枝杈上，然后撕食其内脏和肌肉等柔软部分，剩余部分留在树上。

分布 繁殖于东亚。冬季南迁至印度、东南亚及新几内亚。在

3.124a

3.124b

3.125a

井冈山见于罗浮、小溪洞、宁冈等地。

种群状况　在井冈山冬季及迁徙季常见。

Identification　L ca. 20 cm. Mid-sized brown shrike. Adult has rufous-brown or grey-brown upperparts with rufous-brown rump and rounded tail, buff underparts with rufescent wash to breast-sides and flanks. Broad eyestripes and white supercilium from greyish-white forehead, wings dark brown, with rufous to white edges, and no white carpal patch. Female has some fine dark scalloping below and a dark brown mask, less distinct than male. Juvenile has brown forehead, underparts with more scales. Eyes dark brown; bill black; tarsi black.

Habitat and Behavior　Inhabits scrub, forest edge, plantations, parks and gardens, open country, including farmland, prefers open areas. Usually solitary or in pairs, perches on bushes, wires and small trees, chase flying insects or pouncing on small animals on the ground. Often impales prey to tear guts and muscles, and leaves other parts on the tree.

Distribution　Breeds in E Asia. Migrates south in winter to India, SE Asia and New Guinea. At Luofu, Xiaoxidong and Ninggang in Mt. Jinggang.

Status of Population　Common in winter and on migration in Mt. Jinggang.

3.125b

3.126 棕背伯劳 *Lanius schach* Linnaeus, 1758

识别特征　体型略大而尾长的棕灰色伯劳，约 25 cm。成鸟头顶及颈背深灰色，具黑色眼罩，前额至少有狭窄的黑色，背、腰及体侧红褐，翼及尾黑色，翼上具一白斑，外侧尾羽具棕色羽缘，颏、喉、胸及腹中心部位白色。幼鸟色较暗，两胁及背具横斑，头及颈背灰色较重。有时可见黑化型。虹膜暗褐色；嘴及脚黑色。

生境与习性　喜草地、灌丛、茶林及其他开阔地。立于树枝顶端或电线上，俯视四周，伺机捕食。性凶猛。主要以昆虫为食，也捕食蛙、小型鸟类及鼠类。营巢于树上或高灌木的枝杈基部。窝卵数 4-5 枚。雌鸟孵卵，孵化期 12-14 天。雏鸟在巢期 13-14 天，离巢后 5-7 天仍需要亲鸟饲喂。有时模仿其他鸟的叫声。

分布　伊朗至中国、印度、东南亚至新几内亚。在井冈山广泛分布。

种群状况　在井冈山全年常见。

3.126a

Identification　L ca. 25 cm. Large rufous, grey shrike with long black tail. Adult has black mask with at least a narrow black forehead, grey crown and mantle, bright rufous rear mantle, rump and flanks, black wings with white patch at base of primaries, and narrow tail with rufous fringes on outer feathers. Underparts pale rufous with white throat and breast. Juveniles have basic back pattern with narrow, rather widely spaced blackish scaling above. Some individuals show more black. Eyes dark brown; bill black; tarsi black.

Habitat and Behavior　Inhabits forest edges and open country, grassland, scrub, and tea and other plantations. Bold and conspicuous when perched on treetops or wires, solitary or in pairs. Makes darting sallies after flying insects, or more commonly pounces on grasshoppers and beetles on the ground, also frogs, birds and rats. Nests on the branches of trees or bushes. Clutch size is 4- 5 eggs.

3.126b

Female incubates. Incubation period is 12-14 days. Chicks stay in the nest for 13-14 days, and need brooding for 5-7 days after leaving the nest. Sometimes mimics other bird calls.

Distribution Iran to China, India, SE Asia to New Guinea. Widespread in Mt. Jinggang.

Status of Population Common all year round in Mt. Jinggang.

3.127 虎纹伯劳 *Lanius tigrinus* Drapiez, 1828

识别特征 中等体型的伯劳，约 19 cm。较红尾伯劳明显嘴厚、尾短而眼大。雄鸟头顶及颈背灰色，眼罩宽且黑；背、两翼及尾浓栗色而多具黑色横斑；下体白，两胁具褐色横斑。雌鸟似雄鸟但眼先及眉纹色浅，胁部横纹较明显。幼鸟为较暗的褐色，贯眼纹较不明显，嘴粉色具黑色端。虹膜褐色；嘴近黑色；脚黑褐色。

生境与习性 栖息于平原至低山的森林、林缘及灌丛中。喜开阔林地，多单独或成对活动。典型的伯劳习性。以昆虫为主食。

分布 繁殖于东亚。冬季南迁至马来半岛及大巽他群岛。在井冈山见于长谷岭、下庄等地。

种群状况 在井冈山迁徙季常见。

Identification L ca. 19 cm. Mid-sized brown shrike with large head. Heavier bill, bigger eyes, and shorter tail than Brown Shrike. Male has plain blue-grey crown and nape to upper mantle, broad black mask from forehead to ear coverts; mantle, back, wings and tail chestnut with distinct black bars. Underparts white with slight brown bars on flanks. Female duller than male with whitish eyebrow from lores, and distinct bars from chin to vent. 1st winter has yellowish-brown upperparts with more black scaling on wing coverts, lacked black mask; large flesh-pink bill with only small dark tip. Eyes dark brown; bill black; tarsi dark brown.

Habitat and Behavior Inhabits lowland forests, woodland edges, town parks and open country with bushes. Perches inconspicuously in trees or bushes. Typical shrike, feeds on insects.

Distribution Breeds E Asia. Migrates south in winter to Malay Peninsula and Greater Sunda Islands. At Changguling and Xiazhuang in Mt. Jinggang.

Status of Population Common on migration in Mt. Jinggang.

雀形目 PASSERIFORMES
黄鹂科 Oriolidae

3.128 黑枕黄鹂 *Oriolus chinensis* Linnaeus, 1766

识别特征 体型略大的黄色鹂，约 26 cm，黑色过眼纹从眼先延至枕部。雄鸟金黄色，黑色飞羽具黄色边缘，尾黑色，外侧尾羽末端黄色。雌鸟黑色部分较雄鸟不明显，背橄榄黄色。幼鸟背部橄榄色，下体近白而具黑色纵纹。虹膜暗红色；嘴粉红色；脚近黑。

生境与习性 栖息于平原至低山的阔叶林和针阔混交林，也见于农田、荒地、原野及公园的高大乔木上。多在树冠中隐匿，单只或成对活动。飞行略呈波状，振翅缓慢而有力。以昆虫为主食，也食少量浆果和植物种子。

分布 印度、中国、东南亚。北方鸟南迁越冬。在井冈山地区见于双溪口和南风面。

3.127a

3.127b

3.128

种群状况 在井冈山地区迁徙季可见。

Identification L ca. 26 cm. Large yellow oriole with broad black mask from lores to nape. Male golden-yellow, flight feathers black with yellow fringe, tail black with yellow-tiped outer feathers. Female has olive upperparts, duller than male. Juvenile greenish-yellow with black streaks on whitish underparts. Eyes dark red; bill pink; tarsi blackish.

Habitat and Behavior Inhabits lowland broadleaf forest, theropencedrymion, plantations, parkland and open country with big trees. Solitary or in pairs. Unobtrusive in canopy and best found by voice. Undulating flight with slow, powerful wing beating. Feeds on insects, berries and seeds.

Distribution India, China, SE Asia. Northern populations move south in winter. At Shuangxikou and Nanfengmian in the Jinggangshan Region.

Status of Population Visible on migration in the Jinggangshan Region.

3.129a

雀形目 PASSERIFORMES
卷尾科 Dicruridae

3.129 黑卷尾 *Dicrurus macrocercus* Vieillot, 1817

识别特征 体型中等、通体蓝黑色闪金属光泽的卷尾，约 30 cm。嘴小，尾长、末端分叉且向外卷曲。幼鸟下体下部具浅色横纹。虹膜棕红色；嘴及脚黑色。

生境与习性 栖息于有树的原野、耕地、城市公园等。常立于树梢、光枝或电线上，飞翔于空中捕食昆虫。多成对或成小群活动。繁殖期善鸣叫，领域性强，性好斗。多营巢于阔叶树的枝杈上。窝卵数 3-4 枚，雌雄亲鸟均参与孵卵和育雏。

分布 伊朗至印度、中国、东南亚、爪哇（印尼）及巴厘岛。在井冈山见于罗浮、湘洲、下庄、黄坳、新城区等地。

种群状况 在井冈山夏季及迁徙季常见。

3.129b

Identification L ca. 30 cm. Mid-sized metallic blue-black drongo with long deeply-forked tail, and tips curve outwards. 1st winter has duller wings and tail, underparts with whitish scale. Eyes red-brown; bill black; tarsi black.

Habitat and Behavior Inhabits open country with trees, agricultural fields and urban parks, frequently perches on poles, wires and treetops. Often in pairs or small groups. Sallies to catch insects. Varied ringing calls when breeding, and territorial. Nests on branches of broadleaf trees. Clutch size is 3-4 eggs. Both parents incubate and brood.

Distribution Iran to India, China, SE Asia, Java and Bali. At Luofu, Xiangzhou, Xiazhuang, Huang'ao and Xinchengqu in Mt. Jinggang.

Status of Population Common in summer and on migration in Mt. Jinggang.

3.130 发冠卷尾 *Dicrurus hottentottus* (Linnaeus, 1766)

识别特征 体型略大的卷尾，约 32 cm。通体黑色具蓝绿色光泽，前额具丝状羽冠，体羽斑点闪烁。尾羽较长，宽阔的尾端向上

3.130a

卷曲。第一年冬羽闪斑较少，腹部和臀具白斑。虹膜暗红褐色；嘴黑色，脚黑色。

生境与习性 栖于中低山的各类林，也见于公园和人工绿地。常单独或成对活动于林冠层，也与其他鸟类混群。主要以昆虫为食，也食少量植物种子。繁殖期雄鸟善鸣叫，鸣声粗犷而嘈杂，也做炫耀式飞行。营巢于高大乔木顶端的枝桠上。窝卵数 3-4 枚。孵化期 15-16 天。雏鸟留巢哺育 20-24 天。雌雄亲鸟共同筑巢、孵卵和育雏。

分布 印度、中国、东南亚。在井冈山广泛分布。

种群状况 在井冈山全年常见。

Identification L ca. 32 cm. Largish green-glossed black drongo with heavy bill. Long filiform plumage from forehead over crown. Highly iridescent wings and hackles on breast and mantle, and long tail with broad upwards-curled tips. 1st winter has few glossy spangles, white spots on belly and vent. Eyes reddish-brown; bill black; tarsi black.

Habitat and Behavior Inhabits lowland forests, and open areas with trees. Solitary or in small flocks, sometimes mixed with other species. Chasing insects in the sky. Male has melodious loud singing with harsh screeches and performs acrobatics when breeding. Untidy nest suspended from outer branch. Clutch size is 3-4 eggs. Incubation period is 15-16 days. Chicks stay in nest for 20-24 days. Both parents nest, incubate and brood.

Distribution India, China, SE Asia. Widespread in Mt. Jinggang.

Status of Population Common all year round in Mt. Jinggang.

雀形目 PASSERIFORMES
椋鸟科 Sturnidae

3.131 八哥 *Acridotheres cristatellus* (Linnaeus, 1758)

识别特征 体型稍大的八哥，约 26 cm。大体黑色，嘴基部簇羽突出。初级飞羽基部白色形成块状翼斑，飞行时甚明显。尾端有狭窄的白色，尾下覆羽具黑白色横纹。虹膜橘黄色；嘴浅黄、嘴基红色；脚暗黄色。

生境与习性 活动于近山矮林、路旁、村庄和农作区，也见于苗圃、公园等生境。性活泼，成群活动，不甚惧人。常在耕牛后啄食犁锄翻出的蚯蚓、昆虫和植物块茎等，或在牛背上啄食牛的体外寄生虫。营巢于树洞中或建筑物洞穴内。窝卵数 3-6 枚，多为 4-5 枚。

分布 中国及中南半岛。引种至菲律宾及加里曼丹岛。在井冈山广泛分布。

种群状况 在井冈山全年常见。

Identification L ca. 26 cm. Large black myna with short bushy crest at base of bill. Mostly black except white patch at base of primaries, white bars on vent and undertail coverts, narrow white tail tip. In flight, broad white wing bars distinct. Eyes orange; bill pale yellow with red base; tarsi dull yellow-orange.

Habitat and Behavior Inhabits open agricultural land, towns and gardens. Forages in small or large groups. Bold, often follow cattle to catch insects, or peck parasites of cattle. Nesting in cavities of walls, eaves or trees. Clutch size is 3-6 eggs, mostly 4-5 eggs.

Distribution China and Indo-China Peninsula. Introduced into the Philippines and Kalimantan Peninsula. Widespread in Mt. Jinggang.

3.130b

3.131a

3.131b

3.132 灰椋鸟 *Spodiopsar cineraceus* (Temminck, 1835)

3.132a

识别特征　中等体型的灰褐色椋鸟，约 24 cm。头近黑色，前额至脸颊白色，具黑色絮状羽；喉至上胸灰黑色并夹杂白色丝状羽；腰、腹中央、外侧尾羽末端、尾下覆羽及二级飞羽翼斑为白色（飞行时较明显），其余部分灰褐色。雌鸟色浅而暗，前额白色不明显。虹膜黑褐色；嘴橘红色且尖端黑色；脚暗橘黄。

生境与习性　常成对或集大群活动于低山丘陵、平原和旷野。喜在农田、河谷等潮湿地上觅食，休息时多栖于电线、电杆上和树木枯枝上。飞行迅速，整群飞行。食性以昆虫为主，秋冬季主要以植物果实和种子为食。营巢在树洞、水泥柱上的空洞和人工巢箱中。窝卵数 5-7 枚。孵卵期 12-13 天。孵卵主要由雌鸟承担，雌雄共同育雏。

分布　西伯利亚、中国、日本、越南北部及缅甸北部、菲律宾。在井冈山地区见于黄坳、南风面等地。

种群状况　在井冈山地区迁徙季及冬季常见。

3.132b

Identification　L ca. 24 cm. Mid-sized grey and brown starling. Head blackish, forehead and cheeks white with some dark streaks, and throat to upper breast greyish-black with some white streaks. Rump, central belly, vent tips of tail feathers, and narrow wing bar on secondaries white (distinct in flight), other parts greyish-brown. Female slightly duller than male, white patch on forehead dim. Eyes dark brown; bill deep orange with black tip; tarsi dark orange.

Habitat and Behavior　Inhabits deciduous forest fringes, groves of trees and habitation, often in urban areas, also agricultural land. Lives and flies in flocks, forages on farmland and valleys. Often perches on wires, telegraph poles and deadwood. Feeds on insects, fruit and seeds. Nests in cavities of trees, telegraph poles and walls. Clutch size is 5-7 eggs. Incubation period is 12-13 days. Female incubate, and both parents brood.

Distribution　Siberia, China, Japan, N Vietnam, N Myanmar and the Philippines. At Huang'ao and Nanfengmian in the Jinggangshan Region.

Status of Population　Common in winter and on migration in the Jinggangshan Region.

3.132c

3.133 丝光椋鸟 *Spodiopsar sericeus* (Gmelin, JF, 1789)

识别特征　体型略小的灰白色椋鸟，约 23 cm。雄鸟头部浅色，头顶及脸颊染褐，颏、喉至上胸白色具丝状羽。两翼及尾辉黑，腰浅灰色，余部青灰色。飞行时初级飞羽基部的白斑明显。雌鸟体羽暗淡偏褐色，头部为灰褐色且颈部丝状羽不明显，腰更浅色。虹膜黑色；嘴红色而尖端黑色；脚暗橘黄。

生境与习性　栖息于开阔平原、农耕区和丛林间，多成对或结群活动。从树丛飞出，到草坡、稻田觅食，不甚畏人。常与其他椋鸟混群。主要取食各类昆虫，也食野生果实和杂草种子等植物性食物。营巢于墙洞或树洞中，以干草、鸡毛等做巢。

分布　中国、越南、菲律宾。在井冈山地区见于七溪岭、宁冈等地。

3.133a

种群状况 在井冈山地区全年常见。

Identification L ca. 23 cm. Smallish grey and white starling. Male has pale head, brownish on crown and cheeks, chin, throat and upper breast white with filiform plumage. Wings and tail glossy-black, rump pale grey, other parts slate-grey. In flight, show white patch at base of primaries. Female duller than male, darker on mantle and paler on rump, filiform plumage on neck undistinct. Eyes black; bill red with black tip; tarsi bright orange to dull orange.

Habitat and Behavior Inhabits lowland agricultural areas and open areas with scrub. Forms large flocks on migration. Bold, forages on grasslands and farmlands. Mixed with other starlings. Feeds on insects, also fruit and seeds. Nests in cavities of walls and trees.

Distribution China, Vietnam and the Philippines. At Qixiling and Ninggang in the Jinggangshan Region.

Status of Population Common all year round in the Jinggangshan Region.

3.133b

3.133c

3.134 黑领椋鸟 *Gracupica nigricollis* (Paykull, 1807)

识别特征 体大的黑白色椋鸟，约28 cm。成鸟头白，眼周裸露皮肤黄色；宽阔颈环及上胸黑色；背及两翼黑色，具多道白色翼斑；尾黑而尾端与外侧尾羽白色；下胸至臀部白色。幼鸟较暗淡，无黑色颈环。虹膜深褐色；嘴黑色；脚灰褐色。

生境与习性 常成对或结小群活动于开阔农田、荒地和河流两侧。常与八哥、其他椋鸟混群栖息与觅食，活动在家畜周围。鸣声单调、嘈杂，常且飞且鸣。杂食性，以动物性食物为主。营巢于大树的树杈或枝梢间。窝卵数4-6枚。

分布 中国南部及东南亚。在井冈山见于黄坳、新城区等地。

种群状况 在井冈山全年常见。

Identification L ca. 28 cm. Large white and black starling. Adult has white head with yellow facial skin from lores to ear coverts, broad black collar, black upperparts with white fringes to most wing feathers, tail has white tips and outer rectrices all-white. Underparts and rump white. Juvenile much duller, lacking black collar. Eyes dark brown; bill black; tarsi pale grey-brown.

Habitat and Behavior Inhabits fields, farmland, grassland and riversides. Forages in pairs or in small groups, sometimes mixes with other starlings and mynas. Often around domestic animals. Call is harsh screeches and whistles. Omnivorous, feeds mainly on animal food. Nests in tree branches. Clutch size is 4-6 eggs.

Distribution Southern China and SE Asia. At Xinchengqu and Huang'ao in Mt. Jinggang.

Status of Population Common all year round in Mt. Jinggang.

3.134a

雀形目 PASSERIFORMES
鸦科 Corvidae

3.135 松鸦 *Garrulus glandarius* (Linnaeus, 1758)

识别特征 体小的粉棕色鸦，约35 cm。翼上具黑色及钴蓝色镶嵌图案。头、上背、下体肉桂色，宽阔的髭纹、翼及尾黑色，腰和臀部白色。虹膜红褐色；嘴灰黑色；脚肉棕色。

3.134b

生境与习性　栖息于针叶林、阔叶林、针阔混交林和林缘灌丛的树冠层。多单独或集家族群，迁徙时集大群。嘈杂而机警，遇惊则穿林而飞，振翅沉重、无规律。性杂食，常见其在树间追逐空中飞行的昆虫或在树上寻食野果，是橡子、松子的主要消费者。营巢于枝叶繁茂的高大树木上。窝卵数 3-8 枚。雌鸟孵卵，雄鸟在巢周围守护。雌雄共同育雏，育雏期 17 天。

分布　欧洲、非洲西北部、喜马拉雅、中东至日本、东南亚。在井冈山广泛分布。

种群状况　在井冈山全年常见。

Identification　L ca. 35 cm. Small pinkish-rufous crow with cobalt-blue barred black on greater coverts and outer secondaries. Plain cinnamon head, mantle and underparts; black malar, wings and tail; white rump and vent. Eyes reddish-brown; bill greyish-black; tarsi grey-brown.

Habitat and Behavior　Inhabits deciduous and evergreen broadleaf forests, mixed forests with conifers, and open country with trees. Often solitary or in family parties, but flocks on migration. Noisy and alert in canopy. Flies with labored irregular beat. Feeds on fruit, bird eggs, carrion and acorns. Builds a moss and stick nest in large trees. Clutch size is 3-8 eggs. Female incubates, and male guards. Both parents brood. Brooding period is 17 days.

Distribution　Europe, NW Africa, Himalaya, Middle East to Japan and SE Asia. Widespread in Mt. Jinggang.

Status of Population　Common all year round in Mt. Jinggang.

3.136 红嘴蓝鹊 *Urocissa erythroryncha* (Boddaert, 1783)

识别特征　体长，具红嘴、长尾的蓝鹊，约 68 cm。顶冠后部至颈背白，头至上胸黑色，上背及两翼蓝灰色，腹部及臀白色，尾楔形，中央尾羽蓝色具白端，外侧尾羽具白色端斑和黑色次端斑。虹膜红色；嘴和脚鲜红色。

生境与习性　栖息于山区各种类型的森林，也见于竹林、林缘和村旁。常成对或集小群活动，性活泼而嘈杂。在树间转移时常由一只带头，其余陆续飞去。飞行时多滑翔，两翅平伸，尾羽展开。受惊时吃力鼓翅向远处逃窜。较凶猛，主动围攻猛禽。杂食性。营巢于树木侧枝上，也在高大的竹林上筑巢。窝卵数 3-6 枚，多为 4-5 枚。雌雄轮流孵卵，雏鸟晚成。

分布　喜马拉雅、印度东北部、中国、缅甸及中南半岛。在井冈山广泛分布。

种群状况　在井冈山全年常见。

Identification　L ca. 68 cm. Large blue magpie with red bill and extremely long tail. Black head to upper breast with white rear crown and nape. Upperparts greyish-blue with white tips of blue flight feathers, and underparts white. Blue tail with broad white tips on central feathers, black and white tips on outer feathers. Eyes red; bill red; tarsi red.

Habitat and Behavior　Inhabits forests, forest edge, scrub and villages. Active, noisy and sociable. Groups fly short distance in follow-the-leader fashion, with quick wing-beats and glides, wedged tail outspreading; sustained flight labored and undulating.

3.135a

3.135b

3.136a

3.136b

Aggressively mobs raptors. Omnivorous, readily visits ground. Feeds on fruit, small birds and eggs, insects and carrion. Builds nest cup fairly low in dense-leafed tree or bamboo. Clutch size is 3-6 eggs, mostly 4-5 eggs. Parents incubate in rotation. Chicks are altricial.

Distribution　Himalaya, NE India, China, Myanmar and Indo-China Peninsula. Widespread in Mt. Jinggang.

Status of Population　Common all year round in Mt. Jinggang.

3.137a

3.137 灰树鹊 *Dendrocitta formosae* Swinhoe, 1863

识别特征　体型略小的褐灰色树鹊，约 36 cm。前额、眼先及喉黑色，眼后浅褐色，后枕青灰色。颈侧、上背灰褐色，两翼黑色具白色斑块（飞行时较明显），腰灰白色，尾黑色。胸染棕色，腹部灰色，臀棕黄色。虹膜红褐色；嘴灰黑色；脚深灰色至黑色。

生境与习性　栖息于中低山的阔叶林、针阔混交林，也见于天然林、人工林和城市公园。多成对或集小群活动于乔木的中上层。性怯懦而吵嚷。杂食性，常以浆果、坚果等植物果实与种子为食，也吃昆虫等动物性食物。营巢于树上或灌木上，窝卵数 3-5 枚。雌雄亲鸟轮流孵卵，雏鸟晚成性。

分布　喜马拉雅、印度东部及东北部、缅甸、泰国北部、中南半岛北部和中国（华中、华南及东南）。在井冈山广泛分布。

种群状况　在井冈山全年常见。

Identification　L ca. 36 cm. Smallish grey-brown treepie. Black lores, forehead and throat, slate-grey nape, dusky brown necksides and mantle, prominent white carpal patch on black wing, grey rump, and black tail. Blackish-brown on chest, grey belly, rufous vent. In flight, white stripe through primary bases contrasts with black wings. Eyes reddish-brown; bill grey-black; tarsi grey-black.

3.137b

Habitat and Behavior　Inhabits broadleaf and secondary forests, urban parks and gardens. In pairs or groups. Mainly canopy but sometimes feeds on ground. Feeds on insects, fruit and flowers. Nests in small tree or bush in forest. Clutch size is 3-5 eggs. Parents incubate in rotation. Chicks are altricial.

Distribution　Himalaya, E and NE India, Myanmar, N Thailand, N Indo-China Peninsula and C, S and SE China. Widespread in Mt. Jinggang.

Status of Population　Common all year round in Mt. Jinggang.

3.138 喜鹊 *Pica pica* (Linnaeus, 1758)

识别特征　体型较大、尾长的黑白色鹊类，约 45 cm。头、颈、胸、上体及臀部黑色，肩部、下腹及两胁白色，双翼及尾黑色具蓝绿色金属光泽。飞行时，初级飞羽大体白色，背部具 "V" 形白斑。虹膜暗褐色；嘴黑色；脚黑色。

生境与习性　栖息于山麓、林缘、农田、村庄、城市公园等人类居住附近。除繁殖期成对活动外，常结 3-5 只小群活动。性机警，觅食时总有一鸟负责警卫。飞行显弱，飞行时尾羽扩展、双翅缓慢鼓动，成波浪式前进。在地上活动时成跳跃式。杂食性。营巢于高大乔木上，有时也营巢于高压电柱上。窝卵数 5-8 枚，有时多至 11 枚。雌鸟孵卵由承担，孵卵期 17-18 天。雌雄共同

3.138

育雏，育雏期 1 个月。

分布 欧亚大陆、北非、加拿大西部及美国加利福尼亚州西部。
在井冈山地区见于宁冈、黄坳、七溪岭等地。

种群状况 在井冈山地区全年常见。

Identification L ca. 45 cm. Large black and white magpie with long glossy black tail. Black head, breast and upperparts, glossy blue-purple wings with white scapular patch, white flanks and center of belly, and black vent. In flight, white inner webs of primaries show as large white patch, and V-shaped white patch on back. Eyes dark brown; bill black; tarsi black.

Habitat and Behavior Inhabits open woodland, thickets, agricultural land with scattered trees, suburban and urban parks and gardens. Usually in pairs or small loose parties. Alert, there is always a bird as guard when flock is foraging. Flight appears weak, with rapid fluttery wing beats and swooping glides. Often forages on ground by hopping, perches on buildings, and rides domestic animals. Omnivorous. Building large stick nest atop trees, bushes, poles, and pylons. Clutch size is 5-8 eggs, occasionally 11 eggs. Female incubates. Incubation period is 17-18 days. Both parents brood. Brooding period is about one month.

Distribution Eurasia, N Africa, W Canada and W California, USA. At Ninggang, Huang'ao and Qixiling in the Jinggangshan Region.

Status of Population Common all year round in the Jinggangshan Region.

3.139a

3.139b

3.139 大嘴乌鸦 *Corvus macrorhynchos* Wagler, 1827

识别特征 体大的鸦类，约 50 cm。全身黑色具蓝色光泽，嘴甚粗厚，前额隆起。飞行时，双翼及尾显圆。虹膜褐色；嘴黑色；脚黑色。

生境与习性 栖息于各种森林类型中，尤以疏林和林缘地带较常见。喜在河谷、农田、村庄、沼泽和草地上活动，常跟在家畜身后。非繁殖期成群活动，迁徙时集大群，也与其他鸦类混群。性机警、好斗，攻击猛禽、甚至靠近巢的行人。杂食性。营巢在高大的树杈上，通常为针叶树。窝卵数 3-5 枚。孵卵期 15-19 天，通常 18 天。雏鸟留巢期 26-30 天。雌雄共同孵卵和育雏。

分布 伊朗至中国、东南亚。在井冈山广泛分布。

种群状况 在井冈山全年常见。

Identification L ca. 50 cm. Large purple-glossed black crow with distinct heavy arched bill. Head large with abruptly high and rounded forehead. In flight, show more rounded wings and tail. Eyes dark brown; bill black; tarsi black.

Habitat and Behavior Inhabits forests, prefers sparse woodland and forest edges, also in cultivated land, urban, rural areas. Usually in small groups, but large flocks on migration, sometimes mix with other crows. Often following domestic animals. Alert and aggressive, harass raptors and even pedestrians near nests. Omnivorous. Commonly moves on ground using bouncing hop, also walks. Building a large nest close to trunk of fairly large tree, commonly a conifer. Clutch size is 3-5 eggs. Incubation period is 15-19 days, mostly 18 days. Chicks stay in the nest for 26-30 days. Both parents incubate and brood.

Distribution Iran to China and SE Asia. Widespread in Mt. Jinggang.

Status of Population Common all year round in Mt. Jinggang.

3.140 白颈鸦 *Corvus torquatus* Lesson, R, 1831

识别特征 体型较大的黑白色鸦，约 54 cm。除枕部、上背、颈侧及胸带为白色外，其余部分辉黑色。虹膜深褐色；嘴黑色；脚黑色。

生境与习性 栖息于于平原、丘陵和低山，也见于高至海拔 2500m 左右的山地。通常在清晨飞到开阔的农田、河滩和河湾等地活动和觅食，至晚上才返回村落附近或林缘树上过夜。性机警，比其他鸦类更难接近。栖止时，多伸颈鸣叫。杂食性，主要食物因季节不同而不同。营巢于高大树上，窝卵数 3-7 枚，通常 3-4 枚。

分布 中国（华中、华南及东南），至越南北部。在井冈山见于黄洋界、下庄、厦坪等地。

种群状况 在井冈山全年常见。

Identification L ca. 54 cm. Large black and white crow. White nape, upper mantle, necksides and breast band contrast with other glossy-black parts. Eyes dark brown; bill black; tarsi black.

Habitat and Behavior Inhabits plains and hills, also on mountains up to 2500 m a.s.l. Sometimes seen in mixed flocks with Large-billed Crow. Usually roosts in trees near villages or forest edges at night, forages on cultivated fields or river beds by day. Alert, harder to approach than other crows. Omnivorous, main food changes according to season. Nests in large trees. Clutch size is 3-7 eggs, mostly 3-4 eggs.

Distribution C, S and SE China to N Vietnam. At Xiaping, Xiazhuang and Huangyangjie in Mt. Jinggang.

Status of Population Common all year round in Mt. Jinggang.

3.140a

3.140b

雀形目 PASSERIFORMES
河乌科 Cinclidae

3.141 褐河乌 *Cinchus pallasii* Temminck, 1820

识别特征 体型略大、通体深褐色的河乌，约21 cm。嘴相对较长，体无浅色胸围。有时眼上的白色小块斑明显。幼鸟偏灰，全身具深色鳞状纹。虹膜褐色；嘴深褐；脚深褐。

生境与习性 栖息于山区溪流与河谷沿岸，从不到河流两岸树上停落。常单独或成对活动。多站立在河边或河中露出水面的石头上，尾上翘，头和尾不时上、下摆动。飞行迅速，一般沿河流水面直线飞行，如遇惊扰能迅速折转向相反方向飞去。善于潜水，亦能游泳及在水底行走。全年以昆虫和昆虫幼虫、小鱼虾等动物性食物为主，也偶食植物叶子和种子。营巢于河流两岸石隙间、岩石凹陷处、树根下或瀑布后面的石隙。窝卵数4-5枚，也见3枚和6枚。孵卵期15-16天。雌雄共同育雏，育雏期21-23天。

分布 南亚、东亚、喜马拉雅及中南半岛北部。在井冈山广泛分布。

种群状况 在井冈山全年常见。

Identification L ca. 21 cm. Entirely brown dumpy bird with relatively long bill. Sometimes a small white patch above eyes. Juvenile more greyish, coarsely scaled blackish overall. Eyes brown; bill dark brown; tarsi dark brown.

Habitat and Behavior Inhabits montane rocky streams, shallow rivers or valleys, never perches on trees. Usually solitary or in pairs. Perches on boulders, bobbing and raising tail, jerking occasionally. Fast-moving, flight along streams is low and whirring, turns around immediately when disturbed. Often dives, swims and walks under water, taking various aquatic invertebrates and larvae. Usually nests in crevices in rock-faces, or in artificial situations above water, sometimes behind waterfalls. Clutch size is 4-5 eggs, also seen 3 and 6. Incubation period 15-16 days. Both parents brood, and brooding period is 21-23 days.

Distribution S and E Asia, Himalaya, and N Indo-China Peninsula. Widespread in Mt. Jinggang.

Status of Population Common all year round in Mt. Jinggang.

3.141a

3.141b

雀形目 PASSERIFORMES
鸫科 Turdidae

3.142 蓝短翅鸫 *Brachypteryx montana* Horsfield, 1821

识别特征 中等体型的短翅鸫，约 15 cm。雄鸟通体深蓝色，醒目的白色眉纹有时被掩盖，翼具狭窄的白色肩斑，翼、尾、下腹略带青灰色，下腹中部至尾下覆羽灰白色。雌鸟上体暗橄榄褐色，眼圈、眼先及两翼和尾下覆羽红棕色，下体偏灰褐色。虹膜褐色；嘴黑色；脚深褐色。

生境与习性 栖于中高山的常绿阔叶林和山顶林缘灌丛、竹林中，多靠近溪流或山涧。性隐匿且羞怯，善鸣叫。常单独活动于植被覆盖茂密的地面，有时见于开阔林间空地，甚至山顶多岩的裸露斜坡。取食地上的昆虫等无脊椎动物。营巢于树上或岩石苔藓中。窝卵数常 3 枚。

分布 喜马拉雅至中国南部、东南亚。在井冈山地区见于荆竹山、南风面等地。

种群状况 在井冈山地区全年可见。

Identification L ca. 15 cm. Mid-sized shortwing with long dark legs. Male dark blue above with white supercilium over black lores (sometimes concealed), narrowly white-fringed carpal, slatier on wings, tail and lower underparts, central belly and undertail coverts being almost white. Female dark olive-brown above with pale rufous eyering, lacks white supercilium, rufescent-brown wings and tail, and paler brown underparts with buffy-rufous vent. Eyes dark brown; bill black; tarsi blackish-brown.

Habitat and Behavior Inhabits dense, shady undergrowth and thickets in mature forest, usually near streams or in ravines. Usually solitary and shy, skulks on or near ground. Sometimes comes out into open clearings and even bare rocky slopes of mountain. Feeds on invertebrate on ground. Builds globular nest attached to creepers just above ground. Clutch size is 3 eggs.

Distribution Himalaya to Southern China and SE Asia. At Jingzhushan and Nanfengmian in the Jinggangshan Region.

Status of Population Visible all year round in the Jinggangshan Region.

3.143 蓝歌鸲 *Luscinia cyane* (Pallas, 1776)

识别特征 中等体型的歌鸲，约 14 cm。雄鸟上体青灰蓝色，宽阔的黑色过眼纹延至颈侧和胸侧，颏、喉及下体白色。雌鸟上体橄榄褐，喉及胸褐色并具皮黄色鳞状斑纹，腰及尾上覆羽染蓝。幼鸟颜色同雌鸟，脸和翼褐色更重，腰、尾及尾上覆羽或多或少显蓝色。虹膜褐色；嘴黑色；脚粉色。

生境与习性 栖息于沟谷和溪流两侧的阔叶林、针叶林和针阔混交林下。地栖性，甚隐怯，多单独活动于林下灌丛中。站姿较平，驰走时尾常上下扭动。主要以昆虫及其幼虫为食，还食蜘蛛、多足虫等小型动物。

分布 繁殖于东北亚。冬季迁至印度、中国南部、东南亚。在井冈山见于茨坪等地。

3.142a

3.142b

3.143a

种群状况　在井冈山迁徙季常见。

Identification　L ca. 14 cm. Mid-sized robin. Male has deep blue upperparts and white underparts, broad jet black line from bill base to breast-sides. Female has brown upperparts with dark blue rump and uppertail coverts, and white underparts with brown wash on breast and flanks, slight scaling on chest. Juvenile browner on face and wings, but has variable blue on rump, uppertail coverts and tail. Eyes black; bill black; tarsi pallid pink.

Habitat and Behavior　Inhabits broadleaf forests, coniferous forests and mixed forests around gullies and streams. Generally solitary and skulking, close to or on forest floor, often in dwarf bamboo or dense ground cover, sometimes singing at mid-levels. Stance rather horizontal, pumps tail when on ground. Feeds on insects and larvae, also spiders and myriapods.

Distribution　Breeds in NE Asia. Migrant in winter to India, Southern China, SE Asia. At Ciping in Mt. Jinggang.

Status of Population　Common on migration in Mt. Jinggang.

3.143b

3.143c

3.144 红喉歌鸲 *Calliope calliope* (Pallas, 1776)

识别特征　中等体型褐色歌鸲，约16 cm。脸具醒目图案。上体褐色，两胁皮黄，腹部稍白。成年雄鸟具清晰的白色眉纹和髭纹，眼先黑色，颏喉鲜红镶黑色侧缘。雌鸟颜色较暗淡，颏、喉偏白，部分染粉色。虹膜褐色；嘴深褐色；脚粉褐色。

生境与习性　繁殖季栖于北方低海拔泰加林林缘和近海海岛上，冬季见于有灌丛的草地和芦苇丛间，尤喜靠近溪流等近水地方。典型的地栖鸟类，常在林下灌丛或地边草丛中地面奔跑、跳跃。在地面疾走时，常稍停而把尾巴向上展开如扇、双翼下垂。性机警而胆怯，但繁殖期雄鸟会站在灌木枝头或电线上鸣唱，红色的喉部有节奏地搏动。主要以昆虫为食，也食部分植物性食物，常在地上边走边啄食。

分布　繁殖于东北亚。冬季至印度、中国南部及东南亚。在井冈山见于大井和罗浮等地。

种群状况　在井冈山迁徙季常见。

Identification　L ca. 16 cm. Mid-sized brown robin with distinctive face pattern. Adult has mid-brown upperparts, greyish-brown on flanks, and whitish on belly and vent. Male has clear white supercilium and submoustachial, bold black lores, chin and throat brilliant metallic ruby fringed by black malar. Female has less boldly marked face, and white chin and throat, some have pale pink wash to chin. Eyes dark brown; bill dark brown; tarsi pink-brown.

Habitat and Behavior　Inhabits lower altitude taiga-forest edge in the north and on some offshore islands. In winter seem in grassy areas with bushes and near wetlands with reeds, prefers proximity to streams. Skulks on ground and in low bushes, often cocking tail and drooping wings. Shy and vigilant, but male sings on top of the bush or wires when breeding, and red throat pulsates. Pecks insects and some plants when walking.

Distribution　Breeds in NE Asia. Migrates in winter to India, Southern China and SE Asia. At Dajing and Luofu in Mt. Jinggang.

Status of Population　Common on migration in Mt. Jinggang.

3.144a

3.144b

3.145 红尾歌鸲 *Larvivora sibilans* Swinhoe, 1863

识别特征　体小的歌鸲，约 13 cm。上体橄榄褐，眼先上方和眼圈浅色，尾及尾上覆羽红棕色。下体褐灰色，喉侧及胸部具白色鳞形纹。虹膜褐色；嘴黑色；脚粉褐色。

生境与习性　主要栖息于山地针叶林、针阔混交林和阔叶林中，尤以林木稀疏而林下灌木密集的地方较常见。多单独或成对在植被下层活动。性活跃、善藏匿。站姿略直，在地上走动时，常边走边将尾向上竖起。

分布　东北亚。越冬至中国南部。在井冈山地区见于茅坪、南风面。

种群状况　在井冈山地区迁徙季常见。

Identification　L ca. 13 cm. Small rufous-tailed robin. Upperparts plain olive-brown with pale supraloral and eyering, rufous uppertail coverts and tail, underparts brownish-grey with whitish scallops on throat-sides and breast. Eyes brown; bill black; tarsi pinkish-brown.

Habitat and Behavior　Inhabits well-wooded or deciduous and coniferous forests, typically in dense undergrowth, often in gullies or near streams. Solitary or in pairs, rather terrestrial. Active and shy, stance rather erect, often shivers tail.

Distribution　NE Asia. Wintering to Southern China. At Maoping and Nanfengmian in the Jinggangshan Region.

Status of Population　Common on migration in the Jinggangshan Region.

3.145a

3.145b

3.146 北红尾鸲 *Phoenicurus auroreus* (Pallas, 1776)

识别特征　中等体型、颜色鲜艳的红尾鸲，约 15 cm。具明显的白色翼斑及红棕色的腰。雄鸟头顶至颈背银灰色；眼先、头侧、喉、上背及两翼褐黑，仅翼斑白色；体羽余部栗褐，尾红棕色具深黑褐中央尾羽。雌鸟上体红褐色，眼圈色浅，中央尾羽深色。虹膜暗褐色；嘴黑色；脚灰黑色。

生境与习性　主要栖息于山地、森林、河谷、林缘和居民点附近的灌丛、花园、菜地与低矮树丛。常单独或成对活动。动作敏捷，不时从栖息的树枝上急促飞向地面或附近的其他树枝上啄食昆虫，食后又返回原枝上。尾常上下摆动，头亦点动，还常伴着较微弱而单调的叫声。

分布　见于东北亚及中国。迁徙至日本、中国南部、喜马拉雅、缅甸及中南半岛北部。在井冈山广泛分布。

种群状况　在井冈山冬季常见。

Identification　L ca. 15 cm. Mid-sized colorful redstart with white patch on wings and rufous rump. Male has silver-grey crown to hind-neck, black mask extends to upper breast, black mantle, orange-rufous lower breast to underparts, rufous tail with blackish central rectrices. Female warmer brown upperparts, with buffy eyering and darker central rectrices, paler underparts. Eyes dark brown; bill black; tarsi black.

Habitat and Behavior　Inhabits open hillsides, open forests with rocky area, woodland edges, agricultural margins, parks and large gardens. Usually solitary or in pairs. Sits on prominent perch and shivers tail, with short soft whistled. Agile, makes sallies after insects to ground from bushes, and goes back the perches.

3.146a

3.146b

Distribution Resident NE Asia and China. Migrating to Japan, Southern China, Himalaya, Myanmar and N Indo-China Peninsula. Widespread in Mt. Jinggang.

Status of Population Common in winter in Mt. Jinggang.

3.147 红胁蓝尾鸲 *Tarsiger cyanurus* (Pallas, 1773)

识别特征 体型略小、喉白的鸲，约 15 cm。雌雄都有橘黄色两胁，与白色腹部及臀形成对比。雄鸟上体蓝色，眉纹白；幼鸟及雌鸟上体褐色，尾蓝。虹膜褐色；嘴黑色；脚灰色。

生境与习性 多单独或成对活动于丘陵和平原开阔林地或园圃中滋蔓遮蔽物的地方。地栖性。性虽隐匿，但不甚畏人。停歇时尾常上下摆动。以昆虫为主食，还兼吃一些蜘蛛、果实和草籽等。

分布 繁殖于东北亚及喜马拉雅山。冬季迁至中国南部及东南亚。在井冈山广泛分布。

种群状况 在井冈山冬季常见。

Identification L ca. 15 cm. Smallish robin with orange flanks. Male has blue upperparts with white brow in front of eyes, white underparts with white throat and grey breast-band. Female and 1st winter have olive-brown upperparts with pale eyering, prominent, brown breast-band, and duller blue rump and tail. Eyes brown; bill black; tarsi black.

Habitat and Behavior Usually in understory in fairly open forest as opposed to dense thicket and undergrowth. Constantly flicks tail downwards or fans it sharply. Gleans leaves and trunks, fly catches and forages on ground. Feeds on insects and some plants.

Distribution Breeds in NE Asia and Himalaya. Migrates in winter to Southern China and SE Asia. Widespread in Mt. Jinggang.

Status of Population Common in winter in Mt. Jinggang.

3.148 红尾水鸲 *Phoenicurus fuliginosus* Vigors, 1831

识别特征 体小的鸲，约 14 cm。雄鸟腰、臀及尾栗褐色，其余部位深青蓝色。雌鸟上体灰，眼圈色浅；下体白，灰色羽缘形成鳞状斑纹，臀、腰及外侧尾羽基部白色；尾余部黑色；两翼黑色具两道白色翼斑。虹膜黑色；嘴黑色；脚褐色。

生境与习性 栖息于山区溪流、河谷沿岸，尤以多石的林间或林缘地带的溪流沿岸较常见，也见于平原溪流、湖泊、水塘等岸边。常单独或成对活动。多站立在水边或水中石头上、电线上或村边房顶上。停栖时，常不停地扇开尾羽并上下抖动。飞行时靠近水面，边飞边叫。性好斗，有繁殖领域。以昆虫为主食，也食少量植物果实、嫩叶及草籽等。营巢于河岸、溪流边、稻田壁坎的凹陷处、岩石裂缝间或树洞等处，隐蔽甚好。窝卵数 3-6枚，多为 4-5 枚。雌鸟孵卵，雌雄亲鸟共同育雏。

分布 巴基斯坦、喜马拉雅至中国及中南半岛北部。在井冈山见于茨坪、新城区、荆竹山等地。

种群状况 在井冈山全年常见。

Identification L ca. 14 cm. Smallish robin. Male plain dull slate-grey, with deep chestnut tail, rump and vent. Female upperparts pale grey with pale eyering, dark wings with two prominent white wing bars; underparts white, scaled extensively grey from chin to belly;

3.147a

3.147b

3.148a

white rump and out-rectrices contrasts with greyish-black tail. Eyes black; bill black; tarsi dark brown.

Habitat and Behavior Inhabits streams and valleys in mountains, prefers streams with stones in forests or forest edges, also lakes, ponds and rivers on plains. Usually solitary or in pairs. Often perches on boulders, wires and buildings with tail fanned out, up and down. Close to water with trill in flight. Territorial. Mostly feeds on insects, also some plants. Nests in stone crevise and tree hollow by the river side. Clutch size is 3-6 eggs, usually 4-5. Female incubates, and both parents raise chicks.

Distribution Pakistan and Himalaya to China, and N Indo-China Peninsula. At Ciping, Xinchengqu, and Jingzhushan in Mt. Jinggang.

Status of Population Common all year round in Mt. Jinggang.

3.149 白顶溪鸲 *Phoenicurus leucocephalus* Vigors, 1831

识别特征 体型较大的鸲，约19 cm。雄雌同色。头顶及颈背白色，腰、尾大部及腹部栗红色，余部黑色。幼鸟色暗而近褐，头顶具黑色鳞状斑纹。虹膜褐色；嘴黑色；脚灰黑色。

生境与习性 常栖息于山地溪流和河谷沿岸。有垂直迁徙的习性。常单独或成对活动，有时也成 3-5 只的小群。常站在河边或河中露出水面的石头上，不甚怕人。降落时不停地点头且散开尾羽上下摆动。啄食陆生和水生昆虫，并兼食少量蜘蛛、软体动物、野果和草籽等。营巢于山间激流岩岸的裂缝中、石头下、树洞、岸旁树根间，偶尔也在水边或离水较远的树干上。巢隐蔽，不易发现。窝卵数 3-5 枚。雌鸟孵卵，雌雄亲鸟共同育雏。

分布 中亚、喜马拉雅、中国。越冬至印度及中南半岛。在井冈山见于水口、龙潭等地。

种群状况 在井冈山夏季常见。

Identification L ca. 19 cm. Large robin with prominent white crown. Crown to nape white, rump, tail (except black tip) and belly chestnut, the rest black. Juvenile more brown, black scaled on crown. Eyes brown; bill black; tarsi dark grey.

Habitat and Behavior Inhabits close to rushing streams and rivers in rugged terrain. Usually solitary or in pairs, sometimes in small groups. Often perches on boulders boldly. Bobs on landing, and fans, flicks and wags tail. Feeds on insects, spiders and seeds. Nests covertly in the crevices in rock-faces, or tree holes near streams. Clutch size is 3-5 eggs. Female incubate, and both parents raise chicks.

Distribution C Asia, Himalaya, China. Wintering to India and Indo-China Peninsula. At Shuikou and Longtan in Mt. Jinggang.

Status of Population Common in summer in Mt. Jinggang.

3.150 鹊鸲 *Copsychus saularis* (Linnaeus, 1758)

识别特征 中等体型的黑白灰色鸲，约20 cm。雄鸟头、颈、胸及背黑色，黑色翼具白色翼斑，外侧尾羽、腹及臀白色。雌鸟似雄鸟，但暗灰取代黑色。虹膜褐色；嘴及脚黑色。

生境与习性 常单独或成对活动于村落和人家附近的园圃及栽培地带，或树旁灌丛，也常见于城市庭院中。性活泼，大胆好斗。

清晨常高踞树梢、墙脊、屋顶上啼鸣跳跃，鸣声婉转多变。常在粪坑、猪牛圈、垃圾堆或翻耕地里觅食，有时也在草地上猎取昆虫，尾常上翘。几乎全食动物性食物，兼吃少量草籽和野果。营巢于墙缝、树洞中，或树枝的丫杈处。窝卵数4-6枚，多为5枚。孵化期12-13天。雌雄共同孵卵和育雏。

分布 印度、中国南部、东南亚。在井冈山见于茨坪、罗浮、新城区、荆竹山等地。

种群状况 在井冈山全年常见。

Identification L ca. 20 cm. Mid-sized white, black and grey robin with long tail. Male hood and upperparts glossy black, black wing with white wing bar, black tail with white outer feathers, underparts white except black chest. Females duller, slate-grey replacing black parts in male. Eyes dark brown; bill black; tarsi black.

Habitat and Behavior Inhabits woodland, scrub, farmland and parkland, including urban areas. Forages on the ground for invertebrates; highly territorial. Sings a sweet series of rising and falling whistles, usually from an open perch. Tail often held upright. Nests in the crevices of walls, hollows and crotches of trees. Clutch size is 4-6 eggs, mostly 5. Incubation period is 12-13 days. Both parents incubate and raise chicks.

Distribution India, Southern China and SE Asia. At Ciping, Luofu, Xinchengqu and Jingzhushan in Mt. Jinggang.

Status of Population Common all year round in Mt. Jinggang.

3.150a

3.150b

3.151 蓝矶鸫 *Monticola solitarius* (Linnaeus, 1758)

识别特征 体型略大的蓝色或斑驳褐色矶鸫，约23cm。亚种 *pandoo* 雄鸟通体暗蓝灰色；亚种 *philippensis* 雄鸟头、胸、上体暗蓝色，腹部及尾下深栗色。非繁殖季多具深色及浅色鳞状纹。雌鸟上体灰褐色染蓝，下体皮黄而密布黑色鳞状纹和横纹。幼鸟似雌鸟但上体具黑白色鳞状斑纹。虹膜暗褐色；嘴黑色；脚黑色。

生境与习性 栖息于多岩石的低山峡谷及山溪、湖泊等水域附近的岩石山地，也栖于海滨岩石和附近的山林中，有时也到村庄、果园中。常栖于突出位置，如岩石、房屋柱子及死树，冲向地面捕捉昆虫，而后又返回原栖处。食物以昆虫为主，也食少量果实、草籽等植物性食物。

分布 欧亚大陆、非洲东北部及东南亚。在井冈山见于厦坪、小溪洞等地。

种群状况 在井冈山迁徙季及冬季可见。

Identification L ca. 23 cm. Male race *pandoo* entirely glossy blue with black wings and tail, race *philippensis* has dark blue hood and upperparts, and chestnut underparts. White scales overall in autumn and winter. Female upperparts grey-brown with blue wash, and underparts buff with dark scales and bars. Juvenile similar to female, but more white and black scales on upperparts. Eyes dark brown; bill black; tarsi black.

Habitat and Behavior Inhabits rocky areas on coasts, mountains and steppe. Solitary and bold in winter. Usually perches atop a prominent rock, tree or building, and site-faithful. Feeds on insects, also some fruit and seeds.

3.151a

Distribution　Eurasia, NE Africa and SE Asia. At Xiaping and Xiaoxidong in Mt. Jinggang.

Status of Population　Visible in winter and on migration in Mt. Jinggang.

3.151b

3.152 栗腹矶鸫 *Monticola rufiventris* (Jardine & Selby, 1833)

识别特征　体大的矶鸫，约 24 cm。雄鸟头部具黑色脸罩，上体及尾湖蓝色，颏、喉深蓝色，胸及下体鲜艳栗色，看似一只巨型的蓝色鹟类。雌鸟褐色，头具白色下颊纹和白色月牙状耳斑，上体具深色鳞状斑纹，胸、胁部具褐色鳞状斑纹。虹膜深褐；嘴黑色；脚黑褐。

生境与习性　繁殖期主要栖息于中高山的常绿阔叶林、针阔叶林混交林和针叶林中，尤以陡峭的悬崖和溪流深谷沿岸的森林及其林缘地带较常见。秋冬季见于中低海拔的疏林及林缘、村庄等。常单独或成对活动。常停在小乔木树冠上鸣叫，尾上下摆动。性极机警。食物以昆虫为主。

分布　巴基斯坦西部至中国南部及中南半岛北部。在井冈山见于五指峰、笔架山等地。

种群状况　在井冈山迁徙季和冬季常见。

Identification　L ca. 24 cm. A large rock-thrush. Male has deep blue hood and upperparts, crown bright blue contrast with dark cheek; breast to vent and undertail coverts deep chestnut, like a huge flycatcher. Female upperparts brown with pale crescent on rear ear coverts, and underparts buff with heavily dark scales on breast and flanks. Eyes deep brown; bill black; tarsi dark brown.

Habitat and Behavior　Breeds in mid to high montane coniferous and oak forests, descends to lowland forested and rocky areas in winter. Solitary or in pairs. Prefers tree tops or wires, vocal and bold. Stance upright with tail slightly twitching. Feeds on insects.

Distribution　W Pakistan to Southern China and N Indo-China Peninsula. At Wuzhifeng and Bijiashan Mt. Jinggang.

Status of Population　Common in winter and on migration in Mt. Jinggang.

3.152a

3.153 白眉地鸫 *Geokichla sibirica* (Pallas, 1776)

识别特征　中等体型的深色或褐色地鸫，约 23 cm。雄鸟青黑色，白色的长眉纹宽而显著，两胁具白色细鳞纹，下腹和尾下覆羽染白。飞行时现出翼下两条白色带和白色的外侧尾羽末端。雌鸟橄榄褐色，眉纹、颊纹和喉黄白色，腰部和尾羽青灰色，胸腹浅色而具褐色鳞状斑。虹膜褐色；嘴黑色；脚黄色。

生境与习性　主要栖息于林下植物发达的针阔叶混交林、阔叶林和针叶林。迁徙期间常在林缘、道旁两侧次生林、农田及村庄附近的丛林中活动。常单独或成对活动，迁徙时亦成小群。地栖性，主要在地上活动和觅食，善于在地面行走和奔跑。性隐蔽，受惊扰立刻飞到附近树上。主要食物为昆虫。

分布　繁殖于北亚。冬季迁徙经东南亚至大巽他群岛。在井冈山见于五指峰、八面山等地。

种群状况　在井冈山迁徙季常见。

3.152b

Identification L ca. 23 cm. Mid-sized dark or brown thrush. Adult male dark slaty-grey, blackest on head and throat, with long white supercilium. Some narrow white scaling on flanks, lower belly white, vent and undertail coverts slate scale white. In flight, show double white bars on slaty-black underwing. Female warm olive-brown with yellowish-buff supercilium, submoustachial and chin. Upperparts plain brown, underparts pale buff with heavily scaled dark brown. Eyes dark brown; bill black in male, grey-tipped with yellow base in female; tarsi yellow.

Habitat and Behavior Inhabits mixed deciduous or evergreen broadleaf and coniferous montane forest at lower latitudes, migrant in forest edges, farmland and woodland near villages. Usually solitary or in pairs, in small groups on migration. Forages on ground, fly to tree when disturbed. Feeds on insects mostly.

Distribution Breeds in N Asia. Migrant in winter through SE Asia to Greater Sunda Islands. At Wuzhifeng and Bamianshan in Mt. Jinggang.

Status of Population Common on migration in Mt. Jinggang.

3.153

3.154 怀氏虎鸫 *Zoothera aurea* (Holandre, 1825)

识别特征 体大的褐色地鸫，约 28 cm。周身布满金褐色和黑色的鳞状斑纹，外侧尾羽黑色但末端白色。飞行时可见翼下的黑、白横带。虹膜褐色；嘴深褐色，下嘴基部较浅；脚带粉色。

生境与习性 通常栖居茂密森林，尤以溪谷、河流两岸和地势低洼的密林中较常见。地栖性，常见单独或成对活动，多在林下灌丛中或地面觅食。性胆怯，见人即飞。主要以昆虫等动物为食，亦兼食植物果实、种子等。

分布 广布于欧洲及印度至中国、东南亚。在井冈山见于早禾木、八面山、大坝里等地。

种群状况 在井冈山迁徙季及冬季常见。

Identification L ca. 28 cm. Large brown thrush with heavily scaled plumage and robust bill. Crown, nape to most of upperparts brown, covered with bold, black scaling due to black-edged, brown feathers. Underparts off-white and boldly scaled, with undertail coverts whitish. In flight, note two white wing patches on underwing. Eyes brown; bill dark brown with pale at base; tarsi pink-brown.

Habitat and Behavior Occurs in various forest types, woodland, well-wooded edges of farmland and urban parkland. Forages mostly on forest floor; walks rather than hops like other thrushes, has curious double-bobbing creeping gait. Flies to perch in low trees when flushed. Feeds on insects, also fruit and seeds.

Distribution Widely distributed from Europe and India to China, SE Asia. At Zaohemu, Bamianshan and Dabali in Mt. Jinggang.

Status of Population Common on migration and in winter in Mt. Jinggang.

3.154

3.155 橙头地鸫 *Geokichla citrina* (Latham, 1790)

识别特征 体型略小的鲜艳地鸫，约 22 cm。雄鸟头、颈背、胸及上腹橘黄色，脸颊具两道褐色纵纹（有的亚种无）；背部及尾蓝灰色，翼角具白色横纹（有的亚种无）；下腹及尾下覆羽

3.155a

白色。雌鸟似雄鸟，但颜色较暗淡。虹膜棕褐色；嘴灰黑色；脚橘黄至黄褐色。

生境与习性 常栖息于低山丘陵和山脚地带的山地森林中。常单独或成对活动。地栖性，性羞怯，常躲藏在林下茂密的灌丛中。杂食性，食物以昆虫为主。多在地面活动觅食，有时也在树上吃果实。

分布 巴基斯坦至中国南部及东南亚。在井冈山见于龙潭、八面山、西坪等地。

种群状况 在井冈山迁徙季及冬季常见。

Identification L ca. 22 cm. Smallish colored ground thrush. Male has orange-rufous head and chest, face paler orange-buff with two vertical dark blackish-brown subocular and auricular bars (race *innotata* lacks). Mantle, wings and tail bluish-grey with broad white bar on shoulder (race *innotata* lacks). Vent and undertail coverts pale. Female duller with wings warm brown. Eyes dark brown; bill dark grey; tarsi brownish-pink.

Habitat and Behavior Inhabits dense ground cover in moist deciduous and evergreen forests, also thickets, bamboo groves and plantations. Shy, crepuscular. Feeds on insects and fruit.

Distribution Pakistan to Southern China and SE Asia. Some races are migratory. At Longtan, Bamianshan and Xiping in Mt. Jinggang.

Status of Population Common in winter and on migration in Mt. Jinggang.

3.156 灰背鸫 *Turdus hortulorum* Sclater, PL, 1863

识别特征 体型略小的灰色鸫，约 22 cm。雌雄的两胁及翼下覆羽均为橙色。雄鸟上体全灰，喉灰或偏白，胸灰色，腹中心及尾下覆羽白。雌鸟上体褐色较重，颏喉偏白，胸皮黄色具黑色点斑。虹膜黑褐色；嘴黄色；脚肉色至粉褐色。

生境与习性 常栖息于低山丘陵的茂密森林中，以次生阔叶林最常见。常单独或成对活动，迁徙季节多集几只到十几只的小群。地栖性，善在地面行走跳跃。杂食性，主要在地面啄食果实和昆虫、蚯蚓等。

分布 繁殖于西伯利亚东部及中国（东北）。越冬至中国南部。在井冈山见于茅坪、湘洲等地。

种群状况 在井冈山迁徙季及冬季常见。

Identification L ca. 22 cm. Small-sized, greyish thrush. Male mostly grey on head, breast and upperparts, while lower breast and flanks rich orange. Female browner on upperparts and spotted black on throat to upper breast. Eyes dark brown; bill yellow; tarsi pink-brown.

Habitat and Behavior Occurring in broadleaved evergreen and mixed forests, woodland, shrubby areas and urban parkland. Solitary or in pairs, gregarious on migration. Skulks, keeping to dense vegetation, occasionally foraging in open areas during cold weather. Feeds on insects, earthworms and fruit.

Distribution Breeds in E Siberia and NE China. Winters to Southern China. At Maoping and Xiangzhou in Mt. Jinggang.

Status of Population Common on migration and in winter in Mt. Jinggang.

3.157 乌灰鸫 *Turdus cardis* Temminck, 1831

识别特征　体小的鸫，约 21 cm。雄鸟上体纯黑灰，黑色的头及上胸与白色的下体分界清晰，腹部及两胁具黑色点斑。雌鸟头及上体灰褐，下体白色，两胁沾棕褐，胸及两胁具黑色点斑；与灰背鸫雌鸟相似但体小、嘴黑、体侧斑点较多。虹膜黑褐色；嘴黄色（雄鸟），或近黑（雌鸟）；脚肉色。

生境与习性　常栖于中低海拔的山地森林中，秋冬季也出入于林缘灌丛、村寨和农田附近的小林内。常单独或成对活动，迁徙时结小群。地栖性，甚羞怯。主要以昆虫为食，也吃植物果实与种子。

分布　繁殖于日本及中国东部。越冬于中国南部及中南半岛北部。在井冈山见于八面山、荆竹山等地。

种群状况　在井冈山迁徙季节可见。

3.157a

Identification　L ca. 21 cm. Adult male has glossy black head and upperparts with yellow bill and eyering, throat and breast black, strongly contrasts with white, black-spotted belly, vent clear white. 1st winter male dark grey. Female has brown upperparts and white underparts with buffy-orange wash on flanks, black arrowhead spots from neck-sides and breast to flanks. Eyes dark brown; bill yellow (male), or dark (female); tarsi fresh.

Habitat and Behavior　Prefers mature deciduous and evergreen broadleaf forest when breeding. Occurs in forest, woodland and urban parks in winter. Usually solitary and shy, forages on ground. Mostly feeds on insects, also some fruits and seeds.

Distribution　Breeds in Japan and Eastern China. Wintering in Southern China and N Indo-China Peninsula. At Bamianshan and Jingzhushan in Mt. Jinggang.

Status of Population　Visible on migration in Mt. Jinggang.

3.157b

3.158 白眉鸫 *Turdus obscurus* Gmelin, JF, 1789

识别特征　中等体型的灰褐色鸫，约 23 cm。雌雄都有显著的白色眉纹、黑色眼先和眼下白斑。雄鸟头部青灰色，上体褐色，胸及两胁栗褐色，腹中部及尾下覆羽白色。雌鸟颜色较暗，头橄榄褐色并具白色下颊纹，喉部白色具褐色纵纹。虹膜黑褐色；嘴端黑色，下嘴黄色具黑端；脚偏黄至深肉棕色。

生境与习性　主要栖息于中低山的针叶林、阔叶林和针阔混交林，也见于农田、果园、苗圃和公园。迁徙季节成群活动于树冠层，飞行径直而快速。在地面，也在树上取食，常见于有果实的树上。

分布　繁殖于古北区中部及东部。冬季迁徙至印度东北部及东南亚。在井冈山见于龙潭、笔架山等地。

种群状况　在井冈山冬季及迁徙季常见。

3.157c

Identification　L ca. 23 cm. Mid-sized brown thrush with pale supercilium, below eye patch and black lores. Male hood slaty-grey with short white malar and chin streak; upperparts mid-brown, underparts orange-brown with white central belly and vent. Female duller, head brown with white chin and throat. Eyes blackish-brown; bill dark grey above, yellow below except tip; tarsi dull brownish-yellow.

Habitat and Behavior　In summer in dark taiga in lowland and montane areas. On migration and winter in mature deciduous and

3.158a

evergreen forest, open woodland and parks. In shy flocks, flight fast and direct. Feeds both terrestrially and arboreally, often in fruiting trees.

Distribution　Breeds in C and E Palearctic. Migrant in winter to NE Indian subcontinent and SE Asia. At Longtan and Bijiashan in Mt. Jinggang.

Status of Population　Common on migration and in winter in Mt. Jinggang.

3.158b

3.159 白腹鸫 *Turdus pallidus* Gmelin, JF, 1789

识别特征　中等体型的灰褐色鸫，约 24 cm。雄鸟头及喉灰色，上体至尾上覆羽褐色，胸和两胁染浅棕色，下腹至尾下覆羽白色。雌鸟多褐色，具浅色眉纹，喉偏白而略具细纹。飞行时可见外侧尾羽的宽阔白色末端。虹膜褐色；上嘴灰黑色，下嘴黄色；脚浅褐。

生境与习性　主要栖息于中低山的针阔混交林和针叶林中，常见于河谷与溪流两岸的树林间活动。迁徙时集群出没于林缘、耕地和道旁丛林等开阔处。多在林下层和地面活动觅食。主要以昆虫及其幼虫为食，同时也吃植物果实和种子。

分布　繁殖于东北亚。冬季南迁至东南亚。在井冈山见于笔架山、八面山等地。

种群状况　在井冈山迁徙季及冬季常见。

3.159a

Identification　L ca. 24 cm. A drab brown and grey thrush with white undertail coverts. Male has grey hood, mid-brown upperparts and tail, dark grey on wings; underparts pale with greyish-brown wash on breast and flanks. Female more brown with pale supercilium and dark streaks on withish throat. In flight, shows broad white tip to outer rectrices. Eyes brown; bill dark grey above, yellow below; tarsi pale brown.

Habitat and Behavior　Breeds in lower to mid montane forests, prefers forest edge next to stream or valley. Occurs in open woodland, parks and gardens in winter. Gregarious on migration, and forages on ground. Feeds on insects and larvae, fruit and seeds.

Distribution　Breeds in NE Asia. Winters south to SE Asia. At Bijiashan and Bamianshan in Mt. Jinggang.

Status of Population　Common on migration and in winter in Mt. Jinggang.

3.159b

3.160 斑鸫 *Turdus eunomus* Temminck, 1831

识别特征　中等体型的灰褐色鸫，约 25 cm。颊部灰色，眉纹、颈侧及髭纹浅砖红色，颏、喉白色；上体橄榄褐色，两翼红褐色而飞羽黑褐色；下体偏白色，胸胁具黑色菱状斑。虹膜黑褐色；上嘴偏黑，下嘴黄色具黑端；脚褐色。

生境与习性　繁殖期栖于各类森林和林缘灌丛地带，非繁殖期还出现在农田、果园、路边等。除繁殖季外多结松散小群。性活泼而大胆，不甚怯人。主要以昆虫为食，也食多种灌木和草本植物的果实和种子。

分布　繁殖于东北亚。迁徙至喜马拉雅、中国。在井冈山见于松木坪、龙潭等地。

种群状况　在井冈山迁徙季和冬季常见。

Identification　L ca. 25 cm. Mid-sized grey-brown thrush. Face grey with pale brick-orange supercilium, malar and neck sides;

3.160a

chin and throat whitish; upperparts olive-brown with red-brown wings and dark brown flight feathers; underparts whitish with black rhombus on chest and flanks. Eyes black; bill dark grey above, yellow below except tip; tarsi dull brownish-yellow.

Habitat and Behavior　Summer in woodland from taiga to edges of lowland tundra. Winter flocks in mid-elevation to lowland mixed forest, parks, agricultural land and gardens. Often in loose groups in winter, and foraging in open. Vivacious and bold. Feeds on insects and fruit.

Distribution　Breeds in NE Asia. Migrating to Himalaya, China. At Songmuping and Longtan in Mt. Jinggang.

Status of Population　Common on migration and in winter in Mt. Jinggang.

3.160b

3.161 乌鸫 *Turdus mandarinus* Bonaparte, 1850

识别特征　体型略大的黑色鸫，约 29 cm。雄鸟全黑色，嘴及眼圈橘黄色。雌鸟通体黑褐色，颏、喉及上胸具深色纵纹。虹膜黑褐色；雄鸟嘴黄色，雌鸟深褐色；脚黑色。

生境与习性　喜栖于林区外围、林缘疏林、农田及村镇附近的小树丛中，也进到城市公园、绿地活动。结群或单独活动。平时多栖于乔木上，繁殖期间常隐匿于高大乔木顶部枝叶丛中不停鸣叫。于地面取食昆虫、蚯蚓等，也食部分植物性食物。营巢于村寨附近、房前屋后和田园中乔木主干分枝处。窝卵数 4-6枚。雌鸟孵卵，孵化期 14-15 天。雌雄共同育雏。

分布　中国中部及东部。在井冈山见于茨坪、罗浮、新城区等地。

种群状况　在井冈山全年常见。

3.161a

Identification　L ca. 29 cm. Entirely jet-black thrush with yellow bill and eyering. Female browner than male, with some dark streaks and mottling. Eyes dark brown; bill yellow (male), or dark brown (female); tarsi black.

Habitat and Behavior　Inhabits various forests, open woodland, farmland and urban parks. Bold and noisy. Usually forages on open grassland near forest, feeds on insects and earthworms. Sings in covered canopy of tree when breeding. Nests in a tree, often crotch of trunk. Clutch size is 4-6 eggs. Female incubates, and incubation period is 14-15 days. Both parents brood.

Distribution　Eastern and Central China. At Ciping, Xinchengqu and Luofu in Mt. Jinggang.

Status of Population　Common all year round in Mt. Jinggang.

3.161b

3.162 紫啸鸫 *Myophonus caeruleus* (Scopoli, 1786)

识别特征　体大的鸫，约 32 cm。通体蓝紫色且具金属光泽，头、颈、上背和胸具浅色闪光点斑。虹膜红褐色；嘴黄色或黑色；脚黑色。

生境与习性　常栖息于山地森林溪流沿岸，有时也进到村寨附近的灌丛中。单独或成对活动。常栖止在岩石上，展开尾羽上下扭转。受惊时慌忙逃窜并发出尖厉的警叫声。在地面或浅水间觅食，以昆虫、蟹等为主，偶尔也吃少量植物果实与种子。营巢于山溪近旁的岩石上或岩隙间，也在瀑布后面岩洞中和树根间的洞穴中营巢，也见于庙宇的木梁上或树枝分杈处。窝卵数多为 4 枚。雌雄共同育雏。

3.161c

分布　突厥斯坦至印度、中国及东南亚。在井冈山广泛分布。

种群状况　在井冈山全年常见。

Identification　L ca. 32 cm. Large, entirely glossy blue-black thrush with pale spangles on head, neck, chest and mantle. Eyes red-brown; bill black or yellow; tarsi black.

Habitat and Behavior　Inhabits hilly forests near streams and rivers, sometimes shrubland and even urban parkland. Often perches on boulders by rivers with fanning and twisting tail. Escapes rapidly with shrill whistle when disturbed. Forages on ground for worms and arthropods, also fruit and seeds. Nests on boulder or in crevice of rock, sometime in caves or trees. Clutch size is 4 eggs usually. Both parents brood.

Distribution　Turkestan to India, China and SE Asia. Widespread in Mt. Jinggang.

Status of Population　Common all year round in Mt. Jinggang.

3.163 灰背燕尾 *Enicurus schistaceus* (Hodgson, 1836)

识别特征　中等体型的灰白色燕尾，约 23 cm。头顶及背青灰色，前额具宽阔白带延至眼后，下颊、颏、喉黑色，两翼黑色具白色翼斑，胸腹至尾下覆羽及腰白色，长而分叉的黑色尾羽具白色末端。幼鸟头顶及背青灰褐色，胸部具鳞状斑纹。虹膜黑色；嘴黑色；脚粉红。

生境与习性　主要栖息于中低山的森林和林缘疏林地带的山涧溪流与河谷沿岸，冬季也见于山脚、平原的河流溪谷。常单独或成对活动，喜停栖在水边乱石或激流中露出水面的石头上，上下摆动尾巴，遇惊则紧贴水面沿溪飞行并发出尖哨声。主要以水生昆虫、蚂蚁、毛虫、螺类等为食。多营巢于河岸岩石缝隙中。窝卵数 3-4 枚。雌雄轮流孵卵，雏鸟晚成。

分布　喜马拉雅至中国南部及中南半岛。在井冈山广泛分布。

种群状况　在井冈山全年常见。

Identification　L ca. 23 cm. Mid-sized grey and white forktail. Crown and mantle slaty-grey, bold white patch from forehead to rear of eyes; cheek, chin, and throat black; wing black with white wing patch; breast to undertail coverts, and rump white; long black fork tail with white tip. Juvenile has greyish-brown crown and mantle, scales on breast. Eyes black; bill black; tarsi pink.

Habitat and Behavior　Inhabits montane forest streams. Usually solitary or in pairs, perches on rocks in streams, shakes tail vertically. Flies along stream with high-pitched whistles when disturbed. Feeds on aquatic insects, ants and mollusks. Builds nest near water on rock ledge or crevice. Clutch size is 3-4 eggs. Both parents brood. Chicks are altricial.

Distribution　Himalaya to Southern China and Indo-China Peninsula. Widespread in Mt. Jinggang.

Status of Population　Common all year round in Mt. Jinggang.

3.164 白冠燕尾 *Enicurus leschenaulti* (Vieillot, 1818)

识别特征　体型稍大的黑白色燕尾，约 25 cm。前额和顶冠白，额羽有时耸起成小凤头状；头余部、颈、背及胸黑色；腹部、腰

及尾下覆羽白；两翼黑色具白色翼斑，长而分叉的尾黑色具白色末端，最外侧两枚尾羽全白。虹膜黑褐色；嘴黑色；脚粉色。

生境与习性 常栖息于山涧溪流与河谷沿岸，尤喜水流湍急、河中多石头的林间溪流，冬季也见于山脚平原河谷和村庄附近溪流岸边。常单独或成对活动，喜停栖于岩石或在水边行走，寻找食物并不停地展开叉形长尾。飞行近水面，且飞且叫，每次飞行距离不远。性羞怯，被惊扰立即飞入岸边森林。食物以水生昆虫为主，也食少量植物性食物。营巢于激流附近的岩隙间。窝卵数3-4枚。雌鸟孵卵，雌雄共同育雏。

分布 印度北部、中国南部及东南亚。在井冈山广泛分布。

种群状况 在井冈山全年常见。

Identification L ca. 25 cm. Largish white and black forktail. White forehead and crown, sometimes formed crest; rest of head, neck, back and breast black; belly, rump and undertail coverts white; black wings with white patch, long black fork tail with white tip, outer rectrices all white. Eyes dark brown; bill black; tarsi pink.

Habitat and Behavior Inhabits clear montane streams. Shy, flying off into forest if disturbed. Solitary or in pairs, prefers to perch on rocks, and forages by walking near the stream with long tail fanning. Mostly feeds on aquatic insects, also some plants. Builds nest in hollow in bank or tree root, very near water so often wet. Clutch size is 3-4 eggs. Female incubates, and both parents raise chicks.

Distribution N India, Southern China and SE Asia. Widespread in Mt. Jinggang.

Status of Population Common all year round in Mt. Jinggang.

3.165 小燕尾 *Enicurus scouleri* Vigors, 1832

识别特征 体小的黑白色燕尾，约13 cm。前额具大块白色斑，头、胸、上背和两翼黑色，翼上有白色翼斑，腰白色，短而略分叉的尾白色，中央尾羽黑色。幼鸟头至上背灰色。虹膜褐色；嘴黑色；脚粉白。

生境与习性 常单独或成对活动在湍急的山区溪流与河谷沿岸的岩石上，尤其是瀑布周围。甚活跃而大胆，尾有节律地上下摇摆或扇开。主要以昆虫和昆虫幼虫为食。多营巢于溪边岩石缝隙中，尤其是在一些小型瀑布后面的岩壁缝隙中较常见。窝卵数2-4枚，多为3枚。

分布 突厥斯坦及巴基斯坦至喜马拉雅、印度东北部、中国（华南、华中和台湾）、缅甸西部及北部、中南半岛北部。在井冈山广泛分布。

种群状况 在井冈山全年常见。

Identification L ca. 13 cm. Small white and black forktail with short tail. Head, throat and back black with white patch on forehead and wings, rump white, short forked tail white with black central feathers. Juvenile has grey head and mantle. Eyes brown; bill black; tarsi pale pink.

Habitat and Behavior Inhabits cold, fast-flowing streams, rivers and waterfalls in montane areas. Often perches on boulders in stream with bobbing, flicking wings and fanning tail. Feeds on insects and larvae. Nests in crevice of boulders close to stream, frequently behind waterfalls. Clutch size is 2-4 eggs, mostly 3.

3.164a

3.164b

3.165a

Distribution Turkestan and Pakistan to Himalaya, NE India, S and C China, Taiwan China, W and N Myanmar and N Indo-China Peninsula. Widespread in Mt. Jinggang.

Status of Population Common all year round in Mt. Jinggang.

3.166 黑喉石䳭 *Saxicola maurus* (Pallas, 1773)

识别特征 中等体型的黑、白及赤褐色䳭，约 14 cm。繁殖期雄鸟脸及喉黑色，头顶及背黑色而具棕色羽缘、颈侧具白色斑、翼黑色具白色翼斑和浅色羽缘、腰白、胸及两胁棕色，尾羽黑色。非繁殖羽黑色部分略带褐色。雌鸟头褐色，眉纹浅色，上体有棕色纵纹，下体微带褐色。飞行时翼上白斑显著。虹膜黑褐色；嘴黑色；脚近黑。

生境与习性 栖息于低山、丘陵、原野及湖岸间，喜农田、花园及次生灌丛等开阔生境。单独或成对活动。常站立于突出的低树枝或电线以跃下地面捕食猎物。站立时不断急扭或舒展尾羽。主要以昆虫及其幼虫为食，兼食蚯蚓、蜘蛛、少量的杂草种子等。

分布 繁殖于古北区、日本、喜马拉雅及东南亚的北部。冬季至非洲、中国南部、印度及东南亚。在井冈山见于大井、荆竹山、罗浮、湘洲、新城区等地。

种群状况 在井冈山冬季及迁徙季常见。

Identification L ca. 14 cm. Mid-sized chat. Breeding male black-hooded, rear crown to mantle feathers dark brown with pale fringes, white patch on neck-sides, wing and tail blackish with white wing bar, rump white, breast and flanks rufous. Females and non-breeding males strongly streaked and with paler, brownish hood. In flight, show white wing bar. Eyes black brown; bill black; tarsi black.

Habitat and Behavior Inhabits open country, shrubland, scrubby hillsides, farmland and wetland edges. Usually perches on top of shrubs, exposed branches, stakes and wires boldly. Feeds on insects and larvae, also earthworms, spiders and seeds.

Distribution Breeds in Palearctic, Japan, Himalaya and northern SE Asia. Migrate in winter to Africa, Southern China, India and SE Asia. At Dajing, Luofu, Xiangzhou, Xinchengqu and Jingzhushan in Mt. Jinggang.

Status of Population Common in winter and on migration in Mt. Jinggang.

3.167 灰林䳭 *Saxicola ferreus* Gray, JE & Gray, GR, 1847

识别特征 中等体型的䳭，约 15 cm。雄鸟醒目的黑色脸罩，与白色眉纹及白色的颏、喉形成对比；上背青灰色具黑色纵纹，翼黑色具白色翼斑；下体污白，具灰色胸带；尾黑色，外侧尾羽羽缘灰色。雌鸟上体棕褐色，具白色或皮黄色眉纹，脸罩、两翼和尾棕色较深，腰栗红色，胸及下腹皮黄色。幼鸟似雌鸟，但下体褐色具鳞状斑纹。虹膜深褐；嘴灰黑色；脚深灰色。

生境与习性 主要栖于林缘疏林、开阔灌丛、草坡、沟谷及农田等地，有时也进到阔叶林、针叶林林缘和林间空地。多单独或成对活动，有时也结 3-5 只的小群。常停栖在灌木或小树顶枝上、电线或居民点附近的篱笆上，长时间鸣叫且摆动尾。在地面或于飞行中捕捉昆虫及其幼虫，也食少量野果和草籽。多

3.165b

3.166a

3.166b

营巢于草丛中或灌丛中，也在岸边或山坡岩石洞穴、矮土壁上筑巢。窝卵数 4-5 枚。雌鸟孵卵，孵化期 12 天。雌雄亲鸟共同育雏，留巢期约 15 天。

分布 喜马拉雅、中国南部及中南半岛北部。冬季至亚热带低地。在井冈山见于大井、下庄、茅坪、湘洲等地。

种群状况 在井冈山全年常见。

Identification L ca. 15 cm. Mid-sized, grey or brown chat. Male black masked, with white supercilium and throat; upperparts grey with dark streak, white wing bar on black wings; underparts dirty white with grey breast band; tail black, outer feathers fringed grey. Female mostly brown with long white brow and white throat. Juvenile more scales on underparts. Eyes dark brown; bill black; tarsi dark grey.

Habitat and Behavior Inhabits open country on hill sides and mountains, especially scrubby and grassy areas. Perches on exposed site, usually low bushes, stakes, rocks or wires. Often cocks tail. Feeds on insects and seeds. Nests in bushes or grasses, sometimes in hollow of rock face or earth bank. Clutch size is 4-5 eggs. Female incubates, and incubation period is 12 days. Both parents raise chicks. Brooding period is 15 days.

Distribution Himalaya, Southern China and N Indo-China Peninsula. Migrating to subtropical lowland zone in winter. At Dajing, Xiazhuang, Maoping and Xiangzhou in Mt. Jinggang.

Status of Population Common all year round in Mt. Jinggang.

3.167a

3.167b

雀形目 PASSERIFORMES
鹟科 Muscicapidae

3.168 白喉林鹟 *Cyornis brunneatus* (Slater, 1897)

识别特征 中等体型的鹟，约 15 cm。上体橄榄褐色，眼先白色，颏、喉白色略具深色斑纹，胸带浅褐，腹部污白色。看似翼短而嘴长。虹膜黑褐色；上嘴近黑，下嘴黄色；脚粉红至橙黄。

生境与习性 栖于中低海拔的常绿阔叶林、次生林、茂密竹丛及林缘灌丛。常单独活动，性隐匿，多躲藏在森林下层灌丛和竹丛中活动和觅食，常只闻其声而难见其影。

分布 中国南部。冬季南迁至马来半岛及尼科巴群岛。在井冈山见于五指峰等地。

种群状况 在井冈山夏季常见。IUCN 红色名录（2014）：易危。

Identification L ca. 15 cm. Mid-sized flycatcher with longish, slightly hooked bill and short wings. Upperparts grey olive-brown with pale loral patch. Underparts dirty white with pale brown breast band, white chin and throat with slight dark streaks. Eye dark brown; bill dark above, lower mandible fleshy-orange; tarsi pink to orange-yellow.

Habitat and Behavior Inhabits broadleaved evergreen and mixed forests, bamboo forests from hills to mountains. Perches low and secluded, often picking up prey from ground. Mostly hear it but cannot see it.

Distribution Southern China. Migrating south in winter as far as Malay Peninsula and Nicobars. At Wuzhifeng in Mt. Jinggang.

Status of Population Common in summer in Mt. Jinggang. IUCN Red List (2014): VU.

3.167c

3.168

3.169 褐胸鹟 *Muscicapa muttui* (Layard, EL, 1854)

识别特征　体型略小的偏褐色鹟，约 14 cm。头及上体浅褐色，具白色眼先及眼圈，深色的髭纹将白色的颊纹与白色颏及喉隔开，翼羽羽缘红棕色，腰和尾褐色较浓。下体污白色，胸带及两胁茶褐色。虹膜深褐；上嘴色深，下嘴黄色且尖端色深；脚肉黄色。

生境与习性　见于中低海拔的阔叶林、竹林和次生林，多单独或成对活动，性安静而隐蔽。常在树下部茂密的低枝上，长时间不动，有昆虫飞过时，飞到空中捕食然后又飞回原处。

分布　繁殖于印度东北部，中国（西南、西北和华南）。越冬至印度西南部、斯里兰卡；在缅甸北部及东部和泰国的西北部也有记录。在井冈山见于荆竹山等地。

种群状况　在井冈山迁徙季可见。

Identification　L ca. 14 cm. Smallish brown flycatcher. Head and upperparts pale brown with white lore and eyering, dark malar contrasting with white submoustachial stripe and throat. Rufous-edged flight feathers, rump and tail more rufescent. Underparts dirty white with fulvous breast band and flanks. Eyes dark brown; bill dark above, yellow below with dark tip; tarsi pinkish-yellow.

Habitat and Behavior　Inhabits broadleaf forest, secondary forest and bamboo. Mostly solitary or in pairs. Quiet and retiring, generally found at understory level in forest. Chases insects aerially and perches back.

Distribution　Breeds NE India, SW, NW and S China. Winters to SW India, Sri Lanka and recorded N and E Myanmar and NW Thailand. At Jingzhushan in Mt. Jinggang.

Status of Population　Visible on migration in Mt. Jinggang.

3.170 乌鹟 *Muscicapa sibirica* Gmelin, JF, 1789

识别特征　体型略小、嘴显短的烟灰色鹟，约 13 cm。头及上体深灰色，具白色眼圈和淡色眼先，喉中央白，上胸及胸侧的乌灰色延至腹侧，下腹和尾下覆羽白色。翼上具不明显皮黄色斑纹，翼长至尾的 2/3。虹膜深褐；嘴黑色；脚黑色。

生境与习性　栖息于山区针阔混交林、针叶林及亚高山矮曲林等生境。迁徙季和冬季亦见于山脚和平原地带的落叶和常绿阔叶林、次生林和林缘疏林灌丛。除繁殖期成对外，其他季节多单独活动。觅食于植被中上层。常立于突出的树枝上，冲出捕捉过往昆虫。

分布　繁殖于东北亚及喜马拉雅。冬季迁徙至中国南部、巴拉望岛、东南亚及大巽他群岛。在井冈山见于西坪、五指峰等地。

种群状况　在井冈山迁徙季常见。

Identification　L ca. 13 cm. Smallish dark grey flycatcher with short black bill. Upperparts dark grey with grey lores and pale eyering, narrow whitish center of throat, underpart dirty white with grey-brown breast-band and flanks. Primary projection to 2/3 of tail. Eyes dark brown; bill black; tarsi black.

Habitat and Behavior　Summer in montane regions, in mixed taiga with predominance of conifers, but also mature mixed broadleaf forest, in undergrowth and mid-levels. On migration and

3.169a

3.169b

3.170a

3.170b

in winter occurs in various types of woodland. Breeding in pairs, others solitary. Perches on bare branch, and chases flying insects.

Distribution Breeds in NE Asia and Himalaya. Migrates in winter to Southern China, SE Asia, Palawan and Greater Sunda Islands. At Xiping and Wuzhifeng in Mt. Jinggang.

Status of Population Common on migration in Mt. Jinggang.

3.171 北灰鹟 *Muscicapa dauurica* Pallas, 1811

识别特征 体型略小、头显大的灰褐色鹟,约13 cm。上体灰褐,下体偏白;眼圈白色、冬季眼先偏白色;胸侧及两胁褐灰但无纵纹。嘴较乌鹟长,翼尖延至尾的中部。首次度冬的鸟两翅有翼斑和浅色羽缘。虹膜褐色;嘴黑色,下嘴基黄色;脚黑色。

生境与习性 常见于山地溪流沿岸的混交林、针叶林和落叶阔叶林,迁徙季节和越冬期间也见于山脚和平原地带的次生林、林缘疏林灌丛和农田地边小树丛与竹丛中。常单独或成对,偶见3-5只的小群。多停栖在树冠层中下部侧枝或枝杈上,飞起捕食空中的昆虫,后又回至栖处。尾作独特的颤动。

分布 繁殖于东北亚及喜马拉雅。冬季南迁至印度及东南亚。在井冈山见于笔架山、湘洲等地。

种群状况 在井冈山迁徙季常见。

Identification L ca. 13 cm. Dull grey flycatcher with large head and yellowish base below bill. Upperparts greyish-brown, underparts dirty white washed brown on breast sides. Also white eyering, pale lores and white moustache, formed by the dark malars. Bill longer than Darksided flycater. Primary projection to ½ of tail. Eyes dark brown; bill blackish with prominent yellow base to lower mandible; tarsi black.

Habitat and Behavior Inhabits most forest types, forest edges, woodland and parkland. Usually solitary or in pairs, occasionally in small groups. Forages by sallying for insects from a high perch. Often shakes tail.

Distribution Breeds NE Asia and Himalaya. Migrant south in winter to India and SE Asia. At Bijiashan and Xiangzhou in Mt. Jinggang.

Status of Population Common on migration in Mt. Jinggang.

3.172 白眉姬鹟 *Ficedula zanthopygia* (Hay, 1845)

识别特征 体型略小的鹟类,约13 cm。雄鸟具白色眉纹和翼斑,腰、喉、胸及腹部鲜黄色,臀部、尾下覆羽白色,余部黑色。雌鸟上体灰褐,下体浅黄色,腰暗黄,两翼具明显白色翼斑。虹膜褐色;嘴黑色;脚黑色。

生境与习性 主要栖息于中低山丘陵和山脚地带的阔叶林和针阔混交林中,尤其是河谷与林缘地带有老龄树木的疏林中较常见,也出入于次生林和人工林内,迁徙时见于居民点附近的小树丛和果园中。常单独或成对活动,多在树冠下层低枝处活动和觅食,常飞到空中捕食昆虫,后又落于较高的枝头上。

分布 繁殖于东北亚。冬季南迁至中国南部及东南亚。在井冈山见于大井、西坪等地。

种群状况 在井冈山迁徙季常见。

Identification L ca. 13 cm. Bright black and yellow flycatcher. Male has white supercilium, wing bar and undertail, bright yellow

3.171a

3.171b

3.172a

3.172b

rump and underparts, the rest black. Female dull grey-brown on upperparts, with smaller yellow rump patch, pale underparts faintly washed yellow. Eyes dark brown; bill black; tarsi black.

Habitat and Behavior　Inhabits woodland, scrub and parkland, including urban areas on migration. Forages from understory to canopy, usually near water. Often chases insects from perch.

Distribution　Breeds in NE Asia. Migrates south in winter to Southern China and SE Asia. At Dajing and Xiping in Mt. Jinggang.

Status of Population　Common on migration in Mt. Jinggang.

3.173 绿背姬鹟 *Ficedula elisae* (Weigold, 1922)

识别特征　体型略小的偏绿色鹟，约 13 cm。雄鸟头部及上体深橄榄绿色，眉纹明黄色，眼圈、腰部和下体亮黄色，翼近黑色具白色条状翼斑，尾深色。雌鸟上体暗橄榄绿色，眼圈淡黄色，下体浅黄色，尾或尾上覆羽染锈红色，无黄色腰部和白色块状翼斑。虹膜黑褐色；嘴黑色或黑褐色；脚深色。

生境与习性　繁殖于山地阔叶林中，也出现在针阔混交林的林缘地带，迁徙季节亦见于公园。

分布　泰国、马来半岛、新加坡、越南、日本、中国等地。在井冈山见于笔架山等地。

种群状况　在井冈山迁徙季可见。

Identification　L ca. 13 cm. Smallish green flycatcher. Male head dull greenish-olive with yellow supraloral stripe and narrow eyering, rump bright yellow, wings greyish-black with large white coverts patch. Tail greyish-black, underparts plain yellow. Female upperparts dull olive-green with pale yellow eyering, underparts pale yellow, lacks yellow rump and white wing patch. Eyes dark brown; bill dark brown to black; tarsi black.

Habitat and Behavior　Inhabits montane broadleaf forest and mixed forest edge, also in urban park on migration. Prefers shady areas.

Distribution　China, Thailand, Malay Peninsula, Vietnam, Singapore and Japan. At Bijiashan in Mt. Jinggang.

Status of Population　Visible on migration in Mt. Jinggang.

3.174 黄眉姬鹟 *Ficedula narcissina* (Temminck, 1836)

识别特征　体型略小的鹟，约 13 cm。雄鸟上体及尾黑色，具显著的黄色眉纹，腰黄，翼具白色块斑，颏、喉橙红色，胸、上腹鲜黄色，下腹及尾下覆羽白色。雌鸟上体灰橄榄色，腰橄榄绿色，尾红褐色，两翅橄榄褐色且羽缘较浅，下体污白色，胸具不明显褐色纵纹。虹膜黑色；嘴黑色；脚铅蓝至深褐色。

生境与习性　见于各种有林生境。常单独或成对活动，在森林中下层觅食。

分布　繁殖于东北亚。冬季至泰国南部、马来半岛、菲律宾及加里曼丹岛。在井冈山见于水口等地。

种群状况　在井冈山迁徙季常见。

Identification　L ca. 13 cm. Smallish flycatcher. Male upperparts black with prominent yellow supercilium, white wing patch across coverts; underparts and rump rich orange-yellow, lower belly and undertail white. Females olive-brown on upperparts with olive-green

rump and rufous tail, lacks wing patterns. Eyes black; bill black; tarsi blue-grey to dark brown.

Habitat and Behavior Inhabits forests, woodland, scrub and even parkland on migration. Unobtrusive, forages low in the understory. Solitary or in pairs.

Distribution Breeds in SE Asia. Migrates in winter to S Thailand and Malay Peninsula, the Philippines and Kalimantan Peninsula. At Shuikou in Mt. Jinggang.

Status of Population Common on migration in Mt. Jinggang.

3.175 鸲姬鹟 *Ficedula mugimaki* (Temminck, 1836)

识别特征 体型略小的鹟，约 13 cm。雄鸟上体及尾灰黑色，眼后上方有粗白色眉纹；翼上具明显的白斑，外侧尾羽基部白色；喉、胸及腹侧橘黄；腹中心及尾下覆羽白色。未成年雄鸟上体灰褐色，翼带较不明显。雌鸟上体褐色，具两道翼斑，下体似雄鸟但色淡，尾无白色。虹膜黑色；嘴暗角质色；脚深褐。

生境与习性 主要栖息于中低山和平原湿润森林中，非繁殖期也见于林缘疏林、次生林、果园、灌丛中。常单独或成对活动，多在潮湿和林下溪流较多的森林树冠层枝叶间，停栖位置较其余姬鹟高。

分布 繁殖于北亚。冬季南迁至东南亚。在井冈山见于大井、湘洲等地。

种群状况 在井冈山冬季和迁徙季常见。

Identification L ca. 13 cm. Smallish flycatcher. Male upperparts black with small white patch over eye, white patch on wings and outer base of tail; chin to breast washed bright orange, belly to undertail white. Juvenile male upperparts grey-brown with dim wing bars. Female greyish-brown on upperparts, and shows two thin wing bars, lacks white on tail. Eyes black; bill dark horn; tarsi dark brown.

Habitat and Behavior Inhabits forests, woodland, scrub and even parkland while on migration. Solitary or in pairs. Forages higher up than other *Ficedula* flycatchers.

Distribution Breeds in N Asia. Migrates south in winter to SE Asia. At Dajing and Xiangzhou in Mt. Jinggang.

Status of Population Common on migration and in winter in Mt. Jinggang.

3.176 白腹蓝鹟 *Cyanoptila cyanomelana* (Temminck, 1829)

识别特征 体型较大的鹟，约 17 cm。雄鸟脸、喉及上胸近黑，上体至尾青蓝色，下胸、腹及尾下覆羽白色，与深色的胸截然分开。外侧尾羽基部白色。亚种 *cumatilis* 青绿色、深绿蓝色取代黑色。雌鸟上体灰褐，两翼及尾褐色且肩部沾灰蓝色，喉中心及腹部白。虹膜褐色；嘴及脚黑色。

生境与习性 主要栖息于山地阔叶林和混交林中，尤以林缘和较陡的溪流沿岸及附近有陡岩或坡坎的森林地区较常见。单独或成对活动，多在林冠层取食。主要以昆虫和昆虫幼虫为食。

分布 繁殖于东北亚。冬季南迁至中国、马来半岛、菲律宾及大巽他群岛。在井冈山见于五指峰、领袖峰等地。

种群状况 在井冈山冬季和迁徙季常见。

3.174b

3.175a

3.175b

3.176a

Identification L ca. 17 cm. Largish flycatcher. Male upperparts to tail blue; face to breast black, rest of underparts white. Outer base of tail white. Race *cumatilis* dark green-blue instead of black. Female upperparts greyish-brown with grey-blue wash on shoulder, and brown wings and tail. Eyes dark brown; bill black; tarsi black.

Habitat and Behavior Inhabits forests, woodland, scrub and even parkland on migration. Solitary or in pairs. Forages mainly in canopy. Feeds on insects and larvae.

Distribution Breeds in NE Asia. Migrates south in winter to China, Malay Peninsula, the Philippines and Greater Sunda Islands. At Wuzhifeng and Lingxiufeng in Mt. Jinggang.

Status of Population Common on migration and in winter in Mt. Jinggang.

3.177 海南蓝仙鹟 *Cyornis hainanus* (Ogilvie-Grant, 1900)

识别特征 体型略小的深色鹟，约 15 cm。雄鸟上体至尾深蓝色，脸、颏近黑色，前额及肩部色较鲜亮，喉、胸部深蓝色，下体至尾下覆羽灰白色。亚成体雄鸟的喉近白。雌鸟上体褐色，腰、尾及次级飞羽沾棕色，眼先及眼圈皮黄，胸部橘褐色渐变至腹部及尾下的灰白色。虹膜黑褐色；嘴黑色；脚肉褐色。

生境与习性 主要栖息于低山常绿阔叶林、次生林和林缘灌丛。常单独或成对，偶见 3-5 只一起活动和觅食。频繁穿梭于树枝和灌丛间，在树枝上跳来跳去，不时发出 "ti-ti" 的警诫声。主要以昆虫为食。

分布 中国南部及东南亚。在井冈山见于五指峰、龙潭等地。

种群状况 在井冈山夏季常见。

Identification L ca. 15 cm. Smallish blue or brown flycatcher. Male upperparts deep blue with bright blue on forehead and shoulder, lore, cheek and chin black; throat, breast color varies from blue to bluish-grey; underparts cold greyish. Juvenile male has pale throat. Female greyish-brown from head to mantle, warmer brown on wings, tail and uppertail coverts, throat to breast washed orange-brown and belly grey. Eyes dark brown; bill black; tarsi fresh-brown.

Habitat and Behavior Inhabits broadleaf evergreen and mixed forests, including dense woodland. Solitary or in pairs, occasionally in small groups. Forages mostly in lower story, active and vocal. Feeds on insects.

Distribution Southern China and SE Asia. At Wuzhifeng and Longtan in Mt. Jinggang.

Status of Population Common in summer in Mt. Jinggang.

3.178 小仙鹟 *Niltava macgrigoriae* (Burton, 1836)

识别特征 体小的鹟，约 13 cm。雄鸟上体至尾深蓝色，脸颊、颏、喉黑色，前额、眉纹、颈侧及腰辉蓝色；胸腹深蓝灰色，下腹、尾下覆羽灰白色。雌鸟通体棕褐色，颈侧具闪辉蓝色斑块，眼先、颏、喉和尾下覆羽颜色较浅，翼及尾棕红色，虹膜褐色；嘴黑色；脚黑色。

生境与习性 主要栖息于中低山的山地常绿阔叶林和竹林中，尤喜临近溪流等水域的疏林和林缘地带。冬季多栖于山脚平原

地带。常单独或成对活动于植被中下层。性活泼，善鸣叫，频繁地在树枝间飞来飞去。以昆虫和果实为食。营巢于山地森林附近的岩石洞穴中，也在溪边树洞和岸壁洞穴中。窝卵数 3-5 枚，多为 4 枚。雌雄轮流孵卵，孵化期 12 天，雏鸟晚成性。

分布 喜马拉雅至印度东北部、中国南部及东南亚的部分地区。在井冈山见于荆竹山、小溪洞等地。

种群状况 在井冈山夏季常见。

Identification L ca. 13 cm. Small niltava. Male upperparts deep blue, darkest on cheeks, chin and throat. Forehead, supercilium, neck-patch and rump bright blue; breast deep blue to undertail pale grey. Female overall brown with bright blue neck-patch, paler on lores, chin, throat and undertail coverts, wings and tail rufous-brown. Eyes brown; bill black; tarsi black.

Habitat and Behavior Inhabits montane broadleaves forest, bamboo forests, and forest edges, prefers closer to streams. Often in middle or low stories. Active and vocal, fly in branches frequently. Takes small fruit as well as insects. Places moss nest in rock crevices or tree holes. Clutch size is 3-5 eggs, mostly 4. Both parents incubate. Incubation period is 12 days. Chicks are altricial.

Distribution Himalaya to NE India, Southern China and parts of SE Asia. At Xiaoxidong and Jingzhushan in Mt. Jinggang.

Status of Population Common in summer in Mt. Jinggang.

3.179 铜蓝鹟 *Eumyias thalassinus* (Swainson, 1838)

识别特征 体型略大的鹟，约 17 cm。雄鸟通体蓝绿色，眼先黑色。雌鸟似雄鸟但颜色较淡，眼先烟灰色。雄雌尾下覆羽均具白色羽缘。幼鸟灰褐沾蓝绿，具皮黄及近黑色的鳞状纹及点斑。虹膜黑褐色；嘴黑色；脚近黑。

生境与习性 栖息于中高山的常绿阔叶林、针阔叶混交林和针叶林，尤喜开阔森林或林缘空地，迁徙期见于人工林、果园、农田等各种生境。常单独或成对活动，多在高大乔木冠层，也到林下灌木和小树上活动。胆大不甚怕人，由裸露栖处捕食过往昆虫，也吃部分植物果实和种子。常营巢于岸边、岩坡和树根下的洞中或石隙间，也在树洞、废弃房舍墙壁洞穴中营巢。窝卵数 3-5 枚，多为 4 枚。

分布 印度至中国南部及东南亚、苏门答腊、加里曼丹。在井冈山见于八面山、大坝里等地。

种群状况 在井冈山夏季及迁徙季常见。

Identification L ca. 17 cm. Large turquoise flycatcher with pale fringed undertail coverts. Male entirely greenish-blue with black lores, black wings and tail heavily fringed turquoise. Female duller, lores dusky-grey. Juvenile grey-brown with greenish-blue wash, and buff scales. Eyes dark brown; bill blackish; tarsi black.

Habitat and Behavior Inhabits open mixed lowland and low montane forest, including pines; woodland, orchards and farmland on migration. Solitary or in pairs, perches on canopy boldly, sometimes forages in low bushes. Feeds on insects, also fruit and seeds. Nests in crevices of rock-faces, earth banks or root hollows. Clutch size is 3-5 eggs, mostly 4.

Distribution India to Southern China, SE Asia, Sumatra and

3.178a

3.178b

3.179a

Kalimanta Peninsula. At Dabali and Bamianshan in Mt. Jinggang.

Status of Population Common in summer and on migration in Mt. Jinggang.

3.180 方尾鹟 *Culicicapa ceylonensis* (Swainson, 1820)

识别特征 体小的黄绿色鹟，约 12 cm。雌雄体色相似，头、颈至上胸灰色而略具羽冠和浅色眼圈，背、两翼和尾上覆羽橄榄绿色，腹部和尾下覆羽黄色，尾黑灰色。虹膜黑褐色；上嘴黑色，下嘴浅色；脚黄褐色。

生境与习性 栖息于中低山的阔叶林、竹林、混交林和林缘疏林灌丛。常单独或成对活动，也常与其他鸟混群。活动于森林的底层或中层，性喧闹活跃，在树枝间跳跃，不停追逐捕食过往昆虫。站姿直挺，多鸣叫，常将尾扇开，

分布 印度至中国南部、东南亚。在井冈山见于荆竹山、八面山、龙潭等地。

种群状况 在井冈山全年常见。

3.179b

Identification L ca. 12 cm. Small yellow-green flycatcher with grey hood and chest. Rear crown peak accentuated by crest-like feathers, narrow eyering white. Upperparts olive-green, underparts bright yellow, and tail dark grey. Eyes dark brown; bill black above, pinkish below; tarsi darkpink.

Habitat and Behavior Inhabits broadleaf forests, mixed forests, bamboo forests and forest edges. Solitary or in pairs, also in mixed species flocks. Active at lower and mid-levels in pursuit of insects. Stance rather upright with tail fanned frequently. Rather vocal year-round.

Distribution India to Southern China and SE Asia. At Bamianshan, Longtan and Jingzhushan in Mt. Jinggang.

Status of Population Common all year round in Mt. Jinggang.

雀形目 PASSERIFORMES
王鹟科 Monarchinae

3.181 中南寿带 *Terpsiphone affinis* (Blyth, 1846)

识别特征 中等体型的寿带，不计尾部延长约 22 cm。有棕色和白色两种色型。棕色型雄鸟头黑色具辉蓝色光泽，具明显羽冠，上体及尾红褐色，胸至两胁灰黑色，下腹及尾下覆羽白色，中央尾羽延长 20-30 cm。白色型头部与棕色型同，除两翼具黑色羽缘外其余体羽呈白色。两种色型的雌鸟都似棕色型雄鸟但显暗淡，头部染褐而缺少光泽，尾不延长。虹膜褐色；眼周裸露皮肤蓝色；嘴蓝色；脚蓝黑色。

生境与习性 主要栖息于中低山及平原地带的阔叶林、次生阔叶林、竹林，尤喜沟谷和溪流附近的阔叶林。常单独或成对活动，偶见 3-5 只成群。性害怕，常活动在森林中下层茂密的树枝间，也常与其他种类混群。飞行缓慢，常从栖息的树枝上飞到空中捕食昆虫。落地时长尾高举。

分布 突厥斯坦、印度、中国及东南亚。在井冈山见于茨坪等地。

种群状况 在井冈山迁徙季可见。

3.180a

3.180b

Identification L ca. 22 cm (except extended centered tail feathers). Male has two color morphs, brown and white. Brown morph has glossy black hood and nuchal crest with dark grey hind-collar and breast-band; upperparts mainly plain rufous-brown, belly and vent white. Tail long, extremely elongated central feathers of adult male project up to 30 cm. Apart from black hood male white morph is entirely white. Female resembles brown male, but is duller with smaller crest, black of head tinged blue and lacks elongated tail feathers. Eyes dark red with white eyering; bill blue; tarsi blue-black.

Habitat and Behavior Inhabits mixed deciduous broadleaf, evergreen forest and bamboo forest, prefers nearby to streams. Solitary or in pairs, rarely in small groups. Shy, favors lower and middle levels of dense trees. Flies slowly, chase insects from perches, raises tail when landing.

Distribution Turkestan, India, China and SE Asia. At Ciping in Mt. Jinggang.

Status of Population Visible on migration in Mt. Jinggang.

3.181a

3.181b

雀形目 PASSERIFORMES
扇尾莺科 Cisticolidae

3.182 棕扇尾莺 *Cisticola juncidis* (Rafinesque, 1810)

识别特征 体小的莺，约 10 cm。头顶黑褐色，上体棕色具显著的黑褐色纵纹，腰棕褐色，短而凸的尾具白端。与非繁殖期的金头扇尾莺的区别在于白色眉纹较颈侧及颈背明显为浅。虹膜褐色；嘴灰褐色具黑端；脚粉红至近红色。

生境与习性 主要栖息于开阔草地、稻田、灌丛及芦苇塘等生境。繁殖期单独或成对活动、领域性强，冬季多呈松散小群。性活泼，不停活动和觅食。飞行时尾常扇开并上下摆动。主要以昆虫及其幼虫为食，也吃其他小型无脊椎动物和杂草种子等植物性食物。多营巢于草丛中。窝卵数 4-5 枚。雌雄共同孵卵和育雏。

分布 非洲、南欧、印度、中国、日本、东南亚及澳大利亚北部。在井冈山地区见于宁冈、七溪岭、南风面等地。

种群状况 在井冈山地区全年常见。

Identification L ca. 10 cm. Small, brown bird with dark crown. Upperparts brown with bold dark streaks, and rufous rump; short and graduated tail with white tip. Similar to non-breeding Golden-headed Cisticola, but has paler supercilium. Eyes brown; bill grey with blackish tip; tarsi flesh-pink to red.

Habitat and Behavior Inhabits grasslands and reed beds, typically at wetland margins, but also in rice and sugarcane fields. In loose groups in winter. Active, tail typically fanned in flight. Feeds on insects and larvae, also seeds. Nests in grasses. Clutch size is 4-5 eggs. Both parents incubate and brood.

Distribution Africa and S Europe to India, China, Japan, SE Asia, and N Australia. At Ninggang, Qixiling and Nanfengmian in the Jinggangshan Region.

Status of Population Common all year round in the Jinggangshan Region.

3.182a

3.182b

3.183 纯色山鹪莺 *Prinia inornata* Sykes, 1832

识别特征　体型中等、略单调的偏棕色鹪莺，约 14 cm。繁殖羽具浅色眉纹，皮黄色眼先及浅褐色耳羽；上体暗灰褐，下体淡皮黄色至偏红；楔形的尾显长，且具浅色末端。冬羽羽色浅而平淡，尾更长。虹膜浅褐色；嘴近黑；脚粉红。

生境与习性　栖息于中低山和平原的农田、果园、灌丛、草丛及沼泽中。常结小群活动，在灌木下部和草丛中跳跃觅食。尾时常竖起，飞行呈波浪式，常边飞边叫。主要以昆虫及其幼虫为食，也吃少量其他小型无脊椎动物和杂草种子等。常营巢于芒草丛间。窝卵数 4-6 枚。雌雄共同孵卵，孵化期 11-12 天。

分布　印度、中国、东南亚及爪哇（印尼）。在井冈山见于大井、湘洲、厦坪等地。

种群状况　在井冈山全年常见。

Identification　L ca. 14 cm. Mid-sized, plain brown prinia. Breeding adult upperparts greyish-earth brown with short, pale buff supercilium, buff lores and buffish-brown ear coverts; underparts buff; tail long, graduated, with pale tips. Non-breeding more flat with longer tail. Eyes pale brown; bill black; tarsi pink.

Habitat and Behavior　Inhabits tall grasses, reeds, wetlands and crops, and adjacent scrub and isolated trees. Noisy and conspicuous, usually forages in bushes or grasses. Often cocks tail, wave-like flight with calling. Feeds on insects and seeds. Nests in grasses. Clutch size is 4-6 eggs. Both parents incubate, and incubation period is 11-12 days.

Distribution　India, China, SE Asia and Java. At Dajing, Xiangzhou and Xiaping in Mt. Jinggang.

Status of Population　Common all year round in Mt. Jinggang.

3.184 黄腹山鹪莺 *Prinia flaviventris* (Delessert, 1840)

识别特征　体型中等而尾甚长的橄榄绿色鹪莺，约 13 cm。繁殖期头顶和头侧暗石板灰色，上体橄榄褐色，喉及胸白色，胸及腹部黄色。冬羽颜色稍浅，尾较夏羽长。虹膜浅褐色；嘴夏季黑色，冬季上嘴深灰下嘴浅色；脚橘黄。

生境与习性　栖息于山脚和平原地带的芦苇沼泽、高草地及灌丛等。多在灌丛或草丛下部及地上活动和觅食。飞行有力，常发出"啪、啪"的振翅声响。活动时尾常上下摆动，或垂直翘到背上，并不时发出猫一样的叫声。主要以昆虫及其幼虫为食，也吃植物果实和种子。多营巢于杂草丛间或低矮的灌木上。窝卵数 3-6 枚，多 4-5 枚。雌雄共同孵卵和育雏。孵化期约 15 天。

分布　巴基斯坦至中国南部及东南亚。在井冈山见于湘洲、宁冈、新城区等地。

种群状况　在井冈山全年常见。

Identification　L ca. 13 cm. Slender, olive-green, very long, loose-tailed prinia. Breeding adult head dark slaty-grey, throat and breast white; upperparts brownish to olive-green and underparts yellow. Non-breeding adult paler with longer tail. Eyes reddish-brown; bill black (summer) or blackish-grey above paler below (winter); tarsi orange-yellow.

Habitat and Behavior　Inhabits tall grasses, reeds and scrub.

3.183a

3.183b

3.184a

3.184b

Forages in lower parts of scrub or grasses, and on the ground. Flight powerful with voice on wing beating. Often cocks tail, and nasal mewing. Feeds on insects, fruit and seeds. Nests in grasses or bushes. Clutch size is 3-6 eggs, mostly 4-5. Both parents incubate and brood. Incubation period is 11-12 days.

Distribution　Pakistan to Southern China and SE Asia. At Xinchengqu, Xiangzhou and Ninggang in Mt. Jinggang.

Status of Population　Common all year round in Mt. Jinggang.

3.185 山鷦莺 *Prinia crinigera* Hodgson, 1836

识别特征　体型略大的褐色鷦莺，约16.5 cm。具形长的凸形尾，翼显短。上体栗褐色具黑色及深褐色纵纹；下体污白，两胁、胸及尾下覆羽染黄，胸部黑色纵纹明显。虹膜浅褐；嘴黑色（冬季褐色）；脚偏粉色。

生境与习性　多栖于高草地及灌丛，也见于耕地。常单独或成对活动，也见3-5只的小群。雄鸟于突出处鸣叫。飞行振翼显无力。以昆虫及其幼虫为食。营巢于草丛中或灌丛下部。窝卵数4-6枚。孵化期10-11天。雌雄亲鸟共同孵卵和育雏，雏鸟晚成。

分布　阿富汗至印度北部、缅甸、中国南部（含台湾）。在井冈山地区见于大坝里、南风面、武功山等地。

种群状况　在井冈山地区全年常见。

3.185a

Identification　L ca. 16.5 cm. Large, brownish prinia with long, graduated tail, and relatively short wings. Upperparts rufous-brown with dark streaking on head, mantle and scapulars. Upperparts off-white, washed buff on flanks with faint streaking. Eyes pale brown; bill black (brown in winter); tarsi pink.

Habitat and Behavior　Inhabits scrub, thickets and grassy clearings in forest in hills and above, also scrubby margins of farmland. Solitary or in pairs, sometimes in small groups. Male sings on exposed perches. Flight weak. Feeds on insects and larvae. Nests in lower scrubs or grasses. Clutch size is 4-6 eggs. Incubation period is 10-11 days. Both parents incubate and brood. Chicks are altricial.

Distribution　Afghanistan, N India, Myanmar and Southern China (include Taiwan), At Dabali, Nanfengmian and Wugongshan in the Jinggangshan Region.

Status of Population　Common all year round in the Jinggangshan Region.

3.185b

雀形目 PASSERIFORMES
莺科 Sylviidae

3.186 鳞头树莺 *Urosphena squameiceps* (Swinhoe, 1863)

识别特征　体小而尾极短的树莺，约10 cm。头顶具鳞状斑纹；上体棕褐色，具显著长的浅色眉纹和深色贯眼纹；下体近白，两胁及臀皮黄色。虹膜黑褐色；上嘴褐色，下嘴肉色；脚粉红色。

生境与习性　主要栖于中低山的森林和林缘地带，尤喜林中河谷溪流沿岸及僻静的密林深处。冬季见于较开阔的多灌丛环境。常单独或成对活动于林下地面或近地面处。行动灵活，尾常上翘。主要以昆虫为食。

3.186a

分布　繁殖于东北亚。越冬于东南亚。在井冈山见于茨坪、茅坪等地。

种群状况　在井冈山迁徙季可见。

Identification　L ca. 10 cm. Tiny brown warbler with prominent short tail. Dark scales on crown; upperparts warm brown with buffy-white brow and dark eyestripe (extends to nape); underparts off-white with brown wash on flanks and undertail coverts. Eyes dark brown; bill brown above, and fresh below; tarsi pink.

Habitat and Behavior　Inhabits broadleaved and mixed forest, also woodland, prefer near stream and deep in the forest. Often skulks in dense areas. Solitary, usually seen foraging quietly on forest floor with tail cocking. Feeds on insects.

Distribution　Breeds in NE Asia. Wintering in SE Asia. At Ciping and Maoping in Mt. Jinggang.

Status of Population　Visible on migration in Mt. Jinggang.

3.186b

3.187 强脚树莺 *Horornis fortipes* Hodgson, 1845

识别特征　体型略小的暗褐色树莺，约 12 cm。上体橄榄褐色，具模糊的皮黄色眉纹；下体偏白而染褐黄，尤其是胸侧、两胁及尾下覆羽。幼鸟黄色较多。虹膜褐色；上嘴深褐，下嘴黄色；脚肉棕色。

生境与习性　栖息于中低山常绿阔叶林和次生林及林缘灌丛、竹丛与高草丛中。常单独或成对活动，性胆怯善藏匿，易闻其声但难将其赶出一见。主要以昆虫及其幼虫为食，也吃少量植物果实、种子和草籽。常营巢于灌丛或茶树丛下部靠近地面的侧枝上，也营巢于草丛中。窝卵数 3-5 枚，多为 4 枚。孵卵主要由雌鸟承担。雏鸟晚成性。

分布　喜马拉雅至中国南部及东南亚。在井冈山地区见于大坝里、南风面等地。

种群状况　在井冈山地区春夏季常见。

3.187a

Identification　L ca. 12 cm. Smallish, dusky-brown bush warbler lacking contrast between upperparts and underparts. Upperpart dusky brown with indistinct buff supercilium. Throat to upper belly off-white, increasingly buffy-brown towards lower belly. Flanks to undertail coverts washed rich brown. Juvenile more yellow. Eyes dark brown; bill dark brown above, and yellow below; tarsi brownish-pink.

Habitat and Behavior　Inhabits forest edges at middle elevations, woodland, scrubby hill sides, farmland and thickets. Descends lower in winter. Solitary or in pairs. Sings frequently in spring, but hard to flush. Feeds on insects and larvae, also fruit and seeds. Nests in lower scrubs or grasses. Clutch size is 3-5 eggs, mostly 4. Female incubates. Chicks are altricial.

Distribution　Himalaya to Southern China and SE Asia. At Dabali and Nanfengmian in the Jinggangshan Region.

Status of Population　Common in spring and summer in the Jinggangshan Region.

3.188 远东树莺 *Horornis borealis* (Campbell, CW, 1892)

识别特征　中等体型、头顶偏棕红的褐色树莺，约 15.5 cm。上

3.187b

体棕褐色，有显著的皮黄色眉纹和较弱的深褐色过眼纹，颊部色稍浅；下体浅色，喉白，胸侧、两胁和尾下覆羽染皮黄。似日本树莺，但下体更浅色。虹膜褐色；上嘴褐色，下嘴肉褐色；脚粉灰色。

生境与习性　栖于低山丘陵和山脚平原地带的林缘疏林、次生灌丛、农田、公园等。常隐匿在浓密的灌丛中，难以见到。通常尾略上翘。

分布　繁殖于东亚。越冬至印度东北部、中国南部（含台湾）、东南亚。在井冈山地区见于大坝里、南风面、武功山等地。

种群状况　在井冈山地区迁徙季常见。

Identification　L ca. 15.5 cm. Mid-sized, drab brown warbler with rufous-washed crown. Upperparts rufous-brown with buff supercilium, weak eyestripe, and pale face; underparts pale grey, but paler on chin and throat, washed buff on breast sides, flanks and undertail coverts. Similar-looking to Japanese Bush Warbler but paler on underparts. Eyes brown; bill brown above, yellowish-brown below; tarsi pink.

Habitat and Behavior　Inhabits woodland, scrub, farmland and well-wooded parks. Difficult to see well, often skulks in dense vegetation. Frequently cocks tail.

Distribution　Breeds in E Asia. Winters to NE India, Southern China (include Taiwan), and SE Asia. At Nanfengmian, Dabali and Wugongshan in the Jinggangshan Region.

Status of Population　Common on migration in the Jinggangshan Region.

3.189 东方大苇莺 *Acrocephalus orientalis* (Temminck & Schlegel, 1847)

识别特征　体型略大、嘴长而厚重的苇莺，约19 cm。上体褐色，具显著的皮黄色眉纹；下体污白，喉白色，胸微具深色纵纹，两胁及臀褐色。虹膜褐色；上嘴深灰褐色，下嘴偏粉褐具黑端；脚灰褐色。

生境与习性　栖息于中低山的丘陵和平原，喜芦苇地、稻田、沼泽、水塘及河流沿岸附近的低地次生灌丛。常单独或成对活动，迁徙时亦见几只结小群。繁殖期间雄鸟常站在巢附近的芦苇顶端或附近的小树枝头高声鸣叫，遇人立刻落入苇丛或灌丛中。主要以昆虫为食，也吃少量其他无脊椎动物。

分布　繁殖于东亚。冬季迁徙至印度及东南亚，偶尔远及新几内亚及澳大利亚。在井冈山见于宁冈、黄坳等地。

种群状况　在井冈山迁徙季常见。

Identification　L ca. 19 cm. Large reed warbler with white supercilium extends beyond eye, and long, rather thick bill. Upperparts greyish-brown, darker on wings; underparts dirty white with indistinctly dark streaked breast, and brownish wash on flanks and vent. Eyes brown; bill dark grey-brown above, pinkish-brown below with dark tip; tarsi grey-brown.

Habitat and Behavior　Inhabits wet grassland, marshes, reed beds, riversides and farmland with wet areas. Generally shy, but breeding birds sing from exposed reed perches. Feeds on insects, also other invertebrates.

Distribution　Breeds in E Asia. Migrating south in winter to India

3.188a

3.188b

3.189a

3.189b

and SE Asia and occasionally as far as New Guinea and Australia. At Ninggang and Huang'ao in Mt. Jinggang.

Status of Population　Common on migration in Mt. Jinggang.

3.190a

3.190 黑眉苇莺 *Acrocephalus bistrigiceps* Swinhoe, 1860

识别特征　体型略小的苇莺，约 13 cm。上体棕褐色，有较显著的黑色侧冠纹、细贯眼纹及皮黄色眉纹，腰偏棕红色；下体皮黄色。虹膜褐色；上嘴褐色，下嘴色浅；脚粉色。

生境与习性　栖于中低山丘陵和平原地带的湖泊、河流、水塘、沼泽等水域岸边的灌丛和芦苇丛中。常单独或成对活动，性机警而活泼。繁殖期站在开阔草地上的小灌木顶端或高的草茎梢上鸣叫。主要以昆虫及其幼虫为食，也吃蜘蛛等其他无脊椎动物。常营巢于灌丛和草丛的基部，距地面 0.5-1.5 m。窝卵数 4-6 枚，多为 5 枚。孵化期 14 天左右。

分布　繁殖于东北亚。冬季至印度、中国南部及东南亚。在井冈山地区见于宁冈、黄坳、七溪岭等地。

种群状况　在井冈山地区夏季常见。

3.190b

Identification　L ca. 13 cm. Small reed warbler with marked face pattern. Upperparts mid-brown with prominent black lateral crown-stripe, eyestripe, and buff supercilium; warmer brown on rump; underpart buff. Eye brown; bill brown above, and paler below; tarsi brownish-pink.

Habitat and Behavior　Inhabits reed beds, scrubby grassland, woodland fringes near wetlands and rivers. Usually solitary or in pairs, active and alert. Feeds on insects and spiders. Male sings on exposed sites such as top of scrubs, tall grass stems when breeding. Nests in reeds or adjacent scrubland, not necessarily over water, usually 0.5-1.5 m above the ground. Clutch size is 4-6 eggs, mostly 5. Both adults care for young in the nest for approximately 14 days.

Distribution　Breeds in NE Asia. Migrates in winter to India, Southern China, and SE Asia. At Ninggang, Huang'ao and Qixiling in the Jinggangshan Region.

Status of Population　Common in summer in the Jinggangshan Region.

3.191a

3.191 钝翅苇莺 *Acrocephalus concinens* (Swinhoe, 1870)

识别特征　体型略小、翼短尾长的苇莺，约 14 cm。上体橄榄褐色，腰及尾上覆羽偏棕红色；浅色的眉纹仅及眼，其上不镶黑边，颏、喉偏白，下体皮黄色。虹膜褐色；上嘴褐色，下嘴肉色；脚粉褐色。

生境与习性　栖息于中高山山顶矮灌及草丛。常单独或成对活动，性隐秘、行动敏捷，常隐匿在芦苇和草丛中。繁殖期多站在芦苇和草茎顶端鸣叫。主要以昆虫为食。常营巢于水边或山边苇丛、灌丛与草丛中。窝卵数 3-4 枚。

分布　中国、印度次大陆及东南亚。在井冈山地区见于南风面、江西坳等地。

种群状况　在井冈山地区夏季常见。

Identification　L ca. 14 cm. Smallish plain, unmarked reed warbler with short wings, short primary projection, and long tail. Upperparts

3.191b

olive-brown with more rufous on rump and uppertail coverts, short pale supercilium disappears behind eyes, and not bordered above by black. Chin and throat white, lower underparts yellowish-buff. Eyes brown; bill dark brown above, fresh-brown below; tarsi pinkish-brown.

Habitat and Behavior Inhabits grassland and scrub on mountains. Solitary or in pairs. Shy, but sings on top of reeds or shrubs when breeding. Feeds on insects. Nest is a neat cup of fine fibers lined with down and wool attached to upright plant stems. Clutch size is 3-4 eggs.

Distribution China, Indian subcontinent and SE Asia. At Jiangxi'ao and Nanfengmian in the Jinggangshan Region.

Status of Population Common in summer in the Jinggangshan Region.

3.192 厚嘴苇莺 *Iduna aedon* (Pallas, 1776)

识别特征 体型较大、不具纵纹、翼短尾长的苇莺，约 19 cm。上体棕褐色，嘴粗短，上喙较弯曲，无深色眼线，亦无浅色眉纹，眼先和眼周色浅，颏、喉白，下体皮黄色，尾长而凸。虹膜褐色；上嘴褐色，下嘴色浅；脚灰褐色。

生境与习性 栖息于中低山丘陵和山脚平原，喜林缘、湖边或河谷两岸的小片丛林、灌木林、荆棘丛和高草丛。常单独活动，性机敏而灵巧，多藏匿于茂密的灌丛和草丛中活动和觅食，也到灌木或草茎上部栖息和捕食。主要以昆虫及其幼虫为食，也吃其他小型无脊椎动物。

分布 繁殖于古北区北部。越冬至印度、中国南部及东南亚。在井冈山见于茅坪、黄坳等地。

种群状况 在井冈山迁徙季可见。

Identification L ca. 19 cm. Large, plain, unstreaked reed warbler with short wings and long tail. Upperparts rusty-brown, slightly paler on lore and cheek, lack eyestripe and supercilium, bill short and thick with curved upper mandible; chin and throat whitish, underparts buff; tail long and graduated. Eyes brown; bill dark brown above, and pinkish-horn below; tarsi grey-brown.

Habitat and Behavior Inhabits woodland, forest edge, dense thickets, scrub and bushy areas. Usually solitary, skulks in dense vegetation, sometimes upper. Feeds on insects and other invertebrates.

Distribution Breeds N Palearctic. Winters India to Southern China and SE Asia. At Maoping and Huang'ao in Mt. Jinggang.

Status of Population Visible on migration in Mt. Jinggang.

3.193 高山短翅莺 *Locustella mandelli* (Brooks, WE, 1875)

识别特征 体略小、尾长而凸的褐色莺，约 13 cm。上体单调的深棕褐色，略具皮黄色眉纹；颏、喉浅灰色具不清晰的细纵纹；下体污白色，胸及两胁染褐色；尾下覆羽深褐色具浅色末端。虹膜褐色；嘴深灰色，下嘴基角质色；脚偏粉色。

生境与习性 栖息于低山至中山的林缘、茶园、竹丛、灌丛、茂密蕨类及高草地中。性隐匿，不易见。昼夜发出虫鸣般的叫声。主要以昆虫及其幼虫为食。繁殖期领域性甚强。营巢于草丛几近地面处。窝卵数常 2 枚。

3.192a

3.192b

3.193a

分布　喜马拉雅至东南亚、中国南部。在井冈山地区见于八面山、大坝里、江西坳、南风面等地。

种群状况　在井冈山地区全年常见。

Identification　L ca. 13 cm. Small brown bush-warbler with long graduated tail. Upperparts plain dark chestnut-brown with slight pale grey-brown supercilium, chin and throat pale grey with fine streaks. Underparts grey-white with some brown on breast and flanks, dark brown undertail coverts with markedly paler tip. Eyes brown; bill blackish-grey, pale horn at base; tarsi brownish-pink.

Habitat and Behavior　Skulks at edges of forest and tea gardens, bamboo, bushes, dense ferns, tall grasses and scrub, lower to mid-mountain. Shy and overlooked, usually found by unique insect-like song day and night. Feeds on insects and larvae. Territorial, strongly when breeding. Nests in grasses near ground. Clutch size is 2 eggs.

Distribution　Resident in E Himalaya to SE Asia and Southern China. At Bamianshan, Dabali, Jiangxi'ao and Nanfengmian in the Jinggangshan Region.

Status of Population　Common all year round in the Jinggangshan Region.

3.194 长尾缝叶莺 *Orthotomus sutorius* (Pennant, 1769)

识别特征　体小、嘴细长、尾长而凸的鸟，约 12 cm。额及前顶冠红棕色，颊部、枕部和颈背灰色；上体橄榄绿色，翼及尾羽染深褐色；下体灰色，颏、喉较白。繁殖期雄鸟的中央尾羽较长。虹膜橘黄色；上嘴褐色，下嘴偏粉色；脚粉红色。

生境与习性　栖息于山脚和平原地带的小树丛、人工林和灌木丛。性活泼，常在树枝叶间或灌丛间不停跳动，不时发出单调重复的尖叫声，也到地上活动和觅食。尾常垂直翘到背上，有时飞行时亦如此。主要以昆虫及其幼虫为食，也吃其他小型无脊椎动物和少量植物果实和种子。常营巢于树丛和灌丛，巢多用一片或数片树叶、灌木叶或草叶缝合而成。窝卵数 3-5 枚，孵化期约 12 天。

分布　印度至中国、东南亚及爪哇（印尼）。在井冈山见于厦坪、黄坳、宁冈等地。

种群状况　在井冈山全年常见。

Identification　L ca. 12 cm. Slender bird with narrow, long bill and long, graduated tail. Forehead and crown rufous, cheek, nape and hind neck grey; upperparts olive-green with dark brown wash on wings and tail; underparts grey with white chin and throat. Breeding male has extended central tail feathers. Eyes black with orange orbital ring; bill dark above, pinkish below; tarsi pink.

Habitat and Behavior　Inhabits lower vegetation and thick brush in gardens and secondary forest. Active and vocal. Usually holds tail erect, even in flight. Feeds on insects and worms, also some plants. Nests on bushes with stitched leaves and grasses. Clutch size is 3-5 eggs. Incubation period is 12 days.

Distribution　India to China, SE Asia and Java. At Xiaping, Huang'ao and Ninggang in Mt. Jinggang.

Status of Population　Common all year round in Mt. Jinggang.

3.195 金头缝叶莺 *Phyllergates cuculatus* (Temminck, 1836)

识别特征　颜色鲜艳、嘴显著长的莺类，约 12 cm。头顶红棕色，具狭窄的白色眉纹和深色过眼纹；颏、喉白色，颈背至前胸灰色；上体橄榄绿色，腹部鲜黄色。虹膜褐色；上嘴褐色，下嘴基浅色；脚粉红色。

生境与习性　栖于常绿阔叶林、混交林及林缘的灌丛中，也见于竹林、稀树草坡、村寨周围的次生灌木丛等开阔地带。性活泼好动。繁殖期常成对活动，其余时候也加入混合鸟群。常于空中捕食飞行性昆虫。较隐蔽，但易以鸣声分辨。不以树叶营袋形巢。

分布范围　印度北部至中国南部及东南亚。在井冈山见于荆竹山等地。

种群状态　在井冈山夏季及迁徙季常见。

Identification　L ca. 12 cm. Bright yellow warbler with bright rufous forecrown and notable long bill. Head and neck dark grey with thin white brow and dark eyestripe; upperparts olive green; underparts bright yellow with white throat and grey breast, undertail white. Eyes brown; bill dark brown above, pale at base below; tarsi pink.

Habitat and Behavior　Inhabits undergrowth, mixed bush, bamboo and scrub inside broadleaved evergreen and mixed forests, forest edge. Usually in pairs, also joins mixed flocks. Feeds on insects. Hard to see, but easily recognised by songs. Unlike true tailorbird, does not make a leaf-purse nest.

Distribution　N India to Southern China and SE Asia. At Jingzhushan in Mt. Jinggang.

Status of Population　Common in summer and on migration in Mt. Jinggang.

3.195a

3.195b

3.196 褐柳莺 *Phylloscopus fuscatus* (Blyth, 1842)

识别特征　中等体型的褐色柳莺，约 11 cm。上体深灰褐色，无顶冠纹及翅斑，眉纹在眼前方为白色，眼后方为皮黄色，有深色过眼纹；下体浅褐色，喉至胸污白色。虹膜褐色；上嘴黑褐色，下嘴基黄色；脚淡褐色。

生境与习性　繁殖期栖于低地泰加林，及沼泽、溪流沿岸的疏林与灌丛；非繁殖期见于灌丛、芦苇田、草地等生境，尤喜水域附近。常单独或成对活动，多在林下、林缘和溪边灌丛与草丛活动，较隐蔽。性活泼好动，不断发出重复的叫声。翘尾并轻弹尾及两翼。

分布　繁殖于北亚、西伯利亚、蒙古北部、中国北部和东部。冬季迁徙至中国南部、东南亚、印度次大陆及喜马拉雅。在井冈山见于五指峰、荆竹山、湘洲等地。

种群状况　在井冈山冬季及迁徙季常见。

Identification　L ca. 11 cm. Mid-sized brown leaf warbler. Upperparts dark greyish-brown, lack crown-stripe and wing bar; supercilium whitish before eye, buff behind, contrasting with long black eyestripe which extends from bill-base; underparts pale brown with dirty-white throat and breast. Eyes dark brown; bill blackish-brown above, yellow at base below; tarsi pale brown.

3.196a

Habitat and Behavior In summer favors low, sparse taiga, scrub or bushes near swamps or edges in lowlands. On migration or in winter favors reeds or tall grasses with low trees near water. Often solitary or in pairs, skulks in dense vegetation, but sometimes sallies after insects. Active and vocal, often flicking tail and wings. Gives a short, repeated call.

Distribution Breeds N Asia in Siberia, N Mongolia, Northern and Eastern China. Migrating south in winter to Southern China, SE Asia and Indian subcontinent. At Wuzhifeng, Xiangzhou and Jingzhushan in Mt. Jinggang.

Status of Population Common on migration and in winter in Mt. Jinggang.

3.196b

3.197 巨嘴柳莺 *Phylloscopus schwarzi* (Radde, 1863)

识别特征 体型较大的褐色柳莺，约 12.5 cm。头较大，嘴厚重，脚也粗壮。宽阔的长眉纹于眼前为皮黄色，眼后成乳白色，过眼纹深褐色；上体深橄榄褐色；下体污白，喉浅色，胸及两胁沾皮黄；尾下覆羽浅橘黄色。虹膜深褐色；上嘴褐色，下嘴基部黄褐色；脚黄褐色。

生境与习性 夏季栖于低地落叶林或混交林及林缘灌丛，迁徙时见于混交林、林缘草地、果园和地边灌丛。常单独或成对活动于下层灌木丛和草丛中，并取食于地面。尾及两翼常抽动。主要以昆虫为食。

分布 繁殖于东北亚。越冬于中国南部、缅甸及中南半岛。在井冈山见于五指峰、茅坪等地。

种群状况 在井冈山迁徙季可见。

3.197a

Identification L ca. 12.5 cm. Large brown warbler with large head, rather stout bill and tarsi. Broad, long supercilium buffy-brown before and over eye, white behind; eye stripe dark brown; upperparts dark olive-brown; underparts dusky with off-white throat, buffier on breast and flanks, cinnamon-buff undertail coverts. Eyes dark brown; bill dark brown above, yellowish at base of lower mandible; tarsi yellow brown.

Habitat and Behavior In summer favors low mixed and deciduous thickets, scrubs, edge and taiga with dense undergrowth near water; in winter occurs in mixed forests, forest edges, woodland and scrub, often in dense shrubby margins. Often solitary or in pairs, feeds on insects on ground. Flicks tail and wings.

Distribution Breeds NE Asia. Winters Southern China, Myanmar and Indo-China Peninsula. At Wuzhifeng and Maoping in Mt. Jinggang.

Status of Population Visible on migration in Mt. Jinggang.

3.198 棕腹柳莺 *Phylloscopus subaffinis* Ogilvie-Grant, 1900

识别特征 中等体型的棕黄色柳莺，约 10.5 cm。与褐柳莺近似，但眉纹全为皮黄色；上体橄榄棕色，双翼偏橄榄绿；下体浅皮黄色。虹膜褐色；上嘴深褐色，下嘴基部黄色；脚褐色。

生境与习性 主要栖息于中高山的山地针叶林和林缘灌丛，非繁殖期见于低山丘陵和山脚平原地带的针叶林或阔叶疏林、灌

3.197b

丛和草甸。常单独或成对活动，秋冬季成松散小群。不安时两翼下垂并抖动。主要以昆虫为食。常营巢于幼树中下部枝杈上，或离地面不高的草丛中。窝卵数约 4 枚。

分布 华中及华东。越冬于中国南部、缅甸北部及中南半岛北部的亚热带地区。在井冈山地区见于八面山、大坝里、南风面等地。

种群状况 在井冈山地区夏季及迁徙季常见。

Identification L ca. 10.5 cm. Mid-size brownish-yellow warbler. Resembling Dusky Warbler but supercilium completely yellowish-buff, and upperparts mostly olive-brown, wings more strongly olive-green; chin to belly yellowish-buff. Eyes dark brown; bill dark brown above, yellow at base of lower mandible; tarsi dark brown.

Habitat and Behavior Inhabits mixed forests, montane scrub and forest edges in middle to high mountains; winters to lowland. Often solitary or in pairs, in loose groups in winter. Droops and shakes wings when nervous. Feeds on insects mostly. Nests in short trees, or grasses over ground. Clutch size is 4.

Distribution C and E China. Wintering in subtropics of Southern China, N Myanmar and N Indo-China Peninsula. At Bamianshan, Dabali and Nanfengmian in the Jinggangshan Region.

Status of Population Common in summer and on migration in the Jinggangshan Region.

3.198a

3.198b

3.199 黄眉柳莺 *Phylloscopus inornatus* (Blyth, 1842)

识别特征 体型较小的柳莺，约 9 cm。上体橄榄绿色，下体偏白。眉纹长，几延至颈背，在眼先为黄色，眼后为白色；黑色过眼纹较模糊；后顶冠纹几乎不可见。三级飞羽黑色具白色羽缘，通常具两道翼斑，后一道较宽并具黑色边缘。虹膜褐色；上嘴深灰色，下嘴基黄色；脚褐色。

生境与习性 夏季栖于阔叶林或泰加林林缘，迁徙期间和冬季出现于各种森林类型、农田、城市公园。性活泼好动，频繁扇动翅和尾。常加入混合鸟群。

分布 繁殖于亚洲北部。冬季南迁至印度及东南亚。在井冈山广泛分布。

种群状况 在井冈山冬季及迁徙季常见。

Identification L ca. 9 cm. Small warbler. Upperparts olive-green, and underparts pale. Long supercilium yellowish-buff before eye, grading to white towards nape; weak eyestripe black; rear crown-stripe poorly-defined. Tertials black broadly fringed whitish; two wing bars, the latter usually broader with dark border. Eyes dark brown; bill blackish-grey above, paler yellower-horn at base below; tarsi pinkish-brown.

Habitat and Behavior Summer in broadleaf and edges of coniferous taiga; migration and winter in most forest types, woodland, farmland and even urban parks. Very active, frequently flicks wings and tail. Often joins mixed-species flocks.

Distribution Breeds in N Asia. Migrant south in winter to India and SE Asia. Widespread in Mt. Jinggang.

Status of Population Common on migration and in winter in Mt. Jinggang.

3.199a

3.199b

3.200 黄腰柳莺 *Phylloscopus proregulus* (Pallas, 1811)

识别特征　体型较小的柳莺，约 9 cm。上体橄榄绿色，具柠檬黄色的粗眉纹和浅黄色顶冠纹；腰柠檬黄色，具两道黄色翼斑，三级飞羽羽缘浅色；下体灰白，尾下覆羽沾浅黄。虹膜褐色；嘴黑色，嘴基橙黄；脚淡褐色。

生境与习性　夏季栖于针叶林和针阔叶混交林，也见于阔叶林；迁徙季和冬季常见于林地、灌丛等广阔生境。性活泼敏捷，常悬停捕食。

分布　繁殖于亚洲北部。越冬在中国南部、印度、中南半岛北部。在井冈山广泛分布。

种群状况　在井冈山迁徙季和冬季常见。

Identification　L ca. 9 cm. Small warbler with distinctive pale yellow coronal stripe and lemon-yellow rump. Upperparts olive-green, and underparts greyish-white, pale yellow on vent. Lemon-yellow supercilium contrast with long black eyestripe and dark crown side. Wings brownish-grey with two deep yellow wing bars; tertials blackish broadly fringed white. Eyes brown; bill blackish-brown above, pale orange at base below; tarsi brown.

Habitat and Behavior　Summer in coniferous and mixed taiga; migration or winter wider range of scrub, woodland. Active, often hovers.

Distribution　Breeds N Asia. Winters in Southern China; India and N Indo-China Peninsula. Widespread in Mt. Jinggang.

Status of Population　Common on migration and in winter in Mt. Jinggang.

3.201 双斑绿柳莺 *Phylloscopus plumbeitarsus* Swinhoe, 1861

识别特征　体型较大的柳莺，约 12 cm。上体暗灰绿色，下体偏白。无顶冠纹，黄白色的长眉纹从嘴基延至颈侧，深色过眼纹明显。具两道浅色翼斑，三级飞羽无浅色羽缘。虹膜深褐色；上嘴深褐色，下嘴黄色；脚淡褐色。

生境与习性　夏季栖于山地针叶林和针阔叶混交林、灌丛中；迁徙季及冬季见于低山至山脚的各类森林及林地。

分布　繁殖于中国（东北），东北亚。越冬至泰国及中南半岛。在井冈山见于草坪、荆竹山等地。

种群状况　在井冈山迁徙季常见。

Identification　L ca. 12 cm. Largish leaf-warbler. Upperparts dull grey-green, underparts whitish. Long yellowish-white supercilium from bill base to nape, contrasting with dark eyestripe. Two pale wing bars, to tertials lack pale edges. Eyes dark brown; bill dark brown above, yellow below; tarsi brown.

Habitat and Behavior　Summer in montane taiga, in mixed deciduous and conifer forests and thickets; migration or winter in most forest types and woodland from low elevations to foothills.

Distribution　Breeds NE China, NE Asia. Winters to Thailand and Indo-China Peninsula. At Caoping and Jingzhushan in Mt. Jinggang.

Status of Population　Common on migration in Mt. Jinggang.

3.202 黑眉柳莺 *Phylloscopus ricketti* (Slater, 1897)

识别特征 中等体型的鲜艳柳莺，约 11 cm。上体亮橄榄绿色，下体鲜黄色。鲜黄而长的眉纹与绿黑色的侧冠纹及过眼纹形成对比，顶冠纹橄榄黄色；通常可见两道黄色翼斑。虹膜深褐色；上嘴褐色，下嘴黄色；脚褐粉色。

生境与习性 栖于山地阔叶林和混交林中，也见于针叶林、林缘灌丛和果园。除繁殖期单独或成对活动外，余时常加入混合鸟群。性活泼，鸣声响亮。主要以昆虫和昆虫幼虫为食。常营巢于林下或森林边土岸洞穴中。窝卵数 6 枚。

分布 繁殖于华中、华南及华东。越冬至中南半岛。在井冈山广泛分布。

种群状况 在井冈山全年常见。

Identification L ca. 11 cm. Mid-sized, bright yellow leaf-warbler with distinctive crown pattern. Broad bright yellow supercilium contrasts thick black lateral crown-stripe and eyestripe, median crown-stripe olive-yellow. Upperparts and wings olive-green with two yellow wing bars; underparts bright lemon-yellow. Eyes dark brown; bill dark brown above and yellow below; tarsi brownish-pink.

Habitat and Behavior Inhabits broadleaf evergreen and mixed forests on mountains. Solitary or in pairs when breeding, rest of the time mostly in mixed-species flocks. Active and vocal. Feeds on insects and worms. Nests in mud hollows in forests, or at forest edges. Clutch size is 6 eggs.

Distribution Breeds C, S and E China. Winters to Indo-China Peninsula. Widespread in Mt. Jinggang.

Status of Population Common all year round in Mt. Jinggang.

3.203 极北柳莺 *Phylloscopus borealis* (Blasius, JH, 1858)

识别特征 体型较大、显修长的柳莺，约 12 cm。上体橄榄绿色，下体污白色，胸部染黄，臀部白色。无顶冠纹，黄白色长眉纹前端不到嘴基处，过眼纹近黑。两道翼斑，但前一道通常不明显，三级飞羽无浅色羽缘。初级飞羽超出三级飞羽的部分较长。虹膜深褐色；上嘴深褐，下嘴基橙黄色，颜色对比较双斑绿柳莺明显；脚浅橘褐色。

生境与习性 夏季栖于潮湿的针叶林和针阔叶混交林及其林缘灌丛地带，迁徙期间见于阔叶林、灌丛、果园、庭院和宅旁小林等各种生境。除繁殖期单独或成对活动外，余时常与其他鸟类混群。

分布 繁殖于欧洲北部、亚洲北部及阿拉斯加。冬季南迁至中国南部及东南亚。在井冈山见于五指峰、荆竹山等地。

种群状况 在井冈山迁徙季常见。

Identification L ca. 12 cm. Large, slim-looking leaf warbler. Upperparts dark olive-green, underparts off-white, yellow wash on breast, vent white. Head appears long, with long white supercilium only to lores (not to bill base). Two thin wing bars, upper bar sometimes obscure; tertials lack pale edges. Long primary projection. Eyes dark brown; bill dark brown above, orange at base below; tarsi dull orange-brown.

Habitat and Behavior Summer in most forest from tundra edge

3.202

3.203a

3.203b

and taiga to temperate zone; migration or winter in wooded habitat from broadleaf evergreen forests to scrub and parkland. Solitary or in pairs when breeding. Sometimes joins mixed-species flocks in winter.

Distribution Breeds N Europe, N Asia and Alaska. Migrant south in winter to Southern China and SE Asia. At Wuzhifeng and Jingzhushan in Mt. Jinggang.

Status of Population Common on migration in Mt. Jinggang.

3.204

3.204 云南柳莺 *Phylloscopus yunnanensis* （La Touche, 1922）

识别特征 中等体型的柳莺, 约10 cm。上体灰橄榄色, 下体污白, 腰淡黄色。浅色的顶冠纹在前段几近消失, 眉纹白色、过眼纹黑色。具两道翼斑, 三级飞羽羽缘浅色, 次级飞羽基部无黑色斑块。虹膜褐色; 上嘴深褐色, 下嘴基橙黄色; 脚褐色。

生境与习性 栖于中低海拔的山地阔叶林或针阔混交林中。常活动于森林上层。

分布 主要分布于东南亚和中国西南山地, 也记录于南岭和井冈山地区。

种群状况 在井冈山迁徙季可见。

Identification L ca. 10 cm. Mid-sized leaf warbler with pale yellow rump. Upperparts greyish-green, underparts dirty-white. Pale median crown-stripe absent on forecrown, white supercilium and black eyestripe. Two wing bars, pale edge on tertials, and lack black patch at base of secondaries. Eyes dark brown; bill dark brown above, orange at base below; tarsi brown.

Habitat and Behavior Inhabits montane mixed, low deciduous, broadleaf and conifer forests. Usually active in upper forest levels.

Distribution Mainly distributed in mountainous areas of Southwestern China and SE Asia, also recorded in Nanling Mountains and the Jinggangshan Regions.

Status of Population Visible on migration in Mt. Jinggang.

3.205a

3.205 淡脚柳莺 *Phylloscopus tenellipes* Swinhoe, 1860

识别特征 中等体型的柳莺, 约11 cm。上体橄榄绿色, 与灰色的头顶对比明显, 下体污白色。无顶冠纹, 至少具一道皮黄色的翼斑, 但有时磨损不见。虹膜深褐色; 嘴褐色; 脚浅粉红。

生境与习性 夏季栖于中低山的阔叶林、混交林和针叶林的密林深处, 尤喜河谷两岸的森林; 迁徙季及冬季见于低地有林生境。以特殊的方式向下弹尾。

分布 繁殖于东北亚及日本。越冬于中国 (华东、华南)、东南亚。在井冈山见于茅坪、八面山、龙潭等地。

种群状况 在井冈山迁徙季常见。

Identification L ca. 11 cm. Mid-size leaf warbler with distinctly pale legs. Upperparts olive-green contrast with grey crown, underparts dirty-white. Lack crown-stripe. At least one wing bar, but sometimes abrased. Eyes dark brown; bill dark brown; tarsi whitish-pink.

Habitat and Behavior Summer in thickets, riparian woods and sparse forest; migration or winter in lowland wood habitats. Often wags tail downwards.

3.205b

Distribution Breeds NE Asia and Japan. Winters E and S China, SE Asia. At Maoping, Bamianshan and Longtan in Mt. Jinggang.

Status of Population Common on migration in Mt. Jinggang.

3.206 华南冠纹柳莺 *Phylloscopus goodsoni* Hartert, 1910

识别特征 中等体型、较鲜艳的柳莺，约 10.5 cm。上体鲜橄榄绿色，下体白色，胸部及两胁染黄。黄色顶冠纹在后方更宽阔明显，眉纹鲜黄色。具两道黄色翼斑，三级飞羽无浅色羽缘。虹膜褐色；嘴和脚橘黄色。

生境与习性 栖于山地常绿阔叶林、针阔混交林、针叶林和林缘灌丛地带，秋冬季多下到低山和山脚平原地带。常单独或成对活动，冬季多加入混合鸟群。常在树干上觅食，双翅轮番鼓动。营巢于中山林缘和林间空地等开阔地带的岸边陡坡岩穴或树洞中。窝卵数 4-5 枚。雌鸟孵卵。

分布 中国华中、华南及华东特种。在井冈山广泛分布。

种群状况 在井冈山全年常见。

Identification L ca. 10.5 cm. Mid-sized, brightly colored leaf warbler. Upperparts bright olive-green, underparts off-white with yellow wash on breast and flanks. Median crown-stripe yellow, more distinct and broader on rear crown, supercilium bright yellow. Two pale yellow wing bars, lack pale edge on tertials. Eyes brown; bill and tarsi orange-yellow.

Habitat and Behavior Inhabits broadleaf evergreen and mixed forests from foothills to mid-elevations. Often forages by creeping on tree branches. Frequently flicks a wing in turn. Solitary or in pairs in breeding, often joins mixed-species flocks in winter. Nests in tree hollows or caves at forest edges. Clutch size is 4-5. Female incubates.

Distribution Endemic to E, C and S China. Widespread in Mt. Jinggang.

Status of Population Common all year round in Mt. Jinggang.

3.207 棕脸鹟莺 *Abroscopus albogularis* (Moore, F, 1854)

识别特征 体小、色彩亮丽的莺，约 8 cm。颊部栗红色，与绿色的头顶、黑色的侧冠纹及黑色的喉部形成显著对比。上体绿，下体白，胸部和臀部染黄。虹膜黑褐色；嘴深灰色；脚粉褐色。

生境与习性 栖于中低山的阔叶林和竹林中。冬季常与其他小鸟混群。鸣声单调清脆易识别。主要以昆虫为食。多营巢于枯死的竹子洞中。窝卵数 3-6 枚。

分布 尼泊尔至中国南部（含台湾）、缅甸、中南半岛北部。在井冈山见于八面山、荆竹山等地。

种群状况 在井冈山全年常见。

Identification L ca. 8 cm. Tiny, olive-green warbler with distinct rufous-brown face. Crown olive-green, black lateral crown-stripes broadening at nape. Rest upperparts, wings and tail olive-green, underparts whitish with black throat, yellow wash on breast and vent. Eyes blackish-brown; bill dark grey; tarsi pinkish-brown.

Habitat and Behavior Inhabits bamboo thickets and scrub in broadleaved and mixed evergreen forests. Active, often joins

3.206a

3.206b

3.206c

mixed-species flocks in winter. Gives high-pitched whistles, easily recognised. Feeds on insects. Often nests in hollows in dead bamboo. Clutch size is 3-6 eggs.

Distribution　Nepal to Southern China (include Taiwan), Myanmar and N Indo-China Paninsula. At Bamianshan and Jingzhushan in Mt. Jinggang.

Status of Population　Common all year round in Mt. Jinggang.

3.207a

3.207b

3.208 栗头鹟莺 *Seicercus castaniceps* (Hodgson, 1845)

识别特征　体型较小、颜色鲜艳的莺，约 9 cm。顶冠栗红色，侧顶纹黑色，眼圈白，脸颊及颈部灰色，喉白。上体橄榄绿色，具两道黄色翼斑，腰及下体鲜黄色。虹膜褐色；上嘴黑褐色，下嘴橘黄色；脚褐灰色。

生境与习性　栖于中低山及山脚地带阔叶林与林缘疏林灌丛。繁殖期常单独或成对活动，非繁殖期常加入混合鸟群。多活动在林下灌木丛和竹丛中，常悬停取食昆虫和昆虫幼虫，也吃少量杂草种子等植物性食物。常营巢于阔叶林中树根下的土坎上或溪岸和岩边洞穴中。窝卵数 4-5 枚。雌雄轮流孵卵和育雏。

分布　喜马拉雅至中国南部及东南亚。在井冈山广泛分布。

种群状况　在井冈山全年常见。

Identification　L ca. 9 cm. Bright colored warbler. Chestnut crown contrast with black lateral crown-stripes (from above eye to nape), grey cheek and neck; throat white. Upperparts olive-green with two yellow wing bars, rump and underparts bright yellow. Eyes brown with white eyering; bill dark brown above, orange below; tarsi dark grey.

Habitat and Behavior　Inhabits undergrowth and bamboo thickets in broadleaf evergreen and mixed forests. Solitary or in pairs in breeding, joins mixed-species flocks in winter. Usually keeps to lower and mid-storeys. Often hovers when foraging for insects. Nests in cavities of earth bank or cliff. Clutch size is 4-5 eggs. Both parents incubate and brood.

Distribution　Himalaya to Southern China and SE Asia. Widespread in Mt. Jinggang.

Status of Population　Common all year round in Mt. Jinggang.

3.208a

3.209 淡尾鹟莺 *Seicercus soror* Alström & Olsson, 1999

识别特征　中等体型的橄榄绿色鹟莺，约 11.5 cm。头顶灰色，黑色的顶冠纹和侧冠纹明显，但到前额时非常模糊。多数个体翼斑不明显。下体柠檬黄色，两枚外侧尾羽白色，但内侧一枚白色区域较小。似比氏鹟莺但嘴稍长、尾较短，头顶图案不甚清晰。虹膜褐色；上嘴黑色，下嘴黄色；脚黄褐色。

生境与习性　繁殖于中山的湿润常绿阔叶林中，常见于混合鸟群中。

分布　中国中部及东南部、东南亚。在井冈山见于大坝里等地。

种群状况　在井冈山夏季及迁徙季常见。

Identification　L ca. 11.5 cm. Mid-sized, olive-green warbler. Crown grey with black median crown-stripe and lateral crown-stripe (indistinct on forehead). Yellow wing bar often indistinct or absent.

3.208b

Underparts lemon-yellow, two outer tail-feathers white, but inner one less white part. Similar to Bianchi's Warbler, but bill slightly larger, tail shorter and crown pattern less distinct. Eyes brown; bill black above, and yellow below; tarsi yellowish-brown.

Habitat and Behavior Inhabits broadleaved evergreen and mixed forests from low elevations to 1400 m a.s.l. Joins mixed species flocks during non-breeding period.

Distribution Central and Southeast China, and SE Asia. At Dabali in Mt. Jinggang.

Status of Population Common on migration and in summer in Mt. Jinggang.

3.209

3.210 白眶鹟莺 *Seicercus affinis* (Moore, F, 1854)

识别特征 体小的黄绿色莺，约11 cm。上体橄榄绿色，下体柠檬黄色；顶冠纹和眉纹灰色，黑色的侧冠纹从头顶前部延至枕部，头侧及耳羽灰色，白色眼圈上方有一缺口（亚种 *intermedius* 眼圈黄色）；常具一道黄色翼斑。虹膜褐色；嘴上嘴色深，下嘴黄色；脚黄色。

生境与习性 栖息于中海拔潮湿茂密的常绿阔叶林中，冬季见于低山和山脚地带的次生林、混交林和林缘灌丛且加入混合鸟群。性活跃大胆，活动在林中树木枝叶间，也到林下灌木上活动和觅食。主要以昆虫和昆虫幼虫为食，也吃蜘蛛等。常营巢于常绿阔叶林或松林中的岩坡或沟谷边地上，也见于倒木或树桩上、苔藓植物中。窝卵数4-5枚。

分布 喜马拉雅、中国南部及中南半岛。在井冈山见于八面山、西坪等地。

种群状况 在井冈山全年常见。

3.210

Identification L ca. 11 cm. Compact-looking, olive-green warbler. Upperparts olive-green, and underparts lemon-yellow. Black lateral crown-stripe from fore crown to nape, median crown-stripe and supercilium grey; eyering white and broken above (Race *intermedius* has yellow eyering). Usually has one yellow wing bar. Eyes brown; bill dark above, and yellow below; tarsi yellow.

Habitat and Behavior Inhabits broadleaved evergreen and mixed forests, often in adjoining bamboo thickets; descends lower in winter. Joins mixed species flocks in non-breeding periods. Active and bold, forages in trees and bushes. Feeds on insects and spiders. Nests on the ground under forest, also on fallen tree, stump or bryophyte. Clutch size is 4-5 eggs.

Distribution Himalaya, Southern China and Indo-China Peninsula. At Bamianshan and Xiping in Mt. Jinggang.

Status of Population Common all year round in Mt. Jinggang.

雀形目 PASSERIFORMES
画眉科 Timaliidae

3.211 黑脸噪鹛 *Garrulax perspicillatus* (Gmelin, JF, 1789)

识别特征 体型略大的灰褐色噪鹛，约30 cm。头、颈至胸灰褐色，具黑色脸罩；上体及尾深灰褐色；下体皮黄，尾下覆羽黄褐色。

3.211a

虹膜褐色；嘴近黑，嘴端较淡；脚红褐色。

生境与习性 栖居于平原和丘陵的矮灌木、竹林中，也见于庭院、农田地边和村寨附近的疏林和灌丛内。常成对或成小群在荆棘丛中或灌丛的下层穿梭跳动。性活泼嘈杂，但隐怯。杂食性，以昆虫为主，也吃其他无脊椎动物、植物果实、种子和部分农作物。常营巢于距地面数米高的灌丛、竹林或茂密的树篱间。窝卵数 3-4 枚。雏鸟孵出后，由亲鸟哺育至体羽长成而止。雏鸟出飞后，仍随亲鸟活动。

分布 留鸟于中国（华东、华中及华南）、越南北部。在井冈山见于宁冈、黄坳等地。

种群状况 在井冈山全年常见。

Identification L ca. 30 cm. Large, greyish laughingthrush with black mask. Head to breast greyish-brown with black forehead and cheek; upperparts to tail dark grey-brown, underparts dull buff with rufous-brown undertail coverts. Eyes brown; bill dark with pale tip; tarsi reddish-brown.

Habitat and Behavior Inhabits woodland, thickets, scrubby hillsides, grassy areas and even urban parkland. Forages mostly on ground, sometimes moving to higher branches. Gregarious and usually in small groups. Active and vocal, but shy. Feeds on insects, other invertebrates, fruit, seeds and corn. Nests in bushes and bamboo, several meters above ground. Clutch size is 3-4 eggs. Chicks follow parents forming family group after leaving the nest.

Distribution Resident in E, C and S China and N Vietnam. At Ninggang and Huang'ao in Mt. Jinggang.

Status of Population Common all year round in Mt. Jinggang.

3.212 黑领噪鹛 *Garrulax pectoralis* (Gould, 1836)

识别特征 体型略大、眼先浅色的棕褐色噪鹛，约 30 cm。上体红棕色，具显著的白色眉纹，白色耳羽镶黑色边缘，与黑色或灰色的胸带相接；颏、喉、上胸白色；下体偏白，胁部棕红色；尾羽具白色端斑及黑色次端斑（除中央尾羽外）。虹膜栗色；上嘴黑色，下嘴灰色；脚蓝灰。

生境与习性 栖息于中低山的丘陵和山脚平原地带的阔叶林中，也见于林缘疏林和灌丛。喜集群。多在林下茂密的灌丛或竹丛中活动和觅食。性机警，附近稍有声响便喧闹起来，鸣叫时两翅扇动，并不断点头翘尾。飞行时，一只接一只鱼贯前行。主要以昆虫为食，也吃果实等植物性食物。常营巢于林下灌丛、竹丛或幼树上。窝卵数 3-5 枚，通常 4 枚。

分布 喜马拉雅东段、印度东北部，东至中国（华中、华东），南至泰国西部、老挝北部及越南北部。在井冈山见于五指峰、龙潭、大船、小溪洞等地。

种群状况 在井冈山全年常见。

Identification L ca. 30 cm. Large bulky laughingthrush with pale lores, large white and black tail-tips. Upperparts rufous with distinct white supercilium from pale lores, streaked white ear coverts fringed with black eyestripe and moustachial stripe; broad breast band black or grey; underparts pale with rufous on flanks. Eyes rufous; bill black above and grey below; tarsi blue-grey.

Habitat and Behavior Forages on ground in montane forests

edge, and bushes. Often in small noisy groups, especially being disturbed,and stays out of sight when perching. Flies in long glides one by one. Performs dancing displays with birds hopping about bowing and spreading wings while calling. Feeds on insects, also fruit. Nests in bushes, bamboo and scrubs. Clutch size is 3-5 eggs, mostly 4.

Distribution E Himalaya, NE India, east to C and E China and south to W Thailand, N Laos and N Vietnam. At Wuzhifeng, Longtan, Dachuan and Xiaoxidong in Mt. Jinggang.

Status of Population Common all year round in Mt. Jinggang.

3.213 灰翅噪鹛 *Garrulax cineraceus* (Godwin-Austen, 1874)

识别特征 体型略小而具醒目图纹的褐色噪鹛，约22 cm。头顶、枕部、眼后纹、髭纹及颈侧细纹黑色。眼先及颊部白色；上体灰褐色，初级飞羽羽缘灰色，初级覆羽黑色，次级飞羽、三级飞羽及尾羽具白色端斑和黑色次端斑。下体褐色，颏、喉浅色。虹膜灰白色；上嘴暗灰，下嘴黄；脚暗黄。

生境与习性 主要栖息于中低山的常绿阔叶林、落叶阔叶林、针阔叶混交林、竹林和灌木林等各类森林中。性隐匿，多活动于林下灌丛和竹丛间，有时也在林下地面落叶层活动和觅食。

分布 印度东北部及缅甸北部至中国（华东、华中、东南）。在井冈山见于早禾木等地。

种群状况 在井冈山罕见。

Identification L ca. 22 cm. Smallish marked brown laughingthrush with prominent head pattern. Forehead to nape dark, lores and cheeks white, broad black malar extend to necksides; upperparts grey-brown with black primary coverts, pale blue-grey edged primaries, tertials and secondaries terminally black with narrow white tips; tail brown with broad black subterminal band and white tip. Underparts brown with pale chin and throat. Eyes pale grey; bill dark grey above, yellow below; tarsi dull yellow.

Habitat and Behavior Inhabits scrub, bamboo thickets and broadleaf forest and plantations. Shy and overlooked. Often forages in bushes or bamboo, also on the ground.

Distribution NE India and N Myanmar to E, C and SE China. At Zaohemu in Mt. Jinggang.

Status of Population Rare in Mt. Jinggang.

3.214 棕噪鹛 *Garrulax berthemyi* (Oustalet, 1876)

识别特征 体型略大的棕褐色噪鹛，约28 cm。眼周裸露皮肤蓝色。上体橄榄褐色，眼先、颏黑色，双翼及尾栗红色。腹部灰色，与白色的臀形成对比。虹膜褐色；嘴角质色，嘴基蓝色；脚蓝灰。

生境与习性 主要栖息于中低山的山地常绿阔叶林中，尤以林下植物发达、阴暗、潮湿和长满苔藓的岩石地区较常见。喜结小群活动。性羞怯而嘈杂、善隐藏，多活动在林下灌木丛间地上。常闻其声难见其影。

分布 中国（华中、东南及台湾）。在井冈山见于五指峰等地。

种群状况 在井冈山全年可见。

Identification L ca. 28 cm. A large brown and rufous laughingthrush

3.213

3.214

with bare blue skin around eyes. Upperparts and upper breast grey-brown with black lores, lower forehead, and chin, wings and tail chestnut. Underparts slivery grey, contrasting with white undertail coverts. Eyes brown; bill horn, blue-grey at base; tarsi pale grey.

Habitat and Behavior Inhabits mid-montane forests, descending in winter. Gregarious and noisy, often forages on ground. Hard to see.

Distribution C to SE China and Taiwan China. At Wuzhifeng in Mt. Jinggang.

Status of Population Visible all year round in Mt. Jinggang.

3.215 白颊噪鹛 *Garrulax sannio* Swinhoe, 1867

识别特征　中等体型的灰褐色噪鹛，约 25 cm。前额至枕部深红褐色，深色的眼后纹将浅色眉纹与下颊部隔开；上体及翼灰褐色，下体浅褐色，胸及尾下覆羽染棕红；尾暖褐色。虹膜深褐色；嘴深灰色；脚灰褐色。

生境与习性　栖息于中低山丘陵和山脚平原的矮树灌丛和竹丛，也见于农田和村庄附近的灌丛、芦苇丛和稀疏草地，以及城市庭院等。除繁殖期成对活动外，余时多成小群活动。性活泼而嘈杂，频繁在灌丛间穿梭。性杂食，以昆虫及其幼虫为主。常营巢于柏树、竹和荆棘灌丛中。窝卵数约 4 枚。孵化期为 15-17 天，育雏约 12 天。雌雄亲鸟共同孵卵和育雏。

分布　中国中部及南部。印度东北部、缅甸北部及东部、中南半岛北部。在井冈山见于宁冈、新城区及黄坳等地。

种群状况　在井冈山全年常见。

Identification　L ca. 25 cm. Brown laughingthrush with bold pale brow and cheeks. Forehead to nape dark earth-brown; dark eyestripe behind eye separates broad creamy supercilium from creamy under-cheek. Mantle to rump mid grey-brown, graduated tail warmer. Underparts warm brown to grey-brown from throat to belly, cinnamon on vent. Eyes dark brown; bill dark grey; tarsi greyish-brown.

Habitat and Behavior　Inhabits scrub, forest edge, clearings and bamboo thickets at mid elevations, also urban parks. In small noisy groups when non-breeding. Feeds on insects and larvae. Nests in cypress, bamboo and bushes. Clutch size is 4 eggs. Incubation period is 15-17 days. Brooding period is 12 days. Both parents incubate and brood.

Distribution　Central and Southern China. NE India, N and E Myanmar, and N Indo-China Peninsula. At Ninggang, Xinchengqu and Huang'ao in Mt. Jinggang.

Status of Population　Common all year around in Mt. Jinggang.

3.215

3.216 画眉 *Garrulax canorus* (Linnaeus, 1758)

识别特征　体型略小的棕褐色鹛，22 cm。通体深褐色，白色眼圈在眼后延伸成狭窄的眉纹、顶冠、颈背及上胸具深色纵纹。虹膜棕褐色；嘴偏黄；脚黄褐色。

生境与习性　主要栖息于中低山丘陵和山脚平原地带的矮树丛和灌丛中，也见于农田、村落附近的竹林或庭园中。多成对或结小群活动。性机敏胆怯，常隐匿在浓密的杂草及树枝间跳动鸣叫。歌声悠扬婉转，富于变化，有时也模仿别的鸟叫。性杂食，主要以昆虫为食，也吃野生植物果实、种子及部分农作物。多营巢于灌木上。窝卵数 3-5 枚，常 4 枚。

分布　中国中部及南部、中南半岛北部。在井冈山见于茨坪、八面山、小溪洞等地。

种群状况　在井冈山全年常见。CITES 附录 II。

Identification　L ca. 22 cm. Smallish, brown laughingthrush with distinctive facial patch. White eyering connecting supercilium extends to nape; entirely dark brown with fine dark streaking from head to mantle, and breast. Eyes rufous-brown; bill yellow; tarsi yellow-brown.

Habitat and Behavior　Inhabits broadleaved evergreen and

3.216a

3.216b

mixed forests, forest edges, shrubby thickets, woodland and scrub. Skulks in dense vegetation, usually in pairs or small groups. Highly vocal, sometimes mimics other birds. Feeds on insects and fruit, seeds and corn. Nests in bushes. Clutch size is 3-5 eggs, mostly 4.

Distribution Central and Southern China, N Indo-China Peninsula. At Ciping, Bamianshan and Xiaoxidong in Mt. Jinggang.

Status of Population Common all year round in Mt. Jinggang. CITES App. II.

3.217a

3.217 华南斑胸钩嘴鹛 *Pomatorhinus swinhoei* David, 1874

识别特征 体型略大、嘴长而下弯的钩嘴鹛，约 24 cm。前额及耳羽栗红色，眼先浅色，无浅色眉纹和深色髭纹；上体及尾棕褐色，下体灰白，胸部具浓密的黑色纵纹。虹膜黄至栗色；上嘴深褐色，下嘴浅色；脚肉褐色。

生境与习性 多栖息于丘陵至高山的灌丛、树木、竹丛间，也见于农田地边和村寨附近的小树林和灌木丛。多单独或成对活动，有时松散地结成小群在灌木下层或地上活动。叫声清晰而洪亮，常有互相应叫的习性。主要以昆虫及其幼虫为食，也吃植物种子等。常营巢于灌丛中。窝卵数 3 枚。雌雄亲鸟轮流孵卵，雏鸟晚成。

分布 中国南部特有种。在井冈山广泛分布。

种群状况 在井冈山全年常见。

3.217b

Identification L ca. 24 cm. Large rufous-brown scimitar babbler with distinct long decurved bill. Upperparts to tail rufous-brown with chestnut forehead and ear coverts, lores pale, lacks white supercilium and black moustachial. Underparts grey-white with black streaks on breast. Eyes yellow to chestnut; bill blackish-brown above, pale below; tarsi fresh-brown.

Habitat and Behavior Inhabits undergrowth of montane forests, forest edges and scrub. Usually skulks on the ground, and forages in defoliation. Solitary or in pairs, sometimes in loose groups. Feeds on insects and larvae, also seeds. Nests in bushes. Clutch size is 3 eggs. Both parents incubate.

Distribution Endemic resident of Southern China. Widespread in Mt. Jinggang.

Status of Population Common all year round in Mt. Jinggang.

3.218 棕颈钩嘴鹛 *Pomatorhinus ruficollis* Hodgson, 1836

识别特征 体型略小的深褐色钩嘴鹛，约 19 cm。头深褐色具白色长眉纹、宽阔的黑色过眼纹及栗色的颈圈；喉白，胸具白色纵纹，下体暗褐色。虹膜深褐色；上嘴黑，下嘴黄；脚铅褐色。

生境与习性 栖息于低山和山脚平原地带的阔叶林、次生林、竹林和林缘灌丛，也出入于村寨附近的茶园、果园、路旁丛林和农田地灌木丛间。常单独、成对或成小群活动，有时与其他鸟类混群。多靠近地面，攀爬树干或树枝。主要以昆虫和昆虫幼虫为食，也吃植物果实与种子。常营巢于灌木上。窝卵数 4 枚。

分布 中国中部及南部。喜马拉雅、中南半岛北部、缅甸北部及西部。在井冈山广泛分布。

3.218a

种群状况 在井冈山全年常见。

Identification L ca. 19 cm. Small dark brown scimitar babbler. Head dark brown with distinct long white supercilium, broad dark eyestripe, and rufous nape. White chin and throat contrast with dark reddish-brown streaked breast and upper belly. Upperparts, lower belly and vent dark brown. Eyes dark brown; bill dark above, yellow below; tarsi grey-brown.

Habitat and Behavior Inhabits lower and mid-level mixed and evergreen forest, secondary growth, scrub, tall grasses, bamboo thickets, parks and gardens. Forages on or near ground, and climbs about tree trunks and branches. Sometimes participates in mixed-species flocks. Feeds on insects, fruit and seeds. Nests in bushes. Clutch size is 4 eggs.

Distribution Central and Southern China. Himalaya, N Indo-China Peninsula, N and W Myanmar. Widespread in Mt. Jinggang.

Status of Population Common all the year round in Mt. Jinggang.

3.219 小鳞胸鹪鹛 *Pnoepyga pusilla* Hodgson, 1845

识别特征 体型极小、看似无尾的鹛，约 9 cm。上体灰褐色，具细小的浅色点斑；翼偏棕色，覆羽具浅色斑点；下体茶褐色，满布深色鳞状纹。虹膜深褐色；嘴黑褐色，嘴基黄褐；脚粉红至褐色。

生境与习性 栖息于山区森林，尤喜茂密、林下植物发达、地势起伏的阴暗潮湿森林。单独或成对活动。性隐匿，常在稠密灌木林或竹根间的地面跳来跳去，也在森林地面急速奔跑，形似老鼠。受惊即潜入密丛深处，从不远飞。频繁发出清脆响亮的特有叫声。杂食性，以植物的叶、芽及昆虫等为食。巢见于林下岩石间或长满苔藓植物的岩石壁上。窝卵数约 2 枚。

分布 尼泊尔至中国南部、东南亚。在井冈山见于荆竹山、小溪洞等地。

种群状况 在井冈山全年常见。

Identification L ca. 9 cm. Tiny wren-babbler with tail-less appearance. Upperparts greyish-brown with fine buff speckling, wings richer brown with buff spots on coverts. Underparts buffy to tea-yellow with scaling extending to belly. Eyes blackish-brown; bill dark brown, yellowish at base; tarsi pink to brown.

Habitat and Behavior Inhabits broadleaved evergreen and mixed forest, prefers dense and moist understory. Sedentary and shy, runs on the forest floor like a mouse. Song distinctive. Omnivorous, feeds on insects and plants. Nests between rocks in forest, or on rock-face with moss. Clutch size is 2 eggs.

Distribution Nepal to Southern China and SE Asia. At Xiaoxidong and Jingzhushan in Mt. Jinggang.

Status of Population Common all the year round in Mt. Jinggang.

3.220 丽星鹩鹛 *Elachura formosa* (Walden, 1874)

识别特征 体小而尾短、似鹪鹛的鸟，约 10 cm。上体灰褐满布白色小点斑，两翼及尾具棕色及黑色横斑，下体皮黄褐色而

多具黑色和白色小点斑。虹膜深褐；嘴角质褐色；脚角质褐色。

生境与习性　主要栖息于中高山的山地森林中，尤以林下灌木和草本植物发达的阴暗潮湿的常绿阔叶林和溪流与沟谷林中较常见。地栖性，主要在林下地上灌丛、竹丛和草丛间活动和觅食。善于在地面奔跑，除非迫不得已，一般很少起飞。每次飞行距离亦很短。鸣声响亮。主要以昆虫和昆虫幼虫为食。常营巢于茂密森林中地上，尤喜溪流边和岩石沟谷地区。窝卵数3-4枚。

分布　喜马拉雅东部至中国（西南、华南、东南），缅甸西部及北部和中南半岛北部。在井冈山地区见于南风面、荆竹山等地。

种群状况　在井冈山地区夏季常见。

Identification　L ca. 10 cm. Tiny short-tailed oscine, similar to Wren Babbler. Upperparts grey-brown with heavy small white spots, broad black bars on rufous wings and tail. Underparts peppery buffy-brown with dense black specks and white spots. Eyes dark brown; bill horn-brown; tarsi horn-brown.

Habitat and Behavior　Inhabits understory of broadleaved evergreen forest, scrub and weeds in gullies and rocky areas. Scarce, usually found by unique songs. Runs on the ground, less flies. Feeds on insects and larvae. Nests on the forest floor near stream or stone gully. Clutch size is 3-4 eggs.

Distribution　E Himalaya to SW, S and SE China, W and N Myanmar and N Indo-China Peninsula. At Nanfengmian and Jingzhushan in the Jinggangshan Region.

Status of Population　Common in summer in the Jinggangshan Region.

3.221 红嘴相思鸟 *Leiothrix lutea* (Scopoli, 1786)

识别特征　颜色鲜艳、嘴鲜红色的鹛类，约15.5 cm。黄绿色的头顶、鲜黄色的喉部，与浅色的眼先、灰色的脸及黑色的髭纹相对比；上体橄榄绿，初级飞羽和次级飞羽具黄色和红色的羽缘；下体浅黄，胸橘红色，胁部染灰色。尾近黑而略分叉。虹膜淡红褐色；嘴红色；脚粉红至黄褐色。

生境与习性　栖息于山地常绿阔叶林、竹林和林缘疏林灌丛地带，有时也进到村舍、农田附近的灌木丛中。繁殖季节成对活动，其他季节多成小群活动，也与其他小鸟混群。性机警而喧闹，善鸣叫。主要以昆虫和虫卵等为食，也吃大量植物性食物。常营巢于林下灌木侧枝、小树枝杈上或竹枝上。窝卵数3-4枚。

分布　喜马拉雅、印度东北部、缅甸中部及北部、中国南部及越南北部。在井冈山广泛分布。

种群状况　在井冈山全年常见。CITES 附录 II。

Identification　L ca. 15.5 cm. Mid-sized colorful babbler with bright red bill and slightly notched tail. Yellowish-green crown and yellow throat, contrasting with black malar, pale lores and grey cheeks. Upperparts olive-green with yellow and red outer fringes to primaries and secondaries. Underparts pale yellow with orange on breast and grey on flanks. Eyes reddish-brown; bill red; tarsi pink to yellow-brown.

Habitat and Behavior　Inhabits secondary forests, plantation and montane bamboo scrub. Usually in flocks in dense undergrowth, vigilant and noisy. Often participates in mixed-species flocks in winter. Feeds on insects and their eggs, also vegetables. Nests on branch of

3.220

3.221a

3.221b

scrubs, crotch of bamboo or bushes. Clutch size is 3-4 eggs.

Distribution Himalaya, NE India, C and N Myanmar, Southern China and N Vietnam. Widespread in Mt. Jinggang.

Status of Population Common all year round in Mt. Jinggang. CITES App. II.

3.222 红头穗鹛 *Stachyridopsis ruficeps* (Blyth, 1847)

识别特征 体小的橄榄褐色穗鹛，约 12.5 cm。顶冠棕红色，眼先暗黄；上体暗橄榄色，枕部及头侧沾黄；下体橄榄黄色，喉黄色具黑色细纹。虹膜棕红色；嘴深灰色，嘴基较浅；脚肉褐色。

生境与习性 主要栖息于山地森林林缘。常单独或结小群活动于灌丛中或高草丛，有时也与其他鸟类混群。易被模仿的口哨声吸引前来。主要以昆虫为食，也食少量植物果实与种子。常营巢于茂密的灌丛、竹丛、草丛和堆放的柴垛上。窝卵数 4-5 枚。雌雄共同孵卵和育雏。

分布 喜马拉雅东部至中国中部、南部（含台湾），缅甸北部及中南半岛。在井冈山广泛分布。

种群状况 在井冈山全年常见。

Identification L ca. 12.5 cm. Small olive-brown babbler with orange-red crown and yellowish throat. Upperparts dull olive-green with yellow wash to cheek and nape; underparts olive-yellow with black streaks on chin. Eyes rufous; bill dark grey, pale at base; tarsi fresh-brown.

Habitat and Behavior Inhabits dense undergrowth, scrub, tall grass, bamboo thickets and forest. Usually forages relatively low. Often participates in mixed-species flocks in winter. Easily attracted to imitatation whistles. Feeds on insects mostly, also berries and seeds. Nests in dense bushes, bamboo and grasses. Clutch size is 4-5 eggs. Both parents incubate and brood.

Distribution E Himalaya to Central and Southern China (include Taiwan), N Myanmar and Indo-China Peninsula. Widespread in Mt. Jinggang.

Status of Population Common all year round in Mt. Jinggang.

3.223 红翅鵙鹛 *Pteruthius aeralatus* Blyth, 1855

识别特征 中等体型的鵙鹛，约 17 cm。雄鸟具黑色头顶和过眼纹，宽阔的白色眉纹从眼上方延至颈侧；背及腰灰色；尾黑；两翼黑，初级飞羽羽端白，三级飞羽橘黄；下体灰白。雌鸟色暗，上体灰褐色，翼和尾橄榄绿色。虹膜灰蓝；上嘴蓝黑，下嘴灰；脚粉色。

生境与习性 主要栖息于中高山的落叶阔叶林、常绿阔叶林和针阔叶混交林等茂密的山地森林，冬季下到低山森林及林缘地带。除繁殖期成对活动外，其余季节多单独或成小群活动。性活泼，频繁地在树枝、树干间飞行、跳跃、攀缘。主要以昆虫为食。

分布 巴基斯坦东北部至中国、东南亚及大巽他群岛。在井冈山见于大井和湖洋塔等地。

种群状况 在井冈山夏季可见。

Identification L ca. 17 cm. Sexually dimorphic. Male has black crown and eyestripe, with bold white supercilium from above eye

to nape-sides; mantle to rump grey, wings black with orange tertials and white tips to primaries; tail black; underparts grey-white. Female has grey head, olive-grey mantle and scapulars, olive-green wings and tail; underparts paler. Eyes pale blue-white; bill dark blue-grey above, and grey below; tarsi pink.

Habitat and Behavior　Forages in lower and upper canopy of broadleaf forests and mixed forests, searches for insects on branches and twigs. Mostly in pairs when breeding, solitary or in small group when no-breeding. Active.

Distribution　NE Pakistan to China, SE Asia and Greater Sunda Islands. At Dajing and Huyangta in Mt. Jinggang.

Status of Population　Visible in summer in Mt. Jinggang.

3.224

3.224 淡绿鵙鹛 *Pteruthius xanthochlorus* Gray, JE & Gray, GR, 1847

识别特征　体小的橄榄绿色鵙鹛，约 12 cm。头灰黑色，眼圈白，黑色的嘴粗厚；上体橄榄绿色，具一道浅色翼斑；尾上覆羽墨绿色，尾黑色具狭窄白端；喉及胸白色，胸侧染灰，胁部及臀黄色。虹膜褐色；嘴近黑色；脚灰褐色。

生境与习性　主要栖息于中高山的山地针叶林和针阔叶混交林中，秋冬季也到中低山森林和林缘疏林灌丛地带。繁殖季常单独或成对活动，余时常与其他小鸟混群，多活动在树冠层。看似笨拙的柳莺。主要以昆虫为食，也吃浆果种子等植物性食物。巢常悬吊于树木侧枝枝杈间，用蛛网和枝杈固定，也见于矮灌木和幼树枝杈上。窝卵数 3-4 枚。

分布　巴基斯坦东北部至中国（东南）和缅甸的西部及北部。在井冈山见于龙潭、荆竹山等地。

种群状况　在井冈山夏季及迁徙季可见。

Identification　L ca. 12 cm. Small, robust olive-green forest bird. Head dark grey with white eyering and thick bill; upperparts olive-green with narrow white wing bar; short, broad tail black with narrow white tip. Throat, sides of neck and breast all white; flanks, belly, underwing and undertail coverts lemon-yellow. Eyes mid-brown; bill blackish; tarsi grey.

3.225a

Habitat and Behavior　Inhabits mixed montane broadleaved evergreen and subalpine forest, descending somewhat in winter. Solitary or in pairs when breeding, usually joins mixed-species flocks in winter. Forages on canopy, similar to clumsy leaf warbler. Feeds on insects, also berries and seeds. Builds a pendant nest in trees, scrubs or bushes. Clutch size is 3-4 eggs.

Distribution　NE Pakistan to SE China, W and N Myanmar. At Longtan and Jingzhushan in Mt. Jinggang.

Status of Population　Visible in summer and on migration in Mt. Jinggang.

3.225 金胸雀鹛 *Lioparus chrysotis* (Blyth, 1845)

识别特征　体型略小、色彩鲜艳的雀鹛，约 11 cm。头黑色，具白色的顶冠纹和耳羽；上体橄榄灰色，两翼及尾近黑，飞羽及尾羽具橘黄色羽缘，三级飞羽具白色羽端；下体亮橘黄色，喉黑色。虹膜淡褐；嘴灰蓝；脚偏粉。

3.225b

生境与习性　栖息于中高山的常绿阔叶林、针阔叶混交林和针叶林中，也见于林缘和山坡稀树灌丛与竹林中。常成小群活动，也与其他小鸟混群。性胆怯，常在树枝和竹丛间跳跃，也频繁在林下灌丛间穿梭。主要以昆虫为食。常营巢于常绿阔叶林的林下竹丛和灌丛中。窝卵数 3 枚。

分布　尼泊尔至中国南部、缅甸东北部、越南的北部及中部。在井冈山见于大坝里等地。

种群状况　在井冈山全年可见。

3.226a

Identification　L ca. 11 cm. Small, brightly-colored fulvetta. Head and throat black with white median crown-stripe and silvery-white ear coverts; upperparts olive-grey, wings and tail black with orange-yellow fringed flight-feathers and tail-feathers, white tipped tertials; underparts rich orange-yellow. Eyes brown; bill blue-grey; tarsi pink.

Habitat and Behavior　Inhabits broadleaved evergreen and mixed forests, also forest edges; often skulks in dense bamboo thickets. Regularly participates in mixed feeding flocks. Feeds on insects. Nests in bamboo and bushes. Clutch size is 3 eggs.

Distribution　Nepal to Southern China, NE Myanmar, N and C Vetnam. At Dabali in Mt. Jinggang.

Status of Population　Visible all year round in Mt. Jinggang.

3.226b

3.226 褐顶雀鹛 *Alcippe brunnea* Gould, 1863

识别特征　体型略大的褐色和灰色雀鹛，约 13 cm。前额至尾棕褐色，黑色侧冠纹延至颈侧，脸灰色染些许棕色，下体灰色。无翼斑。虹膜浅褐至棕褐色；嘴深褐或黑色；脚黄褐色或浅褐色。

生境与习性　主要栖息于中低山丘陵和山脚林缘地带的次生林、阔叶林和林缘灌丛与竹丛中。领域性强，常单独在靠近地面或地面觅食。善鸣叫。性羞怯但大胆。主要以昆虫为食，也食部分植物性食物。常营巢于靠近地面的灌丛中。窝卵数 2-3 枚。

分布　中国中部、南部（含台湾）。在井冈山见于茨坪、茅坪、小溪洞等地。

种群状况　在井冈山全年常见。

Identification　L ca. 13 cm. Largish two-tone fulvetta . Forehead to tail dark brown, and underparts dull grey. Black supercilium extend to nape-sides, and brown wash on grey cheeks. Lacks pale wing-patch. Eyes pale brown to rufous-brown; bill dark brown to black; tarsi yellow-brown or pale brown.

Habitat and Behavior　Inhabits scrub layer of evergreen and deciduous forests. Often terrestrial, forages alone close to or on ground amongst leaf litter. Shy, but does not fear people. Vocal. Feeds on insects and some vegetables. Nests in bushes near ground. Clutch size is 2-3 eggs.

Distribution　Central and Southern China (include Taiwan). At Ciping, Xiaoxidong and Maoping in Mt. Jinggang.

Status of Population　Common all year round in Mt. Jinggang.

3.227a

3.227b

3.227 淡眉雀鹛 *Alcippe hueti* David, 1874

识别特征　体型略大的褐色雀鹛，约 14 cm。头灰色，具明显的白色眼圈和不甚清晰的深色侧冠纹；上体为平淡的褐色，下体皮黄色。虹膜红色至栗色；嘴灰色至黑褐色；脚偏粉至暗黄褐。

生境与习性 栖息于中低山山地和山脚平原地带的森林和灌丛。除繁殖期成对活动外，常成小群活动，亦多作为"核心"物种出现在混合鸟群中。性机警，有人靠近立刻发出警诫声。易被"呸……"声吸引。主要以昆虫及其幼虫为食，也吃植物果实、种子、苔藓等植物性食物。常营巢于林下灌丛近地面的枝杈上，也见呈吊篮状以苔藓悬吊于常绿阔叶林下灌木的水平枝上。窝卵数 4 枚。

分布 中国（华南及东南）特有。在井冈山广泛分布。

种群状况 在井冈山全年常见。

Identification L ca. 14 cm. Largish brown fulvetta with dull mid-grey head and pale eyering. Indistinct black lateral crown-stripe extending to nape. Upperparts flat brown, and underparts buff. Eyes red to rufous; bill dark grey to blackish-brown; tarsi pink to dull brown.

Habitat and Behavior Inhabits forest and forest edge at low to mid elevations. Social, vigilant and noisy. Makes an agitated churring when disturbed. Readily attracted by "pishing". Usually nuclear species in mixed-species flocks. Aggressive in mobbing small owls and raptors. Feeds on insects and larvae, also fruit and seeds. Nests in bushes near ground, sometimes builds pendant basket hanging from branch of bush. Clutch size is 4 eggs.

Distribution Endemic to S and SE China. Widespread in Mt. Jinggang.

Status of Population Common all year round in Mt. Jinggang.

3.227c

3.227d

3.228 栗颈凤鹛 *Yuhina torqueola* (Swinhoe, 1870)

识别特征 中等体型、羽冠显著的凤鹛，约 13 cm。羽冠灰色，颊部的栗色延伸成后颈圈，并杂白色纵纹；上体灰褐色，具白色羽轴形成的细小纵纹；下体近白；尾深褐灰具白色羽缘。虹膜浅红褐色；嘴红褐，嘴端深色；脚粉红至褐黄色。

生境与习性 栖息于中低山的常绿阔叶林和针阔叶混交林。非繁殖季节一般结集小群（20-30 只），活动于较高的灌丛顶端或小乔木上。性活泼而嘈杂，常在树枝间跳跃或从一棵树飞向另一棵树。主要以昆虫为食，也兼食植物果实与种子。常营巢于其他鸟类废弃的巢洞或天然洞中。窝卵数 3-4 枚。

分布 印度次大陆东北部、中国南部及东南亚。在井冈山广泛分布。

种群状况 在井冈山全年常见。

Identification L ca. 13 cm. Mid-sized yuhina with prominently rounded crest. Grey head and crown, ear coverts rich chestnut extending as broad chestnut nuchal collar boldly streaked white; upperparts grey-brown with fine pale scapus forming pale streaks; underparts white; tail blackish-brown with white outer fringes. Eyes red-brown; bill dull pink with grey tip; tarsi bright pink to yellow-brown.

Habitat and Behavior Inhabits lower forest canopy. Often in groups (20-30 birds) when non-breeding. Noisy and active. Frequently transfer from one tree to another. Feeds on insects, fruit and seeds. Nests in old nest holes or natural hollows. Clutch size is 3-4 eggs.

Distribution NE Indian subcontinent, Southern China and SE Asia. Widespread in Mt. Jinggang.

3.228a

3.228b

Status of Population　Common all year round in Mt. Jinggang.

雀形目 PASSERIFORMES
鸦雀科 Paradoxornithidae

3.229 棕头鸦雀 *Sinosuthora webbiana* (Gould, 1852)

识别特征　体型纤小、尾长的粉褐色鸦雀，约 12 cm。头棕红色，嘴粗短；翼深栗红色，尾近黑；下体暗灰色，喉略具浅色细纹。虹膜暗褐色；嘴灰或褐色，嘴端色较浅；脚粉灰至铅褐色。

生境与习性　栖息于中低山林缘灌丛地带，也见于疏林草坡、竹丛、矮树丛、高草丛、果园、庭院和芦苇沼泽等生境。平时常集结小群隐匿在灌木荆棘间窜动，较嘈杂。易被 "pishing" 声吸引。杂食性。常营巢于山茶或其他灌木的枝杈及竹丛间。窝卵数 4-5 枚。

分布　中国、朝鲜半岛及越南北部。在井冈山广泛分布。

种群状况　在井冈山全年常见。

Identification　L ca. 12 cm. Small pinkish-brown parrotbill with long tail. Head chestnut with short and stubby grey bill; wings dark chestnut tail nearly black; underparts dull grey with fine pale streaks on throat. Eyes dark brown; bill grey or brown paler on tip; tarsi fresh-grey to grey-brown.

Habitat and Behavior　Inhabits scrub, riparian thickets, woodland edge and fringes of reed beds. Often encountered in noisy, wandering flocks. Easily attracted to soft "pishing" call. Feeds on insects, berries and seeds. Nests in bushes or bamboo. Clutch size is 4-5 eggs.

Distribution　China, Korean Peninsula, and N Vietnam. Widespread in Mt. Jinggang.

Status of Population　Common all year round in Mt. Jinggang.

3.230 灰头鸦雀 *Psittiparus gularis* (Gray, GR, 1845)

识别特征　体大的褐色鸦雀，约 18 cm。头灰色，宽阔的黑色眉纹从前额延至枕侧，眼圈及下颊白色；上背至尾暗褐色；下体皮黄色，喉中心黑色。虹膜红褐；嘴橘黄；脚灰色。

生境与习性　栖息于中低山的常绿阔叶林、次生林、竹林和林缘灌丛中。除繁殖期成对或单独活动外，多成小群活动在林下灌丛和竹丛中。性活泼，较嘈杂。主要以昆虫及其幼虫为食，也吃植物果实和种子。常营巢于林下幼树或竹的枝杈间。窝卵数 2-4 枚，多为 3 枚。

分布　喜马拉雅、印度东北部，至中国南部及东南亚。在井冈山见于湘洲等地。

种群状况　在井冈山全年常见。

Identification　L ca. 18 cm. Large brown parrotbill with large grey head and large orange-yellow bill. Broad black supercilium from forehead to nape, eyering and broad malar white, chin and throat center black; mantle to tail dull brown, and underparts buff. Eyes red-brown; bill orange-yellow; tarsi grey.

Habitat and Behavior　Inhabits scrub, bamboo thickets and forest, from undergrowth to canopy, in hills and low mountains. Social and noisy. Does not fear people. Feeds on insects, fruit and seeds. Nests in scrub and bamboo. Clutch size is 2-4 eggs, mostly 3.

Distribution Himalaya, NE India to Southern China and SE Asia. At Xiangzhou in Mt. Jinggang.

Status of Population Common all year round in Mt. Jinggang.

雀形目 PASSERIFORMES
长尾山雀科 Aegithalidae

3.231 红头长尾山雀 *Aegithalos concinnus* (Gould, 1855)

识别特征 体小、头部图案鲜明的鸟，约 10 cm。头顶及颈背栗红色，宽阔的黑色眼罩从眼先延至颈侧，下颊及颏、喉白色具显著的黑色喉斑；上体灰褐，下体白，胸带及两胁栗色。幼鸟头顶色浅，无黑色喉斑。虹膜浅黄色；嘴黑色；脚红褐色。

生境与习性 栖息于山地森林和灌木林间，也见于果园、茶园等人居附近的小林内。常结小群活动在灌木丛或乔木间，也见与其他小鸟混群。性活泼而嘈杂。杂食性，主要食昆虫等动物性食物，也食少量浆果、杂草种子等植物性食物。巢多见于针叶树上。窝卵数 5-8 枚。由雌雄亲鸟共同担任孵卵工作，但雌鸟坐巢时间较雄鸟长。孵化期约 16 天。雌雄共同育雏。

分布 喜马拉雅、缅甸、中南半岛、中国中部及南部。在井冈山广泛分布。

种群状况 在井冈山全年常见。

Identification L ca. 10 cm. Small tit with striking head pattern. Crown and nape chestnut, black mask from lores to neck sides, under-cheek and throat white with distinct black throat patch. Upperparts greyish-brown. Underparts white with chestnut breast band and flanks. Juvenile duller, lack black throat patch. Eyes pale yellow; bill black; tarsi red-brown.

Habitat and Behavior Inhabits broadleaved evergreen, mixed and pine forests, woodland, scrub and parkland, from lowlands to mountains. Gregarious, forms large flocks of 20 or more. Also in mixed flocks. Noisy and active. Feeds on insects and berries, seeds. Nests in conifers usually. Clutch size is 5-8 eggs. Incubation period is 16 days. Both parents incubate and brood.

Distribution Himalaya, Myanmar, Indo-China Peninsula, Southern and Central China. Widespread in Mt. Jinggang.

Status of Population Common all year round in Mt. Jinggang.

雀形目 PASSERIFORMES
山雀科 Paridae

3.232 远东山雀 *Parus minor* Temminck & Schlegel, 1848

识别特征 体型略大的黑、白、灰色山雀，约14cm。头上部及喉辉黑，与颊部和颈背的白斑强烈对比；上背橄榄灰色，具一条醒目的翼带；下体白色，中央黑带从喉延至尾下覆羽。幼鸟下体黑带较模糊。虹膜褐色；嘴黑色；脚灰褐色。

生境与习性 栖息于山区阔叶林、针叶林和针阔混交林中，也常见于平原地带的林间、庭园、果园和房前屋后。多单独或成对活动，也加入混合鸟群。性活泼大胆，常在树枝间穿梭跳跃寻觅食

3.231a

3.231b

3.231c

物，不甚畏人。飞行略呈波浪状。主要以昆虫为食，也吃少量植物性食物。多营巢于天然树洞、石隙或墙洞间，有时也利用啄木鸟遗弃的树洞。窝卵数多为 6-9 枚，有时多达 12-13 枚。孵卵由雌鸟担任，孵化期约为 14 天。雌雄共同育雏，育雏期 15-17 天。

分布　喜马拉雅、东亚及东南亚。在井冈山广泛分布。

种群状况　在井冈山全年常见。

Identification　L ca. 14 cm. Largish tit with bold black stripe on central belly. Adult head black with white cheek under eyes and white nape patch. Mantle olive-grey, wings black with a distinct white wing bar and pale edging to feathers. Underparts white with black center band (indistinct in juvenile). Eyes brown; bill black; tarsi grey-brown.

Habitat and Behavior　Inhabits broadleaf evergreen, mixed and coniferous forests, woodland, tree plantations, scrub and urban parkland. Usually seen singly or in pairs, occasionally participates in mixed flocks. Wave-like flight. Feeds on insects and some vegetables. Nests in tree hole, crevice of rock face, hollow in walls, or sometimes abandoned hollow of woodpecker. Clutch size is 6-9 eggs usually. Female incubates, and incubation period is 14 days. Both parents brood, and brooding period is 15-17 days.

Distribution　Himalaya, E and SE Asia. Widespread in Mt. Jinggang.

Status of Population　Common all year round in Mt. Jinggang.

3.233 黄腹山雀 *Pardaliparus venustulus* (Swinhoe, 1870)

识别特征　体小而尾短的山雀，约 10 cm。雄鸟头及喉黑色，颊部和颈背具白色斑块；上体深黑灰，翼具两排白色点斑，飞羽羽缘黄色；下体黄色。雌鸟头部灰色，具短的浅色眉纹，喉白，与颊斑之间有灰色下颊纹。虹膜深褐色；嘴灰黑色；脚蓝灰色。

生境与习性　栖息于中低山的各类森林，冬季也见于平原地区的次生林、人工林和林缘疏林灌丛。除繁殖期成对或单独活动外，常 10-30 只结群活动在高大的针叶树或阔叶树上，有时也与其他鸟类混群。杂食性，主要以昆虫为食，有时也吃植物性食物。营巢于天然树洞中。窝卵数 5-7 枚。

分布　中国东南特有种。在井冈山见于五指峰、八面山、笔架山等地。

种群状况　在井冈山全年常见。

Identification　L ca. 10 cm. Small, yellowish tit with short tail. Male head, throat to chest black, cheek and hind neck patch white. Upperparts, wings and tail blackish-grey with two pale bars on wings, and yellow edging to flight feather. Underparts yellow. Female upperparts duller with grey head and white throat. Eyes deep brown; bill blackish-grey; tarsi dark blue-grey.

Habitat and Behavior　Inhabits mixed and deciduous forests, also forest edges, woodland and parkland during winter. Lives in pairs or solitary in summer, forms flocks of 10-30 in winter, sometimes joins mixed flocks. Feeds on fruit and little insects. Nests in tree hollows. Clutch size is 5-7 eggs.

Distribution　Endemic to SE China. At Wuzhifeng, Bamianshan and Bijiashan in Mt. Jinggang.

Status of Population　Common all year round in Mt. Jinggang.

3.232a

3.232b

3.233a

3.233b

3.234 黄颊山雀 *Machlolophus spilonotus* (Bonaparte, 1850)

识别特征 体大的山雀，约14 cm。黑色冠羽显著，眼先黄色。头侧和枕部鲜黄色，黑色过眼纹从眼后延至颈侧。亚种 *rex* 雄鸟上背黑色具蓝灰色点斑，翼黑色具两道白色翼斑；下体灰白色，颏、喉、胸黑色从腹中部延至尾下覆羽。指名亚指图案相似，但上背及下体沾黄。雌鸟色略浅，无黑色腹中线。虹膜暗褐色；嘴深灰至黑色；脚蓝灰至黑色。

生境与习性 主要栖息山地各类森林也见于山边稀树草坡、果园、茶园、溪边和地边灌丛、小树上。性活泼，常结小群活动，有时也同其他鸟类混群。杂食性，主要以昆虫和昆虫幼虫为食，也吃植物果实和种子等植物性食物。营巢于树洞，也在岩石和墙壁缝隙中营巢，有时置于地上。窝卵数 3-7 枚。雏鸟晚成。

分布 喜马拉雅东段至中国南部及中南半岛。在井冈山见于五指峰、八面山、荆竹山、龙潭等地。

种群状况 在井冈山全年常见。

3.234a

Identification L ca. 14 cm. Large tit with conspicuous crest. Male crest black with yellow edge, face to hind neck yellow with black eyestripe extending to neck sides. Race *rex* male upperparts and wings black with blue-grey spotting on mantle and white spotting on wing coverts; large black patch from throat to center belly, contrasting with greyish flanks and undertail. Race *spilonotus* has yellow wash on mantle and underparts. Females similar to male but lack black on underparts, green-yellow instead. Eyes blackish-brown; bill grey to black; tarsi blue-grey to black.

Habitat and Behavior Inhabits broadleaf evergreen, mixed and deciduous forests, also seen orchard, tea garden and bushes. Regularly joins other tits and warblers to form mixed flocks. Omnivorous. Nests in tree hollows, crevices of rock faces or walls, hollows in earth bank, sometimes on ground. Clutch size is 3-7 eggs. Chicks are altricial.

Distribution E Himalaya to Southern China and Indo-China Peninsula. At Wuzhifeng, Bamianshan, Jingzhushan and Longtan in Mt. Jinggang.

Status of Population Common all year round in Mt. Jinggang.

3.234b

3.235 黄眉林雀 *Sylviparus modestus* Burton, 1836

识别特征 体小、具短羽冠的橄榄灰色山雀，约10 cm。形似柳莺。体羽大致橄榄色，具狭窄淡黄色眼圈，浅黄色短眉纹有时被遮盖。虹膜深褐；嘴角质色，基部偏灰；脚蓝灰。

生境与习性 栖息于山地常绿阔叶林、针阔叶混交林、针叶林等各类森林中，也栖于竹林、次生林和林缘疏林灌丛，冬季见于山麓和平原地带的树丛中。常单独或成对活动，也结家族群或与其他鸟类混群。性活泼，行动敏捷，常在树枝和灌丛间跳跃穿梭，边跳边觅食。示警或兴奋时冠羽耸立、浅色眉纹显出。主要以昆虫为食，也吃植物果实和种子等植物性食物。

分布 喜马拉雅、中南半岛及中国南部。在井冈山见于荆竹山、大坝里等地。

种群状况 在井冈山全年常见。

3.235a

Identification　L ca. 10 cm. Small, plain tit with indistinct crest. Pale buff-yellow brow and thin eyering. Mantle, wing and tail uniform olive-yellow; underparts buff-yellow. Eyes dark brown; bill horn, greyish at base; tarsi blue-grey.

Habitat and Behavior　Inhabits broadleaf evergreen, mixed and coniferous forests. Occurs in pairs, small family groups and regularly joins mixed flocks. Active and agility, similar to leaf warbler. Crest erect and brow appears when excited or alert. Feed on insects mostly, also some plants.

Distribution　Himalaya, Indo-China Peninsula and Southern China. At Jingzhushan and Dabali in Mt. Jinggang.

Status of Population　Common all year round in Mt. Jinggang.

雀形目 PASSERIFORMES
绣眼鸟科 Zosteropidae

3.236 暗绿绣眼鸟 *Zosterops japonicus* Temminck & Schlegel, 1845

识别特征　体小的鲜艳鸟类，约 11 cm。上体鲜亮橄榄绿色，具明显的白色眼圈；颏、喉、上胸及臀部柠檬黄色，与白色的腹部对比强烈，下胸及两胁染灰。虹膜红褐色或橙黄色；嘴黑色，下嘴基部稍淡；脚暗铅色。

生境与习性　栖息于阔叶林、针阔混交林、竹林、次生林等各种类型森林中，也栖息于果园、林缘、以及村寨和地边高大的树上。性活泼而嘈杂，除繁殖期成对活动外，常几只至几十只集群活动，也参与混合鸟群。食物以昆虫为主，也吃花蜜、杂草种子、浆果等植物性食物。营巢于阔叶或针叶树及灌木上，呈吊篮式或杯状，多隐藏在浓密枝叶间。窝卵数 3-4 枚，多为 3 枚。

分布　日本、中国、缅甸及越南北部。在井冈山广泛分布。

种群状况　在井冈山全年常见。

Identification　L ca. 11 cm. Small, bright green bird. Upperparts and tail yellow-green, with prominent broad white eyering; bright yellow chin, throat, upper breast and undertail coverts, contrast with grey washed white belly. Eyes brownish-orange; bill dark grey with paler at base below; tarsi grey to black.

Habitat and Behavior　Inhabits deciduous and evergreen broadleaf forests, also built-up areas with trees and scrub. Usually in groups, regularly joins mixed flocks when non-breeding. Active and noisy. Feeds on insects, also nectar, seeds and fruit. Builds pendant basket or cup nest in trees or bushes. Clutch size is 3-4 eggs, mostly 3.

Distribution　Japan, China, Myanmar and N Vietnam. Widespread in Mt. Jinggang.

Status of Population　Common all year round in Mt. Jinggang.

3.237 红胁绣眼鸟 *Zosterops erythropleurus* Swinhoe, 1863

识别特征　中等体型的绣眼鸟，约 12 cm。似暗绿绣眼鸟，但上体灰色较多，两胁栗色（有时不显露），黄色的喉斑较小，初级飞羽翼尖超出三级飞羽较多。虹膜暗褐；嘴灰色，下嘴基黄色；脚灰色。

生境与习性 栖息于中低山的丘陵和山脚平原地带的阔叶林和次生林，迁徙时见于果园、城镇公园及田边和村寨附近的小林内或树上。单独或成对活动，有时也成群。性活泼，行动敏捷。活动时常伴随尖细清脆的叫声。主要以昆虫为食，也吃植物性食物。

分布 中国东部和南部。东亚和中南半岛。在井冈山见于大井、湘洲等地。

种群状况 在井冈山迁徙季可见。

Identification L ca. 12 cm. Small green bird with prominent white eyering. Hood and upperparts bright yellow-green. Yellow chin, throat and vent, and clear white belly, distinct broad chestnut patch on flanks. Primary projection beyond tertials longer than Japanese White-eye, and smaller yellow throat patch. Eyes brown; bill dark grey, pink below and at base; tarsi grey.

Habitat and Behavior Inhabits mature mixed deciduous and coniferous forests and riparian woodland, but on migration occurs in any type of woodland. Solitary or in pairs, sometimes in groups. Active and agility, often gives high-pitch call. Feeds on insects mostly, also some plants.

Distribution Eastern and Southern China. E Asia and Indo-China Peninsula. At Dajing and Xiangzhou in Mt. Jinggang.

Status of Population Visible on migration in Mt. Jinggang.

雀形目 PASSERIFORMES
啄花鸟科 Dicaeidae

3.238 红胸啄花鸟 *Dicaeum ignipectus* (Blyth, 1843)

识别特征 体型纤小的深色啄花鸟，约9 cm。雄鸟上体辉深蓝绿色，黑色脸罩延至上胁；下体皮黄，胸具猩红色斑块，下胸至腹部中央具狭窄黑纹。雌鸟上体橄榄褐色，下体皮黄色。虹膜褐色；嘴及脚黑色。

生境与习性 栖息于中低山和山脚平原地带的阔叶林和次生阔叶林，也常见于开阔村庄、田野附近的次生阔叶林，或溪边树丛间。通常三五只结小群活动于高树顶端，有时也与绣眼鸟等混群。常在盛开花朵的树上觅食。飞行速度快，常边飞边叫。主要以昆虫和植物果实为食，嗜食浆果及寄生在常绿树上的槲寄生果实上的粘物质。巢囊状，多悬挂在细小的树枝梢端，四周有绿叶遮掩。窝卵数多2-3枚。

分布 喜马拉雅、中国南部及东南亚。在井冈山地区见于早禾木、八面山、武功山等地。

种群状况 在井冈山地区全年常见。

Identification L ca. 9 cm. Tiny dark or green flowerpecker. Male upperparts glossy blue-black, mask black extending to upper flanks. Underparts buff with bright crimson patch on breast, and black line on center of breast and belly. Females drab olive-brown, underparts buff. Eyes brown; bill black; tarsi black.

Habitat and Behavior Inhabits broadleaved evergreen and mixed forests from hills to mountains, sometimes descend to lowlands. Usually in small group, also mixed with White-eyes. Forages for small insects and mistletoe fruit in the canopy, difficult to see well.

3.238a

3.238b

Flight fast with high-pitched call. Nests in neat purse-shaped bags suspended from bushes or trees, always covered. Clutch size is 2-3 eggs.

Distribution Himalaya, Southern China, SE Asia and Sumatra. At Bamianshan, Zaohemu and Wugongshan in the Jinggangshan Region.

Status of Population Common all year round in the Jinggangshan Region.

雀形目 PASSERIFORMES
花蜜鸟科 Nectariniidae

3.239 叉尾太阳鸟 *Aethopyga christinae* Swinhoe, 1869

识别特征 体小而短圆的太阳鸟，约10 cm。雄鸟顶冠及颈背金属绿色，脸黑色具辉绿色髭纹；上体深橄榄色，腰黄；尾上覆羽及中央尾羽闪辉金属绿色，中央两枚尾羽延长，外侧尾羽黑色而端白；喉绛紫色渐至上胸的深红色。下体余部污白色。雌鸟较单调，上体橄榄色，下体浅黄绿色，具模糊的浅色眼圈，尾羽无延长。虹膜暗褐色；嘴黑色；脚深褐色。

生境与习性 栖于中低山丘陵和山脚平原地带的阔叶林、混交林及林缘，也见于村寨附近的树丛间。多单独活动，有时成对或结群。性不畏人。喜活动在开花的树冠顶部，不时发出尖细单调的叫声。主要以花蜜为食，也吃昆虫等动物性食物。常营巢于阔叶林中树枝上，巢多呈梨状，系于悬垂的枝叶上。窝卵数 2-4 枚。

分布 中国南部及越南。在井冈山见于夜光山、湘洲等地。

种群状况 在井冈山全年可见。

Identification L ca. 10 cm. Small sunbird. Male iridescent green crown to nape, black facial mask with glossy green malar; upperparts dark olive-green with yellow rump; tail blackish with white tipped outer feathers and two elongated plumes, uppertail coverts and center tail-feathers iridescent green; chin to upper breast maroon, rest of underparts off-white. Female upperparts dull olive-brown with faint eyering, underparts buff, lack elongated plumes. Eyes brown; bill black; tarsi black.

Habitat and Behavior Common resident of broadleaf evergreen and mixed forests, forest edges and woodland, sometimes parkland. Mostly solitary, sometimes in pairs or group. Best seen when feeding at flowering trees. Bold and vocal. Builds pear-shaped nest on suspensory twigs. Clutch size is 2-4 eggs.

Distribution Southern China, and Vietnam. At Yeguangshan and Xiangzhou in Mt. Jinggang.

Status of Population Visible all year round in Mt. Jinggang.

雀形目 PASSERIFORMES
雀科 Passeridae

3.240 山麻雀 *Passer rutilans* (Temminck, 1836)

识别特征 中等体型的麻雀，约 14 cm。雄鸟顶冠及上体栗红色，上背具纯黑色纵纹，喉黑、脸颊及下体污白。雌鸟色较暗，具深色的过眼纹及浅色的长眉纹。虹膜褐色；嘴深灰色（雄鸟），或黄色而嘴端色深（雌鸟）；脚粉褐。

生境与习性 栖于平原至高山的村庄、农田、河谷及其附近的树林、灌丛、荒漠和草甸。喜结群，冬季常与树麻雀混群。杂食性，主要以谷物、草籽等植物性食物为食，也吃昆虫和昆虫幼虫。常营巢于屋檐下和房屋缝隙中，也在岩坡、树上和电杆顶端筑巢。年产 2 窝卵。窝卵数 5-7 枚。孵化期 11-14 天。育雏期 12-15 天。雌雄共同孵卵和育雏。

分布 喜马拉雅、中国（西北、西南、华中、华南及华东）。在井冈山地区见于黄坳、七溪岭等地。

种群状况 在井冈山地区全年常见。

Identification L ca. 14 cm. Mid-sized colored sparrow. Male crown, nape, mantle, scapulars rich chestnut, black streaks on mantle; cheeks dirty white; narrow black throat patch, rest of underparts pale grey. Females dull brown with dark eyestripe and pale supercilium extending to nape. Eyes brown; bill dark grey (male), or yellow with dark tip (female); tarsi pink-brown.

Habitat and Behavior Inhabits forest edges, woodland, scrub and farmland, from lowlands to mountains. Wintering birds often in mixed flocks with Eurasian Tree Sparrow in rural areas. Mainly feeds on corn and seeds, also insects and larvae. Nests under eaves, in crevices of wall, also in trees or top of poles. Clutch size is 5-7 eggs. Incubation period is 11-14 days. Brooding period is 12-15 days. Both parents incubate and brood. Usually breeds twice each year.

Distribution Himalaya, SW, NW, C, S and E China. At Huang'ao and Qixiling in the Jinggangshan Region.

Status of Population Common all year round in the Jinggangshan Region.

3.241 麻雀 *Passer montanus* (Linnaeus, 1758)

识别特征 中等体型的麻雀，约14 cm。成鸟顶冠至枕部暗栗色（不如山麻雀鲜艳），白色的颊部与白色的颈圈相接，颊上具明显黑斑；上体褐色具深色纵纹，下体皮黄色，颏喉中央具黑斑。幼鸟似成鸟但色较黯淡，嘴基黄色，颊部和喉部黑斑不明显。虹膜深褐色；嘴黑色，冬季下嘴基黄色；脚粉褐色。

生境与习性 栖息于居民点或附近的田野。性喜成群，较嘈杂。杂食性，全年主要以各种杂草种子、野果、农作物为食，繁殖期也吃大量昆虫，尤其是雏鸟，几全以昆虫和昆虫幼虫为食。常营巢于村庄的房舍、庙宇、城市建筑物或电杆上，也在土墙洞穴、树洞营巢，或利用废弃的喜鹊巢、烟囱等。窝卵数4-8枚，多为5-6枚。孵化期为10-12天。雏鸟留巢期为15-16天。

分布 欧洲、中东、中亚、东亚、东南亚及喜马拉雅。在井冈山广泛分布。

种群状况 在井冈山全年常见。

Identification L ca. 14 cm. Common sparrow. Adult cap and nape chestnut, white cheeks connecting with white neck collar, distinct black patch on cheek; upperparts warm brown with dark streaking on mantle; underparts off-white with black patch on chin and throat, buff-washed flanks. Juvenile lack black chin and ear coverts patch. Eyes dull brown; bill black, winter yellow at base; legs dull pink.

Habitat and Behavior Inhabits dense urban areas to rural villages. In gardens, parks and agricultural land. Commonly flocks in winter. Feeds on cereal grains, grass and weed seeds, and seed sprouts, switching to insects and spiders when feeding nestlings. Nests in crevice or cavity, often uses buildings and other man-made structures. Clutch size is 4-8 eggs, mostly 5-6. Incubation period is 10-12 days. The brooding period is 15-16 days.

Distribution Europe, Middle East, C, E, and SE Asia, and Himalaya. Widespread in Mt. Jinggang.

Status of Population Common all year round in Mt. Jinggang.

雀形目 PASSERIFORMES
梅花雀科 Estrildidae

3.242 白腰文鸟 *Lonchura striata* (Linnaeus, 1766)

识别特征 中等体型、腰白的文鸟，约11 cm。头及上体深褐色，

3.241a

3.241b

3.242a

眼周及翼较黑，背及翼上覆羽有纤细的白色纵纹；尖形的尾黑色；下体污白，喉、胸、臀及尾上覆羽栗褐并具皮黄色鳞状斑。幼鸟色较淡，腰皮黄色。虹膜红褐色；上嘴黑色，下嘴蓝灰色；脚深灰色。

生境与习性　栖息于中低山丘陵和山脚平原地带，尤以溪流、苇塘、农田和村落附近较常见。性好集群，除繁殖期成对外，余时常成几只到数百只的大群，也与斑文鸟混群。多站在树枝上鸣叫，飞行呈波浪状。性不畏人。以植物性食物为主，也吃少量昆虫等动物性食物。常营巢于溪沟边或庭园内的竹丛、灌丛或树木上，靠近主干的枝叶浓密处。窝卵数多 4-6 枚。孵卵期约 14 天。育雏期约 19 天。雌雄共同孵卵和育雏。

分布　印度、中国南部及东南亚。在井冈山广泛分布。

种群状况　在井冈山全年常见。

Identification　L ca. 11 cm. Mid-sized dark brown munia with white rump. Upperparts mostly brown, wings and face darker brown. Thin streaking over mantle and scapulars. Throat, breast, vent and uppertail coverts brown with buff scales, rest of underparts off-white. Juvenile paler with buff rump. Eyes red-brown; bill black above and blue-grey below; tarsi dark grey.

Habitat and Behavior　Inhabits forest edges, woodland, scrub and farmland, especially with tall grass. In small to large groups when non-breeding, sometimes associated with Scaly-breasted Munia. Wave-like flight. Do not fear people. Phytophagous mainly, but nestlings feed on insects and larvae. Nests in bamboo, bushes or trees, often in dense sites near trunk. Clutch size is 4-6 eggs. Incubation period is 14 days. Brooding period is 19 days. Both parents incubate and brood.

Distribution　India, Southern China and SE Asia. Widespread in Mt. Jinggang.

Status of Population　Common all year round in Mt. Jinggang.

3.243 斑文鸟 *Lonchura punctulata* (Linnaeus, 1758)

识别特征　体型略小的暖褐色文鸟，约 10 cm。上体褐色，喉红褐色；下体灰白，胸及两胁具深褐色鳞状斑。幼鸟下体浓皮黄色而无鳞状斑。虹膜红褐；嘴蓝灰；脚灰黑。

生境与习性　多成群栖息于灌丛、竹丛、稻田及草丛间，也见与白腰文鸟、麻雀等混群。有时数百只聚集在一棵树上，若受惊有一两只飞起，全群随即振翅飞离，并发出呼呼的响声。以吃谷物为主，兼吃少量其他植物种子，较少吃昆虫。常营巢于靠近主干的密集枝杈处。窝卵数 4-8 枚。雌鸟育雏，雏鸟留巢期 20-22 天。

分布　印度、中国南部及东南亚。引种至澳大利亚及其他地区。在井冈山见于茨坪、新城区、罗浮、湘洲、宁冈、黄坳等地。

种群状况　在井冈山全年常见。

Identification　L ca. 10 cm. Small, rather drab finch. Adult upperparts mid-brown, becoming dark red-brown on chin and throat. Underparts off-white with distinctive brown scales on breast and flanks. Juvenile plain brown above, pale buff below lacks scales. Eyes black; bill dark grey; tarsi dark grey.

Habitat and Behavior Inhabits dry glasslands, gardens, cultivated fields and scrub. In small to very large flocks, sometimes mixed with White-rumped Munia and Sparrow. Feeds on cereal grains mainly. Nests in bamboo, bushes or trees, often in dense sites near trunk. Clutch size is 4-8 eggs. Brooding period is 20-22 days. Females brood.

Distribution India, Southern China, SE Asia. Introduced into Australia and elsewhere. At Ciping, Xinchengqu, Luofu, Xiangzhou, Ninggang and Huang'ao in Mt. Jinggang.

Status of Population Common all year round in Mt. Jinggang.

雀形目 PASSERIFORMES
燕雀科 Fringillidae

3.244 金翅雀 *Chloris sinica* (Linnaeus, 1766)

识别特征 体小的黄、灰及褐色雀鸟，约 13 cm。飞行时，可见其亮黄色翼斑。成年雄鸟头及颈背灰色，颊和颏染黄色，背褐色，翼斑、腰、外侧尾羽基部及臀黄色，尾略成叉形，下体暖褐色。雌鸟褐色更浓。幼鸟色淡且多纵纹。虹膜深褐；嘴偏粉，尖端暗色；脚粉褐色。

生境与习性 主要栖息于中低山丘陵或山脚平原的高大树上、苗圃、公园和村寨附近的树丛中。喜栖于裸子植物和电线上。多结群生活。以植物性食物为主，大部分为杂草种子、树木种子、谷物，也吃少量昆虫等。营巢于针叶树幼树枝杈上和杨树、果树、榕树等阔叶树及竹丛中。窝卵数 2-5 枚，多为 4 枚。孵化期 13 天左右。育雏期 14-15 天。雌鸟孵卵，雌雄共同育雏。

分布 西伯利亚东南部、蒙古、日本、中国东部、越南。在井冈山见于茨坪、罗浮、黄坳等地。

种群状况 在井冈山全年常见。

Identification L ca. 13 cm. Small, dark, olive-brown finch with bright yellow wing bar. Male crown and nape grey with greenish-yellow wash on face and chin; upperparts brown with large yellow wing bar, rump and base of outer tail feathers; underparts warm brown. Female browner, and juvenile duller with heavy streaking. Eyes dark brown; bill pink; tarsi pink-brown.

Habitat and Behavior Inhabits mixed and coniferous forests, forest edges and woodland. Also in parks, scrub and farmland during winter, usually in flocks. Often perches on conifer and wires. Phytophagous mainly. Nests in trees and bamboo. Clutch size is 2-5 eggs, 4 mostly. Incubation period is 13 days. Brooding period is 14-15 days. Female incubates, and both parents brood.

Distribution SE Siberia, Mongolia, Japan, Eastern China and Vietnam. At Ciping, Luofu, and Huang'ao in Mt. Jinggang.

Status of Population Common all year round in Mt. Jinggang.

3.245 燕雀 *Fringilla montifringilla* Linnaeus, 1758

识别特征 中等体型、胸棕而腰白的雀鸟，约 16 cm。成年雄鸟头至颈背黑色，上背近黑，两翼及叉形尾黑色，肩斑和翼斑橘红色，初级飞羽基部具白色点斑；喉至胸橘红色，腹部白。非繁殖期雄鸟头及上体偏褐色，下体橘红色较少。雌鸟头更浅

3.244a

3.244b

3.245a

色，颈侧有宽阔的灰色弧形斑块。虹膜褐色；嘴黄色，嘴尖黑色；脚粉褐。

生境与习性 栖息于平原、丘陵到山区的各种森林，迁徙时也到村庄附近的农田中。性喜集群，迁徙时多结成大群，越冬时多成小群活动。杂食性，既在地面上摄食也在树上觅食，主要以草籽、果实等植物性食物为主。

分布 古北区北部。在井冈山见于大井、下庄、黄坳等地。

种群状况 在井冈山秋冬季常见。

Identification L ca. 16 cm. Mid-sized brown and black finch with orange breast and white rump. Breeding male head and nape black, mantle blackish, wings and tail black with orange shoulder patch and wing bar, white spots on base of primaries. Winter male paler, with browner head and mantle, and orange of underparts less bright. Female resembles winter male, but has paler head with broad grey crescent on nape-sides to rear of ear coverts. Eyes brown; bill dull yellow with dark tip; tarsi dull pink.

Habitat and Behavior In summer, mature taiga. In winter, forests, woodlands and agricultural land. Gregarious, in very large groups on migration, and small groups in winter. Feeds on seeds and fruit on ground or in trees.

Distribution N Palearctic. At Dajing, Xiazhuang and Huang'ao in Mt. Jinggang.

Status of Population Common in autumn and winter in Mt. Jinggang.

3.246 普通朱雀 *Carpodacus erythrinus* (Pallas, 1770)

识别特征 体型略小的朱雀，约 15 cm。繁殖期雄鸟上体褐色，头、胸、腰多具鲜亮红色（不同亚种红的程度不同），上背和翼覆羽褐灰色染红色，下体皮黄染粉红色。雌鸟和非繁殖雄鸟无红色，上体青灰褐色，有两条微带褐色的翼带，下体污白，喉、胸及上腹具深色纵纹。虹膜深褐；嘴灰色；脚近黑。

生境与习性 栖息于中高山的针叶林和针阔叶混交林及其林缘地带，迁徙及越冬时见于低海拔的农田、果园、竹林等各种生境。常见单独或成小群活动。树栖性，飞行时略呈波状，两翅扇动迅速。主要以果实、种子、嫩叶等植物性食物为食，繁殖期也吃部分昆虫。

分布 繁殖于欧亚区北部及中亚的高山、喜马拉雅、中国（西北、西南）。越冬南迁至印度、中南半岛北部及中国南部。在井冈山地区见于笔架山、江西坳等地。

种群状况 在井冈山地区冬季及迁徙季可见。

Identification L ca. 15 cm. Breeding male unmistakable with reddish head to breast, and rump; mantle and wing coverts brownish-grey with red wash; underparts washed pink. Female and non-breeding male lack red, upperparts dull greyish-brown with two brownish wing bars, underparts pale with streaks. Eyes dark brown; bill grey; tarsi blackish.

Habitat and Behavior Inhabits woodland, scrub and farmland. Usually single or small flocks but sometimes gather in large flocks at fruiting or flowering trees. Arboreal, and flight wave-like with wings

3.245b

3.246a

3.246b

beating fast. Feeds on fruit, seeds and tender leaves, also insects when breeding.

Distribution Breeds across N Eurasian boreal zone plus high mountains of C Asia, Himalaya, SW and NW China. Winters south to India, N Indo-China Peninsula and Southern China. At Bijiashan and Jiangxi'ao in the Jinggangshan Region.

Status of Population Visible on migration and in winter in the Jinggangshan Region.

3.247a

3.247 褐灰雀 *Pyrrhula nipalensis* Hodgson, 1836

识别特征 中等体型、嘴短而粗壮的雀类，约 16.5 cm。雄鸟头及上体灰色，具狭窄的黑色脸罩，眼下有一白色细纹；飞羽及尾闪辉深绿紫色，翼上具浅色块斑，腰白色，尾长而略凹；下体浅灰褐色，尾下覆羽白。雌鸟多皮黄灰。幼鸟无黑色脸罩。虹膜褐色；嘴绿灰、嘴端黑色；脚粉褐。

生境与习性 栖息于阔叶林和针阔混交林及其林缘，或杜鹃灌丛中。常单独或成对活动，非繁殖期则多成小群在林下灌丛中或树上活动，有时也到地上活动和觅食。性大胆，不甚惧人，活动时频繁发出彼此联络的叫声。飞行迅速而径直。主要以树木、灌木的果实和种子为食，也吃草籽、嫩叶等，偶尔也吃昆虫等动物性食物。

分布 喜马拉雅至中国（西南和东南）、缅甸北部、马来半岛。在井冈山见于松木坪等地。

种群状况 在井冈山全年可见。

3.247b

Identification L ca. 16.5 cm. Mid-sized, greyish finch. Male head and upperparts grey with black mask, and thin white cheek streak under eyes. Flight feathers and tail glossy black with pale wing patch, rump white, and tail slightly notched; underparts pale grey-brown with white undertail coverts. Female more buff. Juvenile lacks black mask. Eyes brown; bill greenish-grey with dark tip; tarsi pink-brown.

Habitat and Behavior Inhabits broadleaved evergreen, mixed and coniferous forests and forest edges. Descends to lower elevations in winter. In pairs or small groups, usually keeping to canopy, sometimes forages on the ground. Feeds on fruit and tree seeds, also weed seeds, tender leaves, and insects.

Distribution Himalaya to SW, SE China, N Myanmar, and Malay Peninsula. At Songmuping in Mt. Jinggang.

Status of Population Visible all year round in Mt. Jinggang.

3.248 黑尾蜡嘴雀 *Eophona migratoria* Hartert, 1903

识别特征 体型略大的雀鸟，约 17 cm。雄鸟通体灰色，具黑色头罩，两翼黑色，飞羽及初级覆羽羽端白色，臀黄褐，尾下覆羽白色。雌鸟褐色较重，无黑色头罩。飞行时，可见浅色腰、白色细小翼斑和白色的翼后缘。虹膜浅红褐色；嘴深黄而端黑；脚粉褐。

生境与习性 栖息于平原的村庄附近、行道树上、公园和苗圃的高树上，也见于丘陵和山区的阔叶林、灌木丛。除繁殖期成对生活外，一般结成小群活动。性活泼，常在树枝上跳跃，并

3.248a

反复从一树转移到另一树上。飞行迅速，微呈波形，群体飞时呼呼作响。食物以植物类为主，兼食昆虫。

分布　西伯利亚东部、朝鲜半岛、日本南部及中国东部。越冬至中国南部。在井冈山见于罗浮、黄坳等地。

种群状况　在井冈山秋冬季常见。

Identification　L ca. 17 cm. Largish finch with chunky yellow bill. Male grey with black cap, wings glossy black with white tips to flight feathers and primary coverts, yellow-brown vent and white undertail coverts. Female browner and lack black cap. In flight, show white small carpal bar, white rear edge to wings, and pale rump. Eyes pale reddish-brown; bill yellow with black tip; tarsi pink-brown.

Habitat and Behavior　Inhabits mixed and deciduous forests, forest edges and woodlands. Also in orchards, wooded farms and parks in winter. Forms small flocks, gathers to feed on pine trees. Active, frequently transfer from one tree to another. Flight fast and sligthy wave-like. Feed on plants mostly, also some insects.

Distribution　E Siberia, Korean Peninsula, S Japan and Eastern China.Wintering to Southern China. At Luofu and Huang'ao in Mt. Jinggang.

Status of Population　Common in autumn and winter in Mt. Jinggang.

雀形目 PASSERIFORMES
鹀科 Emberizidae

3.249 凤头鹀 *Emberiza lathami* Gray, JE, 1831

识别特征　体型较大、具显著细长羽冠的鹀，约 17 cm。雄鸟除两翼及尾栗色外、通体灰黑色。雌鸟羽冠较雄鸟短、上体深橄榄褐色、上背及胸满布纵纹、翼羽色深具栗色羽缘。虹膜深褐；上嘴近黑色，下嘴基粉红；脚肉褐色。

生境与习性　栖息于山麓的耕地和岩石斜坡上，也见于市区和乡村。常单个或成对生活，秋冬季也结群生活。活动取食均多在地面、活泼易见。主要以植物为食，也吃少量昆虫和蠕虫。

分布　印度、喜马拉雅至中国（东南）及中南半岛北部。在井冈山见于大井等地。

种群状况　在井冈山迁徙季常见。

Identification　L ca. 17 cm. Large dark bunting with unique prominent pointed crest. Male black with rufous-brown wings and tail. Female crest shorter, upperparts dark olive-brown with blackish streaks on mantle, chestnut fringes to wing feathers; underparts buff with darkest on steaked chest. Eyes dark brown; bill dusky above, pink at base below; tarsi dull brownish-pink.

Habitat and Behavior　Inhabits farmland, dry grassy and scrubby areas on hills, also seen in urban and rural areas. Usually solitary or in pairs, sometimes gregarious in winter. Active, often forages on ground. Mainly granivorous, also feed on insects and worms.

Distribution　India, Himalaya to SE China and N Indo-China Peninsula. At Dajing in Mt. Jinggang.

Status of Population　Common on migration in Mt. Jinggang.

3.248b

3.249a

3.249b

3.250 蓝鹀 *Emberiza siemsseni* (Martens, GH, 1906)

识别特征 体型较小的鹀,约13 cm。雄鸟通体蓝灰色,仅下腹部、臀及外侧尾羽白色。雌鸟头、上胸红棕色,上体暗褐色具深色纵纹,具两道锈色翼斑和浅色飞羽羽缘,腰灰色,腹部、臀及外侧尾羽白色。虹膜深褐;嘴深灰色;脚粉褐色。

生境与习性 栖息于中低山的次生阔叶林、竹林、针阔混交林和针叶林,非繁殖季节多见于山麓平坝、沟谷和林缘地带,也见于村落附近的灌丛和竹林。常单独、成对或成3-5只的小群,在地上、电线上或山边岩石和幼树上活动、觅食。主要以草籽、种子等植物性食物为食,也吃昆虫等动物性食物。

分布 中国(中部、东南)特有种。在井冈山地区见于茅坪、武功山等地。

种群状况 在井冈山地区迁徙季可见。

Identification L ca. 13 cm. Smallish bunting. Male slaty-blue with white belly, vent and outer tail-feathers. Female has warm rufous-brown hood, dark brown upperparts with heavily blackish streaks on mantle, blackish-brown wings with double wing bar and pale fringes to flight-feathers; rump grey, belly, vent and outer tail feathers white. Eyes dark brown; bill dark grey; tarsi brownish-pink.

Habitat and Behavior Inhabits subtropical valley forests on hills and mountains, favors vicinity of bamboo thickets in secondary forest and scrubby cover of degraded forest. Descends to foothills in winter. Usually solitary or in pairs, sometimes in small groups. Forages on ground. Mainly granivorous, also feeds on small invertebrates.

Distribution Endemic to Central and SE China. At Maoping and Wugongshan in the Jinggangshan Region.

Status of Population Visible on migration in the Jinggangshan Region.

3.251 三道眉草鹀 *Emberiza cioides* (von Brandt, JF, 1843)

识别特征 体型略大的红棕色鹀,约16 cm。雄鸟头部图案显著,具宽阔的白色眉纹、深色眼先和耳羽、白色髭纹和黑色下颊纹,灰色的颈环汇入白色的喉部;头顶至腰红棕色,上背有深色纵纹,尾羽黑色和褐色,外侧尾羽羽缘白色;上胸棕色,渐变为腹部的白色。雌鸟羽色较淡,眉纹及喉皮黄色,下体较暗淡。虹膜黑色;嘴深灰色;脚粉褐。

生境与习性 栖息于低山丘陵和平原的开阔灌丛及林缘地带,尤喜农田、道旁的小树林和灌丛。繁殖期单独或成对活动,雏鸟离巢后多以家族群方式生活,冬季集结成小群。喜停栖在灌木、幼树等高处、电线或电杆上。性胆怯,一见有人便立刻停止鸣叫,或远飞或快速藏匿。繁殖季主要以昆虫为食,非繁殖季主要以草籽、谷粒和嫩叶等植物性食物为食。常营巢于林缘、林下、路边灌丛或草丛中,也见于枝叶茂密的小松树和灌木枝杈上。窝卵数3-6枚,多为4-5枚。孵卵由雌鸟担任,孵化期12-13天。

分布 西伯利亚南部、蒙古、中国北部及东部,东至日本。在井冈山见于茨坪、龙市、黄坳等地。

3.250a

3.250b

3.251a

种群状况　在井冈山全年常见。

Identification　L ca. 16 cm. Large rufous-brown bunting. Male has distinctive pattern on head, bold white supercilium, black lores and ear coverts, white malar and black lateral throat-stripe, grey collar runs to white throat; rufous-brown from crown to rump, black streaks on mantle and back; tail blackish and rufous-brown with white outer fringes to outer feathers. Upper breast orange-brown, then plain dark rufous-brown grading to off-white on belly. Female sandy-brown, supercilium and throat buff, underparts duller than male. Eyes black; bill dark grey; tarsi brownish-pink.

Habitat and Behavior　Inhabits open wooded areas, thickets, cleared forests and agricultural areas in lowlands and low hills. Solitary or in pairs in breeding season, gregarious in winter. Usually perches on poles, wires, tree tops or bushes. Alert, stops singing when sees people, and flies away or hides. Polyphagia. Builds nest in grasses, scrubs or small pines. Clutch size is 3-6 eggs, mostly 4-5. Incubation period is 12-13 by female.

Distribution　S Siberia, Mongolia, Northern and Eastern China to Japan. At Ciping, Longshi and Huang'ao in Mt. Jinggang.

Status of Population　Common all year round in Mt. Jinggang.

3.251b

3.252a

3.252 白眉鹀 *Emberiza tristrami* Swinhoe, 1870

识别特征　中等体型、头具显著条纹的鹀，约 15 cm。繁殖期雄鸟头和喉黑，具白色的顶冠纹、眉纹、髭纹，耳羽后方有一白点；上体灰棕色具深色纵纹，飞羽、尾上覆羽及尾栗红色；下体白色，胸及两胁染暗棕色并具深色纵纹。非繁殖期雄鸟头为灰棕色。雌鸟图纹似繁殖期雄鸟，但较暗淡。虹膜黑色；上嘴深灰色，下嘴偏粉色；脚粉色。

生境与习性　栖息于中低山的阔叶林、针阔混交林和针叶林带，尤喜山溪沟谷、林缘、林间空地和林下灌丛或草丛。仅在迁徙时集结成小群。性胆怯，多在林下灌丛和草丛中活动和觅食，一见有人走过就立刻起飞隐藏于较远的树间或草下。飞行颇快而成直线。主要以草籽等植物性食物为食，也吃昆虫和昆虫幼虫等动物性食物。

分布　中国（东北）及西伯利亚的邻近地区。越冬至中国南部，偶尔在缅甸北部及越南北部有见。在井冈山见于湘洲、大井等地。

种群状况　在井冈山秋冬季常见。

3.252b

Identification　L ca. 15 cm. Smallish bunting with strong facial patterns. Breeding male head black with white crownstripe, supercilium and malar, and white spot on rear ear coverts. Rest of upperparts greyish-brown with dark streaking, flight feathers, uppertail coverts and tail rich chestnut. Throat black, underparts white with dull brown wash on breast and flanks, dark streaked. Non-breeding male head greyish-brown. Females similar to males but less strongly marked. Eyes black; bill deep grey above, and pink below; tarsi deep pink.

Habitat and Behavior　Inhabits shady areas of mixed taiga-forest, particularly beneath stands of fir. Prefers more wooded areas than other buntings in winter. Gregarious on migration. Shy and nervous, forages on forest floor, immediately flies away when people get

3.252c

close. Polyphagia.

Distribution　NE China and adjacent Siberia. Wintering to Southern China and occasionally N Myanmar and N Vietnam. At Xiangzhou and Dajing in Mt. Jinggang.

Status of Population　Common in autumn and winter in Mt. Jinggang.

3.253

3.253 栗耳鹀 *Emberiza fucata* Pallas, 1776

识别特征　体型略大、具显著的栗色耳羽的鹀，约 16 cm。繁殖期雄鸟头顶、枕部灰色，与栗色耳羽、白色髭纹和喉形成对比，黑色下颊纹下延至胸部与黑色纵纹形成的项纹相接；上体灰褐色具深色纵纹，腰红棕色，尾深褐色；下体浅色，具黑、白及栗色胸带，两胁染棕色。雌鸟与非繁殖期雄鸟相似，但色彩较淡而少特征。虹膜深褐；上嘴灰色，下嘴粉灰色；脚粉红。

生境与习性　栖息于低山、丘陵、平原、河谷、沼泽等开阔地带，尤以有稀疏灌木的林缘沼泽草地较常见，也出现在田边、地头和附近的草地灌丛中。繁殖期成对生活，迁徙时结群飞行，并常和其他鹀类混群，但在越冬地多分散地单个活动。性大胆，除非极其接近时才飞离。在地上或草丛与灌丛中觅食。繁殖期主要以昆虫及其幼虫为食，非繁殖期主要以草籽、灌木果实、谷粒等植物性食物为食。

分布　喜马拉雅西段至中国、蒙古东部及西伯利亚东部。越冬至朝鲜半岛、日本南部及中南半岛北部。在井冈山地区见于七溪岭、宁冈、南风面等地。

种群状况　在井冈山地区秋冬季常见。

Identification　L ca. 16 cm. Largish bunting with distinctive chestnut ear-patch. Breeding male crown and nape grey, contrast with chestnut ear coverts, white malar and throat, black lateral throat-stripe extending as necklace of black streaks. Upperparts grey-brown with black streaking, rump rufous-brown, tail blackish-brown. Underparts pale with black, white and chestnut breast-bands, and rufous wash on flanks. Female duller than male. Eyes black; bill grey above, and pinkish-grey below; tarsi brownish-pink.

Habitat and Behavior　Inhabits open grassy habitat with thickets, including rank meadows and wetland fringes; winters in open agricultural land. Inconspicuous and bold, somewhat gregarious, flight described as typically slow and hesitating. Feeds on or near the ground for insects or fallen grass seeds.

Distribution　W Himalaya to China, E Mongolia and E Siberia. Wintering to Korean Peninsula, S Japan and N Indo-China Peninsula. At Qixiling, Ninggang and Nanfengmian in the Jinggangshan Region.

Status of Population　Common in autumn and winter in the Jinggangshan Region.

3.254a

3.254 小鹀 *Emberiza pusilla* Pallas, 1776

识别特征　体小、头部栗色的鹀，约 13 cm。雄鸟头顶、眼先、颊部栗红色，具黑色侧冠纹，耳羽后缘镶黑色；上体灰褐色具黑色纵纹；下体偏白，胸及两胁具浓密黑色纵纹。雌鸟头部暗棕色，侧冠纹不甚清晰。虹膜黑色，具狭窄白色眼圈；嘴深灰色；脚肉褐色。

3.254b

生境与习性　非繁殖季栖息于低山、丘陵和山脚平原地带的灌木丛、村边树林与草地、苗圃、农田中。多结群生活，分散活动于地上。频繁在草丛和灌木间穿梭跳跃，也栖于小树的枝上，见人立刻落下藏匿。主要以草籽、种子、果实等植物性食物为食，也吃昆虫等动物性食物。

分布　繁殖在欧洲极北部及北亚。冬季南迁至印度东北部、中国及东南亚。在井冈山广泛分布。

种群状况　在井冈山秋冬季常见。

Identification　L ca. 13 cm. Small bunting with distinctly chestnut face. Male has chestnut crown, face and lores with black lateral crown-stripes, postorbital line wrapping around rear edge of ear coverts; upperparts greyish-brown with black streaking; underparts white with heavily black streaking on breast and flanks. Female duller on head and lateral crown-stripes less clear. Eyes black with narrow white eyering; bill grey; tarsi dull pink.

Habitat and Behavior　Inhabits woodland openings, gardens, bracken, especially grassy spots and cultivated areas. Gregarious, feeds mostly on the ground, flies into trees when disturbed. Molts before southward migration. Feeds on grass seeds and fruits mostly, also some insects.

Distribution　Breeds in extreme N Europe and N Asia. Migrating in winter to NE India, China and SE Asia. Widespread in Mt. Jinggang.

Status of Population　Common in autumn and winter in Mt. Jinggang.

3.255 黄眉鹀 *Emberiza chrysophrys* Pallas, 1776

识别特征　中等体型、略具羽冠的鹀，约 15 cm。雄鸟似白眉鹀但眉纹前半部黄色，喉白色，下体更白而多纵纹，翼斑也更白，腰更显斑驳。黄眉鹀的黑色下颊纹比白眉鹀明显，并分散融入胸部纵纹中。雌鸟和幼鸟颊部褐色，而非黑色。虹膜深褐；嘴粉色，嘴峰及下嘴端灰色；脚粉红。

生境与习性　非繁殖季栖息于低山丘陵和山脚平原地带的混交林和阔叶林中，尤其于林间路边和溪流沿岸较常见，也见于少树的灌丛草地和农田地带。一般成对或成小群活动，有时亦与其他鹀类混群。在地上草丛或灌木低枝上活动和觅食。性胆怯，善藏匿。主要以草籽等植物性食物为食，也吃少量昆虫。

分布　繁殖于俄罗斯贝加尔湖以北。越冬在中国南部。在井冈山见于大井、罗浮、西坪等地。

种群状况　在井冈山秋冬季可见。

Identification　L ca. 15 cm. Mid-sized bunting with slight crest. Breeding male similar to Tristram's Bunting, but long supercilium yellow at fore and white at rear, wing bar white, underparts paler with more streaks, rump mottled. Black lateral throat-stripe (broader than Tristram's) dispersedly merge into streaks on breast. Female and juvenile are brownish on cheeks, not black. Eyes dark brown; bill pink with grey culmem and tip; tarsi deep pink.

Habitat and Behavior　Inhabits mixed forests, forest edges and woodland, also well-wooded fringes of farmland. Usually in pairs or small groups, sometimes forages with other buntings. Forages in bushes or on ground. Shy, hides immediately when people getting close. Feeds on grass seeds mostly, also some insects.

Distribution　Breeds in Russia (north of Lake Baikal). Winters in Southern China. At Dajing, Luofu and Xiping in Mt. Jinggang.

Status of Population　Visible in autumn and winter in Mt. Jinggang.

3.256 田鹀 *Emberiza rustica* Pallas, 1776

识别特征　体型略小、略具羽冠的鹀，约 14.5 cm。繁殖期雄鸟头黑色具白色眉纹和髭纹，及深色下颊纹，喉白色；颈背、胸带、两胁纵纹及腰棕色。雌鸟与非繁殖期雄鸟相似，头褐色，眉纹和髭纹皮黄色，褐色耳羽镶黑色边缘，其后部通常具一近白色点斑。似小鹀，但腰红褐色，头部无栗色标记。虹膜

深栗褐；嘴深灰，基部粉灰；脚偏粉色。

生境与习性 栖息于平原杂木林、人工林、灌木林和沼泽草甸中，也见于低山区和山麓及开阔田野中。繁殖期间多成对或单独活动，非繁殖期间集结成群，但在越冬地多在草丛、农田中分散或单独活动。性大胆，不甚畏人。在地面取食，主要以各种杂草种子、植物嫩芽、灌木浆果等植物性食物为食，也吃昆虫、蜘蛛等无脊椎动物。

分布 繁殖于欧亚大陆北部的泰加林。越冬至中国。在井冈山见于中井等地。

种群状况 在井冈山迁徙季可见。

Identification L ca. 14.5 cm. Smallish bunting with distinct crest and striking head pattern. Breeding male has black head, with white supercilium from above eye to nape, white malar and dark lateral throat-stripe; hind neck, breast band, streaks on flanks, and rump chestnut. Female and non-breeding male head brown, lacking black, but does have black outline to brownish ear coverts. Resembles Little Bunting, but rump rufous, and lack chestnut on head. Eyes dark brown; bill dark grey, pink grey at base; tarsi dull pink.

Habitat and Behavior Winter inhabits dry lowland woodland, forest on lower slopes, riverine scrub and margins of agricultural land with rank vegetation. Forages in loose groups in winter. Does not fear people. Feeds on weed seeds, berries, tender shoot, and insects, spiders.

Distribution Breeds in N Eurasian taiga. Wintering to China. At Zhongjing in Mt. Jinggang.

Status of Population Visible on migration in Mt. Jinggang.

3.257 黄喉鹀 *Emberiza elegans* Temminck, 1836

识别特征 中等体型、羽冠显著的鹀，约15 cm。雄鸟头具黑色脸罩，黄色的羽冠前缘黑色，颏及喉黄色；后颈环灰色，并具黑色胸兜。雌鸟似雄鸟但色暗，褐色取代黑色，皮黄色取代黄色，无黑色胸兜。虹膜深栗褐；嘴近黑，下嘴基部较浅；脚浅粉褐。

生境与习性 栖息于低山丘陵的次生林、阔叶林、针阔叶混交林的林缘灌丛中，尤喜河谷与溪流沿岸疏林灌丛，也见于有树的山边草坡和农田、道旁及居民点附近的小块次生林内。单独或成小群活动。性胆小，见人即藏匿或飞走。多沿地面低空飞行，觅食亦多在林下灌丛、草丛或地上，有时也到乔木树冠层间觅食。以昆虫和昆虫幼虫为食，也吃植物性食物。

分布 分布不连贯，中国（中部和东北）、朝鲜半岛及西伯利亚东南部。在井冈山见于长谷岭、下庄等地。

种群状况 在井冈山秋冬季可见。

Identification L ca. 15 cm. Mid-sized, crested bunting with bold facial patterns. Breeding males black masked, contrasting with yellow crest black at fore, and yellow throat. Hind necklace grey, and breast patch black. Rest of upperparts and wings chestnut with dark streaks. Female resembles male, but duller; brown instead of black, buff instead of yellow, and lacks breast patch. Eyes dark red-brown; bill dark grey, pink at base of below; tarsi pink-brown.

3.256a

3.256b

3.257a

3.257b

Habitat and Behavior Inhabits open deciduous woodland, town parks, and woodland and forest edges with tall grasses, bordering overgrown agricultural land, often on low hillsides. Singly or in small groups. Shy, hides or flies away immediately when people getting close. Forages on ground, or in bushes and grasses, also up to canopy of trees. Feeds on insects and some seeds.

Distribution Disjunct distribution across Central and NE China, Korean Peninsula and SE Siberia. At Changguling and Xiazhuang in Mt. Jinggang.

Status of Population Visible in autumn and winter in Mt. Jinggang.

3.258a

3.258 黄胸鹀 *Emberiza aureola* Pallas, 1773

识别特征 中等体型、色彩鲜艳的鹀，约 15 cm。繁殖期雄鸟顶冠及颈背栗色，脸及喉黑，栗色胸带将黄色的半颈环与黄色的胸腹部间隔开，翼具较大的白色斑块及狭窄的白色翼斑。非繁殖期雄鸟色彩暗淡，灰褐色的耳羽镶黑边，颏及喉黄色，无栗色胸带。雌鸟及幼鸟上体浅褐色，眉纹浅皮黄色，上背具显著纵纹，下体浅黄。虹膜深褐；上嘴灰色至黑褐，下嘴粉褐；脚淡褐。

生境与习性 栖息于低山丘陵和开阔平原地带的灌丛、草甸、草地和林缘地带，尤喜溪流、湖泊和沼泽附近的灌丛、草地，也见于田间地头。非繁殖期成群活动。白天在地上、草茎或灌木枝上活动和觅食，晚上栖于草丛中。繁殖期主要以动物性食物为主，非繁殖期则主要以植物性食物为主。

分布 繁殖于西伯利亚至中国（东北）。越冬至中国南部及东南亚。在井冈山见于大井等地。

种群状况 在井冈山迁徙季偶见。IUCN红色名录（2014）：濒危。

3.258b

Identification L ca. 15 cm. Mid-sized colorful bunting. Breeding male has chestnut crown and nape, black cheek and throat, yellow half-collar separated with yellow breast and belly by chestnut breast band, carpal with marked white bars. Non-breeding male duller, grey-brown ear coverts bordered narrow black, chin and throat yellow, lack breast band. Female and juvenile has pale supercilium, heavily streaked mantle and flanks, underparts pale yellow. Eyes dark brown; bill grey to dark brown upper, pinkish-brown below; tarsi pale brown.

Habitat and Behavior Inhabits open scrub, wet grasslands and farmland, including dry paddy fields; mostly in lowlands. In groups in winter. Forages in grasses, bushes, or on ground by day, rests in grasses at night. Phytophagous when non-breeding.

Distribution Breeds in Siberia to NE China. Wintering to Southern China and SE Asia. At Dajing in Mt. Jinggang.

Status of Population Rare on migration in Mt. Jinggang. IUCN Red List (2014): EN.

3.259 栗鹀 *Emberiza rutila* Pallas, 1776

识别特征 体型略小、色彩鲜艳的鹀，约 15 cm。繁殖期雄鸟头、上体及胸栗色而腹部黄色。非繁殖期雄鸟头、胸及上背染黄色，有时有一条栗色胸带。雌鸟无甚特色，顶冠、上背、胸及两胁

3.259a

具深色纵纹，腰棕色，无白色翼斑或尾部白色边缘。虹膜深褐；嘴灰褐色；脚淡肉褐色。

生境与习性 栖息于较为开阔的稀疏森林中，也见于农田地边的灌丛草地。非繁殖季多成小群活动。鸣叫时多停在树顶或枝梢上。主要以植物性食物为主，兼食昆虫等。

分布 繁殖于西伯利亚南部至蒙古北部。越冬至中国南部及东南亚。在井冈山见于大井、黄坳、湘洲等地。

种群状况 在井冈山秋冬季常见。

Identification L ca. 15 cm. Smallish colorful bunting. Breeding male hood and upperparts chestnut, and underparts bright yellow. Non-breeding male duller and flecked with yellow. Female upperparts, breast and flanks streaked, rump rufous-brown, lacks white wing bar and white outer rectrices. Eyes dark brown; bill grey-brown; tarsi fresh-brown.

Habitat and Behavior Inhabits open forests, and dry agricultural land with trees or shrubs. Usually in small groups, but large flocks recorded during migration. Phytophagous mainly, also feeds on insects.

Distribution Breeding in S Siberia to N Mongolia. Wintering to Southern China and SE Asia. At Dajing, Huang'ao and Xiangzhou in Mt. Jinggang.

Status of Population Common in autumn and winter in Mt. Jinggang.

3.259b

3.260 灰头鹀 *Emberiza spodocephala* Pallas, 1776

识别特征 体小的褐色鹀，约14 cm。指名亚种繁殖期雄鸟头、颈背及喉灰色，眼先及颏黑；上体余部栗色而具显著黑色纵纹，具两道白色翼斑，外侧尾羽白色；下体浅黄或近白，胁部具深褐色纵纹。雌鸟及冬季雄鸟头橄榄色，具浅黄色眉纹、下颊纹和喉部，耳羽深色，上背和胸、胁多具纵纹。各亚种羽色变化较多。虹膜深栗褐；上嘴近黑并具浅色边缘，下嘴偏粉色且嘴端深色；脚粉褐。

生境与习性 栖息于山区河谷溪流两岸、平原沼泽地的疏林和灌丛中，也见于山边杂林、草甸灌丛、山间耕地及公园、苗圃和果园中。非繁殖期常成小群活动。不断弹尾以显露白色的外侧尾羽。繁殖期主要以昆虫和昆虫幼虫等动物性食物为主，非繁殖期主要以草籽、谷粒等为食。

分布 繁殖于西伯利亚、日本、中国（东北及中部、西部）。越冬至中国南方。在井冈山广泛分布。

种群状况 在井冈山秋冬季常见。

3.260a

Identification L ca. 14 cm. Smallish brown bunting. Male has olive-grey hood with black lores and chin, upperparts brown with dark streaks, two wing bars, and white outer rectrices; underparts pale yellow with dark brown streaks on flanks. Female has pale yellowish supercilium, submoustachial and throat, dark ear coverts, dark streaks on olive breast-band and flanks, rest of underparts off-white. Subspecies varies. Eyes dark brown; bill greyish-pink with grey tip and culmen; tarsi dull brownish-pink.

3.260b

Habitat and Behavior Inhabits forest, woodland edges, scrub, parks, gardens and agricultural land, usually inconspicuous, staying near cover, but entering fields for water grain. Calls and nervous behavior, continually twitches tail revealing white outer feathers. Phytophagous in winter.

Distribution Breeds in Siberia, Japan and NE, Central and Weastern China. Winters to S China. Widespread in Mt. Jinggang.

Status of Population Common in autumn and winter in Mt. Jinggang.

3.260c

第 4 章
井冈山地区哺乳动物区系

Chapter 4

Mammal Fauna of the Jinggangshan Region

基于 2010–2015 年在井冈山地区开展的数次科学考察，并查看井冈山保护区标本馆的标本及已发表的文献资料，共整理记录了井冈山地区哺乳动物 8 目 20 科 54 属 70 种，包括近期发表的 5 种江西省新纪录，6 种中国特有种；1 种国家 I 级重点保护野生动物，14 种国家 II 级重点保护野生动物；列入 CITES 附录 I 7 种，列入 CITES 附录 II 4 种，列入 CITES 附录 III 2 种；IUCN 濒危（EN）等级 1 种，IUCN 红色名录易危 (VU) 等级 6 种。

本书共收录井冈山地区哺乳动物 37 种。

We recorded 70 mammal species belonging to 54 genera, 20 families and eight orders based on continual field surveys conducted from 2010 to 2015 in the Jinggangshan Region, examined specimens from the Museum of Jinggangshan National Nature Reserve and published species records. One of which is listed in China Key List: I, 14 species are listed in China Key List: II, five species were first recorded in Jiangxi, six species are endemic to China; seven are listed in CITES App. I, four in CITES App. II, two in CITES App. III; and one is listed on IUCN Red List as Endangered (EN) and six species as Vulnerable (VU).

We selected 37 mammal species from the Jinggangshan Region for this atlas.

鼩形目 SORICOMORPHA
鼩鼱科 Soricidae

4.1 臭鼩 *Suncus murinus* (Linnaeus, 1766)

识别特征 体型似鼠，但吻部较尖长且具较多口须。体重可达 75 g；头躯长 110-150 mm，尾长 60-90 mm。背毛灰黑色，有银灰色光泽；腹部为淡灰色。尾基部特别粗大，刚毛明显。体侧中部有一腺体，能分泌奇臭的黄色液体。

生境与习性 栖于井冈山地区农田和城镇草地等环境，也进入村落民房，尤其出没于厨房、阴沟、墙角、柴堆及杂物等处。夜间较活跃，受惊后发出"唧唧"的尖叫，并放出臭液。捕食各种小动物，如蝼蛄、蚱蜢、蟑螂、金龟子和鱼虾等，偶见捕食小型鼠类、植物种子和块根。

分布 分布于中国长江以南各省区、江西、贵州、云南、广东、广西、福建、浙江、海南、台湾、香港和澳门。

种群状况 常见种。

Identification Mouse-like, snout long and pointed, with profuse vibrissae; weight up to 75 g, head-body length 110-150 mm, tail length 60-90 mm. Grey-black with silver-grey shine to light grey on the belly hair; thick tail with obvious bristle. Secreting smelly and yellow liquid through glands on the middle of body sides.

Habitat and Behavior Inhabits farmland and grassland in rural and urban areas in the Jinggangshan Region, sometimes appears in village houses, especially in the kitchen, sewers, corner of walls, ricks and sundries. Quite active at night, squeaking and emitting smelly liquid when frightened. Preys on many animals, such as mole crickets, grasshoppers, cockroaches, beetles, fish and shrimp. Occasionally feeds on small rodents, plant seeds and roots.

Distribution South side provinces of the Yangtze River in China including Jiangxi, Guizhou, Yunnan, Guangdong, Guangxi, Fujian, Zhejiang, Hainan, Taiwan, Hong Kong and Macau.

Status of population Common species.

4.2 灰麝鼩 *Crocidura attenuata* Milne-Edwards, 1872

识别特征 体型较小，体重约 10 g。吻端尖长，吻须发达，向后拉超过耳基部；眼小，耳郭明显且露于毛被之外。背毛暗灰褐色，略带金属光泽；腹毛暗灰色，尖端染浅棕色，毛基部暗灰色；尾背腹面均被有稀疏长毛。

生境与习性 栖于井冈山大坝里茶场等山地林区，喜在山溪河谷两岸灌木丛中活动，草丛中也有发现。捕食森林害虫，对林业有益。

分布 西藏、云南、四川、贵州、安徽、浙江、江苏、江西、福建、湖北、湖南、广东、广西、海南和台湾等。

种群状况 常见种。

Identification Small, approximate 10 g of weight. Long and pointed snout with developed beard which reaching base of the ear; eyes small; obvious auricle sticking out from hair. Dorsal pelage dark grey-brown, with metallic luster, ventral hair dark-grey and the

4.1

4.2

tip light brown, basal part dark-grey, both sides of the tail have long and sparse hairs.

Habitat and Behavior Roosts in forest zones in tea plantations of Dabali in the Jinggangshan Region, prefers bushes by side of mountain steams, also found in grassland. Preys on forest pests, beneficial to forestry.

Distribution Xizang, Yunnan, Sichuan, Guizhou, Anhui, Zhejiang, Jiangsu, Jiangxi, Fujian, Hubei, Hunan, Guangdong, Guangxi, Hainan and Taiwan.

Status of population Common species.

鼩形目 SORICOMORPHA
鼹科 Talpidae

4.3 长吻鼹 *Euroscaptor longirostris* (Milne-Edwards, 1870)

识别特征 体中等大小，体重约 30 g。吻尖而裸露无毛，吻背面中央具凹槽。眼小，无外耳郭，外耳道隐于被毛中。四肢粗短，前足掌宽厚并向外翻转，爪又粗又长，适于地下挖土打洞（故俗名地拱子、鼹鼠）。尾短，约等于后足长度。背毛棕褐或暗褐色，腹毛比背毛稍浅。

生境与习性 栖于井冈山朱砂冲等山地、河谷和山间盆地中的次生阔叶林、稀树灌丛、林缘草地及农耕地（如行洲村苗圃）。营地下生活，多选择腐殖质丰富、土层松软地方的栖息。主要以地下土壤中昆虫幼虫和蠕虫等为食。

分布 江西、四川、云南、陕西、青海、广西、广东、福建等。

种群状况 中国物种红色名录：易危。

Identification Medium size, nearly 30 g; pointed snout without hair, middle part of dorsal snout has grooves eye small; external ears absent, the external auditory canal hidden in hair; outward-turned forefeet with thick and long sharp claw for underground digging (also called ground arches, the mole); short tail, equal to hind feet. Dorsal seta brown or dark-brown, darker than ventral hair.

Habitat and Behavior Inhabits secondary broad-leaved forest, bushes, grassland and farmland (e.g. plant nursery in Xingzhou village) between river valleys and intermountain basins near Zhushachong Forest Farm in Mt. Jinggang. Underground life habits, prefers rich humus and soft soil. Mainly feeds on insect larvae and worms in the soil.

Distribution Jiangxi, Sichuan, Yunnan, Shaanxi, Qinghai, Guangxi, Guangdong and Fujian.

Status of population China Species Red List: VU.

翼手目 CHIROPTERA
菊头蝠科 Rhinolophidae

4.4 中华菊头蝠 *Rhinolophus sinicus* K. Andersen, 1905

识别特征 前臂长 45-52 mm，颅全长 19-23 mm。头顶具有复杂鼻叶，其中马蹄叶较大，两侧下缘各具一片附小叶，鞍状叶左右两侧呈平行状，顶端圆，连接叶阔而圆。背毛毛尖栗色，毛基灰白色，腹毛赭褐色。头骨矢状嵴明显，颚桥短。

生境与习性 常栖于自然岩洞、废弃防空洞、坑道、窑洞等，在井冈山水库大坝坑道等地有发现。可集成上百只的群体，可见与皮氏菊头蝠等同栖一洞。特别喜食蚊类。粪便可作肥料。秋末冬初交配，翌年初夏产仔，每年 1 仔，幼仔生长到次年秋达性成熟。

分布 国内分布长江以南地区及陕西、甘肃、西藏。

种群状况 常见种。

Identification Forearm length 45-52 mm, greatest skull length 19-23 mm; noseleaves complex; larger horseshoe leaf, on each side with a piece of supplementary leaflet below; lateral margins of sella in parallel and with round top, broad and rounded connecting process. Tip of dorsal hair maroon, grey hair base, ochre abdominal hairs. Well-developed sagittal crest and short jaw bridge.

Habitat and Behavior Roost in natural grottos, abandoned air-raid shelter, tunnels and caves; has been found in the tunnel at Mt. Jinggang Reservoir Dam. Colony size can be hundreds of

4.3

4.4a

4.4b

4.4c

individuals, sometimes coexist with *Rhinolophus pearsoni*. Prey preferences are mosquitoes, feces can be used for fertilizer. Mating in late autumn or early winter, females give birth to one infant next early summer, juveniles reach sexual maturity following autumn.

Distribution South of the Yangtze River and Shaanxi, Gansu, Xizang.

Status of Population Common species.

4.5 中菊头蝠 *Rhinolophus affinis* Horsfield, 1823

识别特征 体中等大小，前臂长 47-54 mm，颅全长 22-24 mm。具有复杂的鼻叶，其中鞍状叶两侧缘凹入，前面观略似提琴形；连接叶较低圆。第3、第4、第5指掌骨近等长。第3下前白齿小，位于齿列之外。体背毛深暗褐色，腹毛淡。尾短，与股间膜近平行。

生境与习性 常见栖息在潮湿的山洞和废弃矿井、坑道等地，曾在井冈山石溪村观音庙大峡谷捕获，可与大蹄蝠、小菊头蝠、皮氏菊头蝠、中华鼠耳蝠等同穴共栖。以蚊类、蛾类等昆虫为食。秋季交配，翌年5-6月产仔，每胎1仔。具有冬眠习性。

分布 较广泛，在长江以南各省区均见分布。

种群状况 常见种。

Identification Medium size, forearm length 47-54 mm, greatest length of skull 22-24 mm; noseleaf structure complex, the sella concaves on the edges of both sides, fiddle-like from the front view, connecting process lower but round. The third, fourth and fifth metacarpal lengths subequal; the third lower premolar small and lies outside the dentition. Dorsal hair dark-brown, darker than abdominal hair. Short tail, and approximately parallel with tail membrane.

Habitat and Behavior Inhabits wet caves and abandoned mines, tunnels and other places, have been collected in the Grand Canyon near the Buddhist Temple at Shixi village in Mt. Jinggang. Roosts with *Hipposideros armiger*, *Rhinolophus pusillus*, *R. pearsoni*, *Myotis chinensis*. Feeds on insects such as mosquitoes and moths. Mating in autumn, females give birth to off spring next May or June. Hibernates.

Distribution Provinces south of the Yangtze River.

Status of population Common species.

4.6 小菊头蝠 *Rhinolophus pusillus* Temminck, 1834

识别特征 体型小，前臂长约37 mm。具复杂的鼻叶，其中马蹄叶两侧的小副叶退化，鞍状叶上窄下宽，与连接叶之间具明显凹陷；连接叶呈尖三角形，高出鞍状叶顶端。前颌骨为软骨，鼻骨突起，上颌第一前白齿位于齿列中。体背毛呈茶褐色，毛基部灰白色；腹毛肉桂色。

生境与习性 栖于湿度较大的岩洞、坑道（如井冈冲水库附近）和防空洞内。有冬眠习性。捕食蚊、蚋等小飞虫。

分布 安徽、福建、广西、贵州、海南、河北、湖南、江苏、江西、陕西、四川、浙江、广东、云南、西藏及香港。

种群状况 常见种。

Identification Small size, forearm length 37 mm; complex noseleaf, supplementary leaflet below the horseshoe degraded, the bottom of sella wider than the top, with obvious sunken connected with connecting process; triangular connecting process, higher than the top of sella leaf. Premaxilla is cartilage, and nasal bone sticking out, the first premolar in upper jaw lies in the tooth row. Dorsal hair dark-brown, grey bases, abdominal pelage cinnamon.

Habitat and Behavior Lives in high humidity caves, tunnels (such as Jinggangchong Reservoir) and air-raid shelters. Hibernates, feeds on mosquitoes and blackfly among other insects.

Distribution Anhui, Fujian, Guangxi, Guizhou, Hainan, Hebei, Hunan, Jiangsu, Jiangxi, Shaanxi, Sichuan, Zhejiang, Guangdong, Yunnan, Xizang and Hong Kong.

Status of population Common species.

4.7 皮氏菊头蝠 *Rhinolophus pearsoni* Horsfield, 1851

识别特征　体型中等，前臂长 51-60 mm。具有复杂鼻叶，其中马蹄叶宽大，覆盖上唇；鞍状叶两侧近中部各有一凹刻，形成上部窄而基部宽；连接叶自鞍状叶顶端起呈圆弧形向下延伸，其后缘明显低于鞍状叶顶端。头骨颧宽略大于后头宽，矢状嵴较高。上颌第一前臼齿甚小，位于齿列中。体毛长而柔密，背毛暗褐色或棕褐色，腹毛稍淡。

生境与习性　栖息于潮湿岩洞（如井冈山红军游击洞）或人工洞（如水库大坝隧道）。集数只或十余只的小群，但与同洞中其他种蝙蝠不混群；有冬眠习性。食虫，捕食害虫对人类有益。

分布　江西、安徽、福建、广东、广西、贵州、湖北、湖南、江西、浙江、陕西、四川、西藏和云南等。

种群状况　较常见。

Identification　Medium size, forearm length 51-60 mm, with complex noseleaf, large horseshoe covering upper lip; sella sunken near the middle of both sides, and the upper narrower than the base; connecting process in cylindrical arc shape, extending downward from top of sella, rear edge is obviously lower than sella tip. Zygomatic width slightly greater than mastoid width, high sagittal crest. The first upper premolar very small. Long and soft hair, dark-brown or brown, abdominal fur paler.

Habitat and Behavior　Inhabits humid caves (such as Mt. Jinggang Red Army guerrilla hole) or artificial holes (such as dam tunnel). Does not mix with other species in the same cave, and in small colony of a dozen. Hibernates. Preys on harmful insects; beneficial to humans.

Distribution　Jiangxi, Anhui, Fujian, Guangdong, Guangxi, Guizhou, Hubei, Hunan, Jiangxi, Zhejiang, Shaanxi, Sichuan, Xizang and Yunnan.

Status of population　Common species.

4.7

4.8

4.8 大耳菊头蝠 *Rhinolophus macrotis* Blyth, 1844

识别特征　体较小，前臂长 45.2-47.7 mm，颅全长 17-20 mm。耳大。马蹄叶宽，中央具明显缺刻，两侧各具 1 小附叶；鞍状叶较宽，但其宽度小于高度；连接叶低而平滑，起自鞍状叶顶端基部。头骨矢状嵴不明显。背毛毛基灰白色，毛尖暗褐色；腹毛色浅淡。

生境与习性　栖于山洞，数量稀少。可与小菊头蝠、鼠耳蝠等种类栖于相同的洞穴内。

分布　陕西、浙江、江西、四川、贵州、广西、云南、福建、重庆和广东。在井冈山水库大坝、观音庙大峡谷和朱砂冲等地均有本种分布。

种群状况　种群数量较小。

Identification　Small-sized horseshoe bat with forearm of 45-48 mm. Greatest skull length is 17-20 mm. Ear is relatively large. Horseshoe very wide with a deep median emargination and an adhering leaf on each side. Inferior surface of the sella is very broad, and is parallel sided; apex is rounded off and deflected downward.

Sella projects strongly forwards. The sagittal crest of skull is not strong. The color of dorsal fur is grey above and paler below; ventral pelage is paler than dorsal.

Habitat and Behavior　Population size is low. Often co-occurs with other *Rhinolophus* and *Myotis* species.

Distribution　Shaanxi, Zhejiang, Jiangxi, Sichuan, Guizhou, Guangxi, Yunnan, Fujian, Chongqing, Guangdong. Several specimens were collected at the tunnel of Mt. Jinggang Reservoir Dam, Grand Canyon of Guanyin Temple and Zhushachong Forest Farm.

Status of population　Small populations.

翼手目 CHIROPTERA
蹄蝠科 Hipposideridae

4.9 大蹄蝠 *Hipposideros armiger* (Hodgson, 1835)

识别特征　体型大，为国内最大的食虫蝙蝠之一，前臂长 83-98 mm。马蹄叶略呈方形，其两侧各具 4 片小副叶；马蹄叶上方为一横列的中叶，中央具 3 片突起的纵棱。头骨吻部由前向后逐渐升高，呈斜坡状；矢状嵴发达；上颌第一小前白齿位于齿列外。上体棕褐色，下体毛黄褐色。

生境与习性　常栖息于大岩洞或人工洞内，集结数十只或数百只的大群，同洞中可见多种其他种类蝙蝠。在井冈山石溪村观音庙大峡谷和水库大坝坑道采集过本种。具冬眠习性，以昆虫为食，捕食蛾、螟等相对大型的昆虫。

分布　分布于安徽、福建、广东、广西、海南、贵州、湖南、湖北、江苏、江西、陕西、四川、浙江、云南、西藏、香港、澳门及台湾。

种群状况　常见种。

Identification　Large size, one of the largest insectivorous bats, forearm length 83-98 mm, horseshoe leaves slightly square, with 4 pieces of supplementary leaflets on each side; with transverse middle lobes above, which has 3 pieces of probuberant vertical ridge. Rostrum slopes up from anterior to posterior gradually; well-developed sagittal crest; the first upper premolar outside the toothrow. Sepia dorsal hair, yellowish-brown ventral pelage.

Habitat and Behavior　Inhabits large grottos or artificial caves, colonies can be as large as tens of thousands of individuals, co-occurs with other bat species. Has been collected from the Grand Canyon near Buddhist Temple of Shixi village and the tunnel of Mt. Jinggang dam. Hibernates. Preys on relatively large insects such as moths.

Distribution　Anhui, Fujian, Guangdong, Guangxi, Hainan, Guizhou, Hunan, Hubei, Jiangsu, Jiangxi, Shaanxi, Sichuan, Zhcjiang, Yunnan, Xizang, Hong Kong, Macau and Taiwan.

Status of population　Common species.

4.10 无尾蹄蝠 *Coelops frithii* Blyth, 1848

识别特征　体小无尾，前臂长 35-38 mm，颅全长 16 mm 左右。具有特化的鼻叶，有前、中、后鼻叶之分，后鼻叶两侧各有一个小叶。耳大漏斗状，呈半透明状。体毛背腹各异，背毛基部黑褐色，毛尖赤褐色，腹毛基部灰褐色，毛尖灰白色。头骨吻部较短，矢状嵴不发达。

生境与习性　洞穴型蝙蝠（栖于井冈山水库附近废弃隧道中），可与大蹄蝠、大长翼蝠等多种蝙蝠混栖。具冬眠习性，以小型昆虫为食。

分布　分布于海南、广西、福建、重庆、云南和台湾。井冈山标本为江西省新纪录（徐忠鲜等，2013）。

种群状况　种群数量稀少。

Identification　Small size without tail, forearm length 35-38 mm, the greatest skull length 16 mm. Specialized nose leaf, can be divided into anterior, intermediate and posterior noseleaf, the posterior has one leaflet on each side. Big funnel-shaped ears, translucent. Base of dorsal fur black-brown, hair top russet, grey-brown belly fur. Muzzle short, sagittal crest undeveloped.

Habitat and Behavior　Cave roosting bats (inhabits abandoned tunnel near Mt. Jinggang Reservoir), co-exists with *Hipposideros armiger* and *Miniopterus schreibersi*. With hibernation habits. Feeds on small insects.

Distribution　Hainan, Guangxi, Fujian, Chongqing, Yunnan and Taiwan. Collected from Mt. Jinggang as a new occurrence record for Jiangxi in 2013 (Xu *et al.*, 2013).

Status of population　Quite rare.

翼手目 CHIROPTERA
蝙蝠科 Vesperitilionidae

4.11 西南鼠耳蝠 *Myotis altarium* Thomas, 1911

识别特征　体型中等，前臂长 39-45 mm，颅全长约 16 mm。耳壳狭长，前折超过吻端约 7 mm，耳屏尖长，约 10 mm。第 3-第 5 掌骨近等长。体毛较长而柔和，背毛棕褐色，腹毛毛色近似。头骨吻短而鼻额部略凹，矢状嵴和人字嵴均不发达。

生境与习性　栖于海拔 1000 m 以下的岩洞或人工洞（如井冈山水库大坝隧道）中，采集时曾见单只伏于小石凹中。具有冬眠习性。以昆虫为食。

分布　我国特有种（模式产地：四川峨眉山）。分布于四川、云南、贵州、江苏、浙江、广西、江西、广东等地。

种群状况　中国物种红色名录：近危。

Identification　Medium size, forearm length 39-45 mm, greatest skull length 16 mm. Long ear, which exceeds the rostrum about 7 mm, tragus long and narrow, about 10 mm. Third, fourth and fifth metacarpal lengths equal. Long and soft fur, dorsal pelage sepia, similar to belly hair color. Short muzzle, nose and forehead slightly concave, with undeveloped sagittal crest and herringbone crest.

Habitat and Behavior　Roosts in grottos or artificial caves (such as the tunnel of Mt. Jinggang dam) below 1000 m a.s.l., had been seen lying in sunken stone. Hibernation habits. Feeds on insects.

Distribution　Endemic to China (type locality: Emei Mountain in Sichuan). Sichuan, Yunnan and Guizhou, Jiangsu, Zhejiang, Guangxi, Jiangxi and Guangdong.

Status of population　China Species Red List: NT.

4.12 中华鼠耳蝠 *Myotis chinensis* (Tomes, 1857)

识别特征　体型较大的蝙蝠种类之一，前臂长 69-71 mm。无鼻叶。耳郭长且端部较狭窄，向前折可达或接近吻端，耳屏长而尖，约为耳长之半。头骨吻部微上翘，脑颅部近圆形，矢状嵴和人字嵴均明显。上颌有 2 对向中间斜生的小门齿，第 2 前臼齿小，位于齿列中。背部毛褐色，毛尖灰褐色；腹部毛暗灰色。

生境与习性　多栖息于较大型的岩洞。单只或数只高挂在岩洞顶壁，有时与其他鼠耳蝠种类混群。有冬眠习性，食较大型昆虫。

分布　浙江、江苏、福建、广西、广东、海南、安徽、江西、湖南、陕西、四川、贵州、云南及香港。

种群状况　较常见。

Identification　One of the largest species of bats, forearm length 69-71 mm, absence of noseleaf. Ear long, tragus long and sharp, about half of the ear length. Muzzle upwarps, skull rounded, sagittal and lambdoidal crests weak. Two pairs of small incisors in upper jaw inclined forward to the middle, the second premolar small and lies in the toothrow. Dorsal pelage brown, with slightly paler hair tips,

dark-grey ventral pelage.

Habitat and Behavior　Found in large caves, single or a few hanging from the top of caves, sometimes coexist with other myotis species. Hibernates. Preys on large insects.

Distribution　Zhejiang, Jiangsu, Fujian, Guangxi, Guangdong, Hainan, Anhui, Jiangxi, Hunan, Shaanxi, Sichuan, Guizhou, Yunnan and Hong Kong.

Status of population　Common species.

4.13 东亚伏翼 *Pipistrellus abramus* (Temminck, 1840)

识别特征　体型很小，前臂长 31-35 mm；耳小，略呈钝三角形，向前折达眼与鼻孔之间；耳屏小，端部钝圆，外缘基部有凹缺，向前微弯。第 5 指比第 3 或第 4 指长，翼膜止于足基部。尾长，仅尾尖从股间膜后缘穿出，股间膜呈锥状。后足短小。头骨甚宽，吻部鼓出，宽扁，无眶后突，颧弓细；犬齿发达，具后小尖，上下颌第一前臼齿很小。阴茎骨呈 "S" 形。上体毛深褐色，下体毛灰褐色。

生境与习性　多栖息于建筑物（在井冈山行洲村民房发现），特别喜欢栖息在瓦房及天花板之中，通常集数只小群潜伏在房檐下和门窗砖缝内。傍晚飞出，以蚊等小型昆虫为食。有冬眠习性。

分布　广布全国各地城乡，为最常见的蝙蝠种类。

种群状况　常见种。

Identification　Small size, forearm length 31-35 mm; ear small and triangle; tragus small and blunt, outside edges concave and bend forward slightly. The fifth phalanx longer than third and fourth phalanx, wing membrane extends to toe base. Tail long, only the tip of tail sticking out from tail membrane, and the membrane vertebral shaped. Hind feet short. Skull wide, rostrum bulgy, wide and flat. Without postorbital process, zygomatic arch weak; developed canine with a tip behind, the first premolar of the upper and lower jaws very small. The baculum S-shaped. Upper body hair dark-brown, while the lower body hair grey-brown.

Habitat and Behavior　Mainly inhabits buildings (found in a house in Xingzhou village at Mt. Jinggang), especially prefers to roost in tile-roofed houses and ceilings, and usually a small group lurking just under eaves and cracks between door and windows or between blocks. Flying out at dusk, feed on mosquitoes and other small insects. Have hibernation habits.

Distribution　Widely distributed across urban and rural areas and is the most common bat species seen in the evening.

Status of population　Common species.

4.14 褐扁颅蝠 *Tylonycteris robustula* Thomas, 1915

识别特征　体小，体重为 5 g，前臂长 25 mm；拇指基部和足掌具肉垫；耳短小，端部钝圆，耳屏窄而钝。距较长，有窄长的距缘膜。头颅骨甚小而扁平，颅高约为颅宽的 1/2，矢状嵴不发达。体毛暗褐，背毛端部深褐色，腹毛偏淡。

生境与习性　热带型蝙蝠，多栖息于被甲虫蛀食的竹洞中。

分布　2013 年 8 月在江西井冈山朱砂冲林场采集到 1 只雌性个体，是江西省首次记录到该种蝙蝠（张秋萍等，2014）。随着近年研究的深入，该种在我国的分布已经由原来的云南和广西 2 省，扩大到海南、贵州、四川、江西和广东等 7 省区。

种群状况　稀少种类。中国物种红色名录：易危。

Identification　Small-sized bat with an average weight of 4.7 g and an average forearm length of 25.2 mm. Thumbpads fleshy, and footpads trapezoidal in shape. Ear with blunt tip is short and small, while the tragus is narrow and blunt. Its calcar is long with narrow with a narrow membrane. Braincase is extraordinarily flattened and wide, and the height of braincase is approximately half of the breadth of braincase. Sagittal crest is very low. Pelage is dark buff-grey or cinnamon-brown on the back; ventral surface is paler, less dark-brown and more grey.

Habitat and Behavior　Tropical species. Often found inside bamboo chambers gnawed by beetles.

Distribution　In 2013, one *Tylonycteris robustula* specimen was collected from Zhushachong Forest Farm in Jiangxi (Mt. Jinggang).

4.13

4.14

This represents the first recorded specimen in Jiangxi (Zhang *et al.*, 2014). In China, distributed in Yunnan, Guangxi, Hainan, Guizhou, Sichuan, Jiangxi and Guangdong.

Status of population　Rare species. China Species Red List: VU.

4.15 毛翼管鼻蝠 *Harpiocephalus harpia* (Temminck, 1840)

识别特征　为蝙蝠科中体型较大的种类，雌性前臂长为 49-53 mm，雄性为 45-50 mm，鼻部前端呈短管状，耳壳卵圆形，耳屏披针型，较长，且有一基凹。背部毛黄褐色，毛尖褐栗色，翼膜淡黑褐色，后足、股间膜及尾膜密生黄褐色细毛，后足相对较短。头骨较粗壮，矢状嵴与人字嵴明显；颧弓发达；吻端短宽且中央深凹，额骨前端凹陷明显；下颌骨冠状突较高，角突与颌关节突较短且宽。

生境与习性　森林性蝙蝠，以甲虫为食。在相同的环境中还采集到泰坦尼亚彩蝠、大耳菊头蝠和中管鼻蝠等。

分布　于 2013 年 8 月在江西井冈山观音庙大峡谷和朱砂冲，用蝙蝠竖琴网采集到 9 只蝙蝠标本，是江西省翼手目分布新纪录（陈柏承等，2015）。国内分布于台湾、云南、广东、福建、江西和广西等。

种群状况　稀少种类。中国物种红色名录：易危。

Identification　Large-sized species in Vespertilionidae. Forearm length 45-50 mm in females and 49-53 mm in males. Nostrils are short tubes. Ear is oval in shape, and the tragus is long-lanceolated with an obvious depression at the base. Pelages are thick and soft. The hind legs, wing membrane and uropatagium are covered with sparse brown hair. Skull is strong with high sagittal crest and well-developed zygomatic arch. Its snout is short and wide with a deep concave. Mandible coronoid process is high, while condyle and angular process are short and wide.

Habitat and Behavior　Roosts in forests. Feeds mainly on beetles and usually shares the same foraging habitat as *Kerivoula titania*, *Rhinolophus macrotis*, and *Murina huttoni*.

Distribution　In 2013, using harp traps, nine *Harpiocephalus harpia* specimens were collected from Grand Canyon near Guangyin Temple and Zhushachong Forest Farm in Jiangxi (Mt. Jinggang). These represent the first records in Jiangxi (Chen *et al.*, 2015). In China, found in Taiwan, Yunnan, Guangdong, Fujian, Jiangxi and Guangxi.

Status of population　Rare species. China Species Red List: VU.

4.16 泰坦尼亚彩蝠 *Kerivoula titania* Bates, 2007

识别特征　体型较小，前臂长 30-34 mm，颅全长 14-15 mm。无鼻叶，耳郭呈漏斗状，耳屏略呈披针形。脑颅骨表面光滑，略显扁平型，无矢状峭和人字峭。翼膜和尾间膜呈灰色，被覆稀疏浅毛。翼膜后缘末端延伸至后足第一趾基部；尾间膜前缘末端止于脚踝。背毛总体近灰色，毛基部黑色，毛尖端深灰色，腹毛的基部为黑色，尖端稍白而略带灰褐色。

生境与习性　森林性蝙蝠，标本使用蝙蝠竖琴网捕获，采集于毛竹林与常绿阔叶林混交林之中的通道。

分布　2013 年 7 月和 8 月，在江西井冈山石溪村观音庙大峡谷和朱砂冲林场采集到 6 ♂ 6 ♀，为江西省翼手目分布新纪录（李锋等，2015）。中国除海南和台湾已有报道外，江西为中国大陆地区首次分布报道。

种群状况　稀少种类。为近年发表的新种（Bates *et al.*，2007），国内兽类分布新纪录（Wu *et al.*，2012），濒危状况未评价。

Identification　Small-sized bat in Vespertilionidae with forearm length of 30-34 mm and greatest skull length of 14-15 mm. Its nose is sample. Ear is funnel-shaped with a long-lanceolated tragus. Skull is relatively flattened without obvious sagittal crest or herringbone crest. Patagium and membrane among tail are gray and covered by sparse shallow pelage. Patagium is onto the base of first toe of hindfoot, while membrane of tail connects to edge of ankle. Color

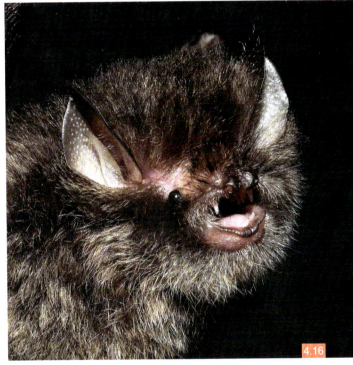

of dorsal fur is almost grey; sometimes hair bases black, while termini dark-grey; ventral hair bases black, while termini slightly white or grey-brown.

Habitat and Behavior　Roosts in forests; collected from mixed zones of bamboo forest and evergreen broad-leaved forest using harp traps.

Distribution　In 2013, twelve *Kerivoula titania* specimens (six male, six female) were collected from Guanyinmiao Grand Canyon, Shixi village and Zhushachong Forest Farm in Jiangxi (Mt. Jinggang). These represent the first recorded specimens in mainland China and Jiangxi (Li *et al.*, 2015).

Status of population　Rare taxon. The first recorded in mainland China by Wu *et al.* (2012). Not listed in China Species Red List.

灵长目 PRIMATES
猴科 Cercopithecidae

4.17 藏酋猴 *Macaca thibetana* (Milne-Edwards, 1870)

识别特征 体型较大，体重 15-20 kg，体长 580-710 mm，尾长 70-90 mm。头较大，有颊囊。雄性脸肉色，眼周白色；雌性脸红色，眼周粉红色。额部和两颊有长而密的毛，故又称毛面短尾猴。头顶和颈部毛褐色，下颌、颊为灰褐色，背、腰至尾基、尾背面毛为黑褐色；四肢外侧同背部毛色，内侧浅灰色。

生境与习性 栖于亚热带常绿阔叶林、落叶混交林、灌木林或多石岩的稀树山坡上。昼行性，多在地面上活动，夜晚在崖壁或山洞中栖息；"猴王"带领的群居生活。杂食性，食野果、竹笋、昆虫、蜥蜴、小鸟等，也盗食农作物。

分布 中国特有种。分布于云南、贵州、四川、福建、浙江、江西、甘肃、湖南和湖北等地。

种群状况 国家 II 级保护野生动物。IUCN 红色名录：近危。CITES 附录 II。中国物种红色名录：易危。

Identification Body size large, weight 15-20 kg, body length 580-710 mm, length of tail 70-90 mm. Head large, has cheek pouch. Male bare face flesh color, edges of eyes white; the female face is red and edges of eyes pink. The forehead and cheek with long and dense hair, which is why this species also called hairy face short tail macaque. The head and neck grey, while the chin and cheek grey-brown; back, waist, until to the base and back of tail dark-brown; the color on back of arms and legs similar to on the back of body; and paler grey ventrally.

Habitat and Behavior Inhabits subtropical evergreen broad-leaf forest, mixed deciduous forest, bush wood or on rare tree stone hillsides. Diurnal animal, mostly moves on the ground, roosts in rock walls or caves at night; social life led by "Monkey King". Omnivorous, eats fruits, bamboo shoots, insects, lizards and birds, sometimes steals crops.

Distribution Endemic species to China. Widespread in Yunnan, Guizhou, Sichuan, Fujian, Zhejiang, Jiangxi, Gansu, Hunan and Hubei.

Status of population China Key List: II. CITES App. II. IUCN Red List: NT. China Species Red List: VU.

兔形目 LAGOMORPHA
兔科 Legoridae

4.18 华南兔 *Lepus sinensis* Gray, 1832

识别特征 中型兔类，体重 1.5-2.0 kg，体长 350-430 mm，尾长 50-60 mm。冬毛背部黄褐色，间杂有黑色针毛，腹毛白色或浅棕黄色。耳郭长，70-80 mm，耳背棕黄色。尾毛短，背面棕灰色，腹毛污白色。

生境与习性 适应性强，栖息环境多种多样。多栖于井冈山的山坡灌丛或草丛，昼夜活动。草食性，以草、嫩枝叶等为食。

4.17

4.18a

4.18b

分布 长江以南各省区。

种群状况 分布较广，但种群密度不大。

Identification Medium sized rabbit, weight 1.5-2.0 kg, body length 350-430 mm, tail length 50-60 mm. Winter pelage on the backside yellowish-brown, mixed with black-tipped hairs, underside white or pale yellow. Long ears, 70-80 mm, claybank hair of the ear back. Tail hair short, grey above and pale white below.

Habitat and Behavior Well-adapted ability, diverse habitat environment. Inhabits bushes and grassland in Mt. Jinggang, active during day and night. Herbivore, feeds on grass and young leaves.

Distribution Provinces south of the Yangtze River.

Status of population Widely distributed but low population density.

食肉目 CARNIVORA
灵猫科 Viverridae

4.19 果子狸 *Paguma larvata* (Hamilton-Smith, 1827)

识别特征 外形较肥胖，体重 4-8 kg。头顶和颜面部有明显的白斑，头额中央具一条显著的白纹（故又称：白鼻心，花面狸）；身上无斑纹，体背和四肢多为灰棕色或棕黄色，腹部浅黄色；尾较长，基部毛同体色，尾端黑色。

生境与习性 夜行性，喜栖于岩洞、石隙、树洞或灌丛中。善攀爬树，取食各种野果（故名：果子狸），也捕食小动物。

分布 西南、华南、华中及华东各省。

种群状况 中国物种红色名录：近危。

Identification Obese, weight 4-8 kg. The top of head and face have obvious white spot, with significant white stripe on central forehead (also called white nose, masked civet); no stripe on other parts of body, back and limbs mostly grey brown or brown, belly light yellow; long tail, the color of basal tail same as body, tail end black.

Habitat and Behavior Nocturnal, roost in caves, stone cracks, tree holes or bushes. Climbs well, feeds on various wild berries and preys on small animals.

Distribution SE, S, C and E China.

Status of population China Species Red List: NT.

4.20 斑灵狸 *Prionodon pardicolor* Hodgson, 1842

识别特征 体型较小，约 500 g。吻部突出，颜面部狭长。从肩至尾基部具排成纵列的棕黑色圆形或卵圆形毛色斑，斑块向体侧逐渐变小且不规则。背部毛较深，为淡褐或黄褐色，腹面变成乳黄或乳白色。尾长而呈圆柱状，有 9-11 个暗色环，尾尖多数灰白色。

生境与习性 栖于阔叶林、稀疏灌丛或高草丛。夜行性，多地栖生活，亦会上树。肉食性，以鼠类、鸟类、蛙和昆虫等小动物为食。

分布 广东、广西、江西、贵州和云南等地。

种群状况 数量稀少。国家 II 级保护野生动物。CITES 附录 I。中国物种红色名录：易危。

4.19a

4.19b

4.20a

Identification Small size, body weighs 500 g. Pointed rostrum, narrow face. Longitudinal brownish-black round or oval spots from shoulders to tail base, with plaque being smaller and irregular to the body sides. Back hair pale-brown or brown, darker than those on the belly, which is yellow fraction or milky white. Tail long and cylindrical, with 9-11 dark rings, almost grey at the tail tip.

Habitat and Behavior Roosts in broad-leaved forest, shrubbery or tall grass. Nocturnal, moves on the ground mostly, sometimes in trees. Carnivore, feeds on rodents, birds, frogs, insects and other small animals.

Distribution Guangdong, Guangxi, Jiangxi, Guizhou and Yunnan.

Status of population Quite rare. China Key List: II. CITES App. I. China Species Red List: VU.

4.20b

食肉目 CARNIVORA
獴科 Herpestidae

4.21 食蟹獴 *Herpestes urva* (Hodgson, 1836)

识别特征 中型兽类，重2-3 kg。颈短、身躯粗壮；尾基部粗大，尾端较细；四肢短，五趾中第三、第四趾爪长而锐利。肛门腺一对，可排放臭气驱敌。体毛像披棕蓑，长而蓬松，呈灰棕褐色，从口角经颈侧直到肩部各有一条白色纵纹；体背针毛黑褐、红棕、灰白等毛色相间杂，腹毛较浅淡，尾端毛尖较多白色。

生境与习性 喜栖居山林沟谷及溪流两旁的丛林，尤其是有山坑的密林。日间活动，晨昏是觅食高峰。能利用灵敏的嗅觉，准确寻找深藏地下的蚯蚓和昆虫幼虫，水边的蛙类、蟹类和螺类等小型动物取食。

分布 云南、福建、广西、浙江、江西、广东和海南等长江以南省区。

种群状况 常见种。中国物种红色名录：近危。

Identification Medium-sized, body weighs 2-3 kg. Neck short; body strong; tail base thick, terminus slender; limbs short; the third and fourth toe with long and sharp claws. One pair of anal glands can drive away enemy with odorous emissions. Body hair like straw or palm-bark rain cape, long and fluffy, greyish-brown, from the mouth through the throat until shoulders have white vertical stripes; dorsal hairs mixed with dark-brown, reddish-brown, gray and white, the belly hair paler, almost white tail top.

Habitat and Behavior Inhabits valleys and jungle between mountains, especially jungles with holes. Diurnal activities, feeding peak time is mostly in the morning and twilight. Good scenting ability could easily locate and hunt the underground worms and insects, frogs, crab and snails near the river.

Distribution Yunnan, Fujian, Guangxi, Zhejiang, Jiangxi, Guangdong, Hainan and other provinces south of the Yangtze River.

Status of population Common species. China Species Red List: NT.

4.21

食肉目 CARNIVORA
鼬科 Mustelidae

4.22 猪獾 *Arctonyx collaris* Cuvier, 1825

识别特征 外形和大小与狗獾极其相似，但猪獾的鼻与上唇间裸露，鼻吻部形成猪鼻状圆形鼻垫。体毛黑褐色，间杂灰白色针毛；从前额到额顶中央，有一条短宽的白色条纹，并向后延伸至颈背；两颊在眼下各具一条污白色条纹；下颌及喉部白色，向后延伸近肩部。

生境与习性 栖于森林、灌丛、草丛等环境。洞栖，洞道长，内分主道、侧道及盲端，有2-3个洞口，冬季洞的结构较复杂。生活习性与狗獾相似，但性情更凶猛，遇敌即前脚低俯，发出似猪的怒吼声，再挺立身躯以牙和利爪回击。杂食性，捕食各种小动物，兼食植物的根、茎和果实。

分布 中国除台湾和海南外，各省均有分布。

种群状况 IUCN红色名录：近危。中国物种红色名录：易危（VU）。

Identification Similar to *Meles meles*, but nose and upper lips naked, the nose and snout forms pig snout-liked nose round pads. Body hair dark-brown, mixed with grey aciulum, short and wide white stripes from forehead and central forehead extend to nape; below the eyes each has white stripes; lower jaw and throat white, extending close to shoulders.

Habitat and Behavior Inhabits forests, shrubs, grassland and other environments. Cave-dwelling, caves with long road and can be divided into main branch, by road and blind side, there are 2-3 holes, the hole structure is more complicated in winter. Habits similar to *Meles meles*, but more fierce, when facing enemies bends down forefeet and issues a roar like pig, then stand upright and fight back with teeth and sharp claws. Omnivorous, feeds on a variety of small animals, and eats plant roots, stems and fruits.

Distribution All of China except Taiwan and Hainan.

Status of population IUCN Red List: NT. China Species Red List: VU.

4.23 鼬獾 *Melogale moschata* (Gray, 1831)

识别特征 体较小，一般不超 1.5 kg。鼻吻突出，颈粗短，耳短圆而直立。趾爪侧扁而弯曲，前爪第二、第三爪特别粗长，适于挖掘。通身毛色为灰褐色，头顶向到后腰有一断续的白色纵纹；前额、眼后、颊和颈侧均有不定型的白斑，喉、胸、腹部的毛污白或浅黄色。

生境与习性 喜栖居山区农田附近的丛林和草丛。善打洞，但洞穴简单。夜行性，多单独活动，常在潮湿的作物区或水溪旁活动，能用强爪和长吻扒挖寻找食物，外出活动通常循一定的路径。杂食性，主食各种小动物，亦食植物的根、茎和果实。

分布 广布于长江以南各省区。

种群状况 中国物种红色名录：近危。

Identification Small-sized badger, generally not exceeds 1.5 kg. Snout prominent; neck stubby; ears short, round and upright. Claws flat and curved; claws of second and third fingers especially thick and long, adapting to digging. Pelage color is grayish-brown, with an intermittent white vertical stripe from the top of the head to the back; forehead, back of eyes, cheeks and neck sides have shapeless white spots, hair of throat, chest and stomach white and light yellow.

Habitat and Behavior Inhabits jungles and grasslands near farmland in mountains. Good at digging holes, but caves are simple. Nocturnal, mostly moves about separately, usually moves near wet crop areas or rivers, can use strong claws and long snout for digging food, usually follows certain paths for outdoor activities. Omnivorous, mainly eats small animals, also eats plant roots, stems and fruit.

Distribution Widely distributed in south side provinces of the Yangtze River.

Status of population China Species Red List: NT.

4.24 黄喉貂 *Martes flavigula* (Boddaert, 1785)

识别特征 属于较大的貂鼬类动物，体长 450-650 mm，体重 2-

3 kg。体型细长，耳短圆，不太突出。体毛短而稀疏，尾部毛不见蓬松。吻部到头顶、颈、背和尾部，以及四肢显示暗棕色或黑褐色，但喉胸部为鲜黄色是其最大的特征，故名黄喉貂。

生境与习性 栖息在各类混交林，筑巢于树洞或者石堆中。以鼠类等小型动物为食，也食野果。还有食蜂蜜的习性，故有"蜜狗"的雅号。

分布 我国大部分省区均有分布。

种群状况 数量较稀少。国家 II 级保护野生动物。中国物种红色名录：近危。

Identification A large, robust, muscular and flexible carnivorous animal with a body length of 450-650 mm and weight of 2-3 kg. Hair is relatively short and sparse, while pledge in tail is bushy. Color of pelage from muzzle to tail and limb is blackish-brown or dark-brown; throat and chest are distinctively bright yellow. According to these characteristics, it is called the Yellow-throated Marten.

Habitat and Behavior Roosts in all kinds of mixed forest. Usually nests in holes in tree or among rocks. Diet consists of rats and small animals, but also wild fruit. Honey has become part of diet.

Distribution Most provinces in China.

Status of population Rare species. China Key List: II. China Species Red List: NT.

4.25 黄鼬 *Mustela sibirica* Pallas, 1773

识别特征 体型细长，四肢短，头小而颈长，体重 500-700 g。鼻垫基部及上、下唇为白色，喉部及颈下常有白斑。耳壳短宽，尾长为体长之半，肛门腺发达。体毛为棕褐色，可因不同地方和季节有深浅变化；腹毛稍浅淡，背腹毛色无明显的分界。

生境与习性 栖息环境多样，平原、丘陵、山区的各种生境均有分布，尤以平原地区数量大。平时栖居在乱石堆和倒木下，在繁殖季节和冬季有较稳定的洞穴。遇险时，会从肛门腺分泌臭液，发出难闻的气味。肉食性，主要捕食鼠类和各种小动物，偶尔伤害家禽。

分布 广泛分布于全国各省区。

种群状况 中国物种红色名录：近危。

Identification Slender body, short legs, small head and long neck, weight 500-700 g. Nose pad base, upper and lower lips white, throat and neck below always has white spot. Short and wide ears, tail length is half of body length, developed anal gland. Brown pelage, may change with environment and season; abdominal hair slightly pale, without obvious boundaries between the dorsal and ventral hair.

Habitat and Behavior Inhabits plains, hills and mountains, especially plains. Usually inhabits rubble and fallen trees, has stable caves in breeding season and winter. In distress secretes smelly liquid from anal gland. Carnivorous, mainly feeds on mice and other small animals, occasionally harms poultry.

Distribution Widely distributed in China.

Status of population China Species Red List: NT.

4.24a

4.24b

4.25

食肉目 CARNIVORA
猫科 Felidae

4.26 豹猫 *Prionailurus bengalensis* Kerr, 1792

识别特征　体长 360-600 mm，尾长 150-370 mm，体重 3-8 kg。背部黄褐至棕灰色。全身布满棕褐至淡褐色斑点、色斑似豹，故名豹猫。头部至肩部有 4 条向后的纵向黑纹。耳背黑色，有一白斑。

生境与习性　栖息于山地森林、居民区等各种类型的环境中。主要营地面生活。主食鼠类、鸟类、野兔等小型脊椎动物。

分布　江西等全国各地均有分布。

种群状况　目前种群数量较小。中国物种红色名录：易危。

Identification　Body length of 360-600 mm and tail length 150-370 mm. Weight 3-8 kg. Color of dorsal fur varies from yellow-brown to brownish-grey. Body decorated with brown or light brown spots, it therefore similar to leopard; four obvious black stripes from head to shoulders. Back of ear is black with a white spot.

Habitat and Behavior　Found in montane forests and residential areas. Mainly feeds on small terrestrial vertebrates such as mice, birds and hares.

Distribution　Jiangxi and Most provinces in China.

Status of population　Rare species. China Species Red List: VU.

4.26

偶蹄目 ARTIODACTYLA
猪科 Suidae

4.27 野猪 *Sus scrofa* Linnaeus, 1758

识别特征　家猪的祖先，外形与家猪相似，但嘴更长，性凶猛。幼仔身上有许多条纹，长成后才逐渐消失。雄性的犬齿形成獠牙状。

生境与习性　适应能力很强，能找到隐蔽环境和食物的地方，就可以生存。喜集群生活，但无固定住地，繁殖时用树枝和杂草堆积成巢。杂食性，吃植物的嫩枝、根和果实，也吃动物尸体和其他小动物、昆虫及地下的软体动物。

分布　广布全国各地。

种群状况　因全国环境状况改变，野猪数量有增加的趋势，有时会危害农作物。

Identification　Ancestors of domestic pigs, similar to hogs, but with longer snout, more fierce. Pups with many stripes, disappear after growing up. Male canine is tusk-like.

Habitat and Behavior　Strong adaptive capacity. Prefers sociality, without fixed residence, in breeding season nests piled by branches and weeds. Omnivorous, eats plant shoots, roots and fruits, also dead animals and other small animals, insects and mollusks underground.

Distribution　Widely distributed in China.

Status of population　Wild boar have increased due to changes in environmental conditions, sometimes damage crops.

4.27a

4.27b

偶蹄目 ARTIODACTYLA
鹿科 Cervidae

4.28 毛冠鹿 *Elaphodus cephalophus* Milne-Edwards, 1872

识别特征 似赤鹿但额顶有一簇黑褐色冠状长毛,故称毛冠鹿。雄性有角短小而不分叉,几乎隐于毛丛中。耳背有一块白斑,无额腺,但眶下腺特别发达。体毛灰褐、赤褐色或灰色;腹部及尾腹面为白色,背面黑褐到黑色。

生境与习性 栖息于海拔较高人迹罕至的高山大岭,有隐蔽场所、有充足的食物和没有人为干扰就可以生活。不结群,多晨昏活动,白天隐于密林和灌丛中。喜食百合科、虎耳草科、蔷薇科和玄参科植物的嫩枝叶,也食野果和种子。

分布 甘肃、青海、陕西、河南、江西和长江以南各省区(除海南)。

种群状况 IUCN红色名录:近危。中国物种红色名录:易危。

Identification Looks like *Muntiacus muntijak*, a cluster of dark-brown coronal hair in the frontoparietal, called Tufted Deer. Male horned, short, small and unbranched, almost hidden in hair. A white spot on the back of ears, without frontal gland, but with very developed infraorbital gland. Body hair grayish brown, auburn or grey, abdomen and tail ventral hair white, dark-brown to black hair on the back.

Habitat and Behavior Roosts in hidden places in mountains at higher elevations, solitary, and mostly moves during morning and dusk, hidden in woods and bushes during the day. Prefers the branches and leaves of Liliaceae, Saxifragaceae, Rosaceae and Scrophulariaceae; also eats berries and seeds.

Distribution Gansu, Qinghai, Shaanxi, Henan, Jiangxi and south of the Yangtze River (except Hainan).

Status of population IUCN Red List: NT. China species Red List: VU.

4.29 赤麂 *Muntiacus muntjak* (Zimmermann, 1780)

识别特征 中型鹿类,雄体体重可达30 kg。雄鹿有角,单叉型,角柄较长,角尖向内弯。脸狭长,额部具明显的"V"形黑纹。体毛光滑细密,体色以棕红为主、可随年龄和季节变为黄褐色深黄色;胸腹部淡黄色至白色;下肢暗褐或黑色。

生境与习性 喜单独活动,有一定的领域。栖息于食物丰富、隐蔽场所、人为干扰少、气候温暖的环境。稍有惊动即迅速藏匿,但会折回原地,家域比较稳定。采食各种植物的嫩枝叶,也食青草、落地的野果和农作物等。发情或者天气变化时常在夜晚发出叫声。

分布 分布广东、广西、湖南、江西、贵州、四川、云南、西藏和海南等地。

种群状况 中国物种红色名录:易危。

Identification Medium-sized deer, weight 30 kg. Males have single fork type antler that emerge from long pedicles with the tips of the inner horn. Face narrow and long, with a distinct V-shaped

4.28a

4.28b

black stripe on the forehead. Body hair smooth, fine and thick. The main color of dorsal body brown, changed from yellowish-brown to dark-brown in difference of ages and seasons; venter and chest hair light yellow and white; the lower limbs dark-brown or black.

Habitat and Behavior　Like traveling alone with distinct territorial boundaries. Inhabits secluded places with rich food, warm climate, and not interfered by human. It is slightly disturbed and quickly hides, but soon turn back. Their home range often is stabilized. Feeds on tender branches and leaves, also eats grass, wild fruits and crops. They often make sounds at night in mating season and climate change.

Distribution　Guangdong, Guangxi, Hunan, Jiangxi, Guizhou, Sichuan, Yunnan，Xizang and Hainan.

Status of population　China Species Red List: VU.

4.30 水鹿 *Rusa unicolor* (Kerr, 1792)

识别特征　大型鹿类之一，体长可达 2 m，体重可达 200 kg。雄体有角，长而粗，角表面粗糙形成纵嵴，角分三叉，第一叉（眉叉）与主干成锐角；鹿角每年冬季脱换，春季刚刚长出的嫩角未钙化，取下干燥后称为鹿茸。耳大而直立，眶下腺发达；体毛粗而厚密，多灰褐色，或棕褐色。尾较长，尾毛黑色，长密而蓬松。

生境与习性　栖息于面积较大的阔叶林、混交林、山地草坡、稀树草原等环境，喜爱在林下有水源及林、灌、草相间的地段生活。生性机警，听觉和嗅觉灵敏，发现险情即飞奔而逃，并发出尖叫声。性喜水，夏天常到山溪中水浴。广泛取食各种植物的嫩茎叶、花和果实。

分布　广布于华南及贵州、四川和青海的部分地区。

种群状况　野外数量不多。国家 II 级保护野生动物。IUCN 红色名录：易危。中国物种红色名录：易危。

Identification　Large deer, body length up to 2 m, weight to 200 kg. Males have horns, long and thick, rough surface with longitudinal ridges, divided into trigeminal, the first fork (brow tine) forms an acute angle with trunk; antlers molt every winter, tender antlers grow in spring and non-calcified (drying after removing called Pilose antler). Large, erect ears, developed infraorbital gland. Body hair thick and dense, mostly taupe or tan. Tail long; tail hair black, long, dense and fluffy.

Habitat and Behavior　Inhabits larger areas of broad-leaved forest, mixed forest, mountain grassland and savanna, likes to live in understory with water and mix of forest, bushes and grass. Alert, with sensitive hearing and smell. Runs away and screaming if danger presenting. Likes water, often bathes in mountain streams in summer. Extensively feeds on tender leaves, flowers and fruits of various plants.

Distribution　Widely distributed in S China and some regions of Guizhou, Sichuan and Qinghai.

Status of population　Rare species. China Key List: II. IUCN Red List: VU. China Species Red List: VU.

4.29a

4.29b

4.30

偶蹄目 ARTIODACTYLA
牛科 Bovidae

4.31 中华鬣羚 *Capricornis milneedwardsii* David, 1869

识别特征　因最早在印尼苏门答腊被发现，又叫苏门羚。体型较大，体重 50-70 kg。两性均有短角，耳长似驴。四肢粗壮，具有偶蹄，蹄间有足腺，能分泌黏液以利攀爬悬岩。体毛大部分黑褐色，杂以棕灰色毛，颈、肩背面有浅棕色长鬣毛（故名鬣羚），尾短，尾基部毛锈棕色。

生境与习性　典型栖息环境为树林繁杂、乔木、灌木交替，间有裸岩、陡壁的环境。单独或 2-3 只一起活动，遇敌即向险峻的悬崖峭壁躲避。以杂草及各种植物嫩枝叶、菌类和松萝为食。

分布　西藏、青海、甘肃、陕西、四川、贵州、云南、湖北、湖南、安徽、江西、浙江、福建、广东和广西等地。

种群状况　国家 II 级保护野生动物。CITES 附录 I。IUCN 红色名录：近危。中国物种红色名录：易危。

Identification　Found in Sumatra at first, also called Sumatran Serow. Larger-sized artiodactyl, weighing 50-70 kg. Present two short horns. Ears long and like of the donkey. Thick limbs has pedal gland between hooves secreted mucus to climb cliff. Pelage color mostly black-brown, mixed with grey; the back of neck and shoulder have brown long mane (hence called Serow); short tail with rust-brown hair bases.

Habitat and Behavior　Typical habitat mixed with trees, shrubs, even bare rock, and steep cliffs. Lives alone or together with 2-3 individuals. Hides in steep cliffs when encountering danger. Feeds on weeds, tender branches and leaves of various plants, fungi and Chinese Usnea.

Distribution　Xizang, Qinghai, Gansu, Shaanxi, Sichuan, Guizhou, Yunnan, Hubei, Hunan, Anhui, Jiangxi, Zhejiang, Fujian, Guangdong and Guangxi.

Status of population　China Key List: II. CITES App. I. IUCN Red List: NT. China Species Red List: VU.

啮齿目 RODENTIA
松鼠科 Sciuridae

4.32 隐纹花松鼠 *Tamiops swinhoei* (Milne-Edwards, 1874)

识别特征　体型较小的松鼠，体长约 130 mm。体背毛棕褐色，从背中央向两侧共有 7 条条纹，分别为黑色、棕褐色、棕黄色、浅黄色或白色，最外侧的条纹不明显。耳尖有束毛，基部黑色，末端白色；腹毛灰黄色；尾毛蓬松呈棕褐色，其两侧有镶棕黄色边缘的黑色纵纹。

生境与习性　为典型的树栖啮齿动物，井冈山多栖于密林，也常见于山区村前屋后的大树上；在树洞或在树枝上筑巢。杂食性，以花、果、嫩叶及昆虫为食。

分布　河北、河南、山西、陕西、甘肃、宁夏长江以南各省区等地。

种群状况　常见种。

4.31

4.32a

4.32b

Identification Small-sized squirrel, body length 130 mm. Dorsal hair brown, with seven longitudinal stripes from the central back to both sides, as following color black, tan, brown, pale yellow or white, the outermost ones unobvious. A tuft of hair on the tip of each ear with black hair bases and white hair termini. Grayish yellow abdominal hair. Tail hair fluffy, brown, with black stripes edged with brownish-yellow on both sides.

Habitat and Behavior Typical arboreal rodent. Inhabits jungles in Mt. Jinggang, also found in large trees near villages; roosts in trees hollows or nests on tree branches. Omnivorous, eats flowers, fruits, tender leaves and insects.

Distribution Hebei, Henan, Shanxi, Shaanxi, Gansu, Ningxia and provinces south of the Yangtze River.

Status of population Common species.

4.33 红腿长吻松鼠 *Dremomys pyrrhomerus* (Thomas, 1895)

识别特征 中型大小松鼠，但吻部较长。两颊染锈红色，体型也较瘦长。体背黄褐色，毛基部灰黑色；腹毛灰白色，臀部及腿部两侧红褐色。尾基部毛黄褐色，毛较长而暗淡，具白色毛尖，形成类似黑、灰色相间毛环。

生境与习性 栖于井冈山针阔混交林及林缘。以树栖为主，也常见在地面活动。白天活动，采食各种野果和植物嫩枝叶，有时也寻找鸟卵及昆虫为食。

分布 云南、四川、安徽、贵州、湖南、湖北、江西、广东、广西和海南等地。

种群状况 较常见。中国物种红色名录：近危。

Identification Medium-sized squirrel, with longer snout and slender body. Rusty-red hair of cheek; dorsal hair brown with greyish-black hair bases; grey hair abdominal hair, buttocks and both sides of leg hair red-brown; tail hair yellowish-brown on the tail base, followed hairs longer and pale, with white tips, forming black, grey hair rings.

Habitat and Behavior Inhabits conifer and broad-leaved mixed forest and forest edges in Mt. Jinggang. Arboreal mainly, but also moves on ground. Moves during the day, feeds on diverse wild fruit and tender branches and leaves, sometimes looks for bird eggs and insects.

Distribution Widely distributed in Yunnan, Sichuan, Anhui, Guizhou, Hunan, Hubei, Jiangxi, Guangdong, Guangxi and Hainan.

Status of population Common species. China Species Red List: NT.

4.34 帕氏长吻松鼠 *Dremomys pernyi* (Milne-Edwards,1867)

识别特征 体重约 230 g，体长 180-220 mm、体型大小似赤腹松鼠，但吻部较长，体也较瘦长。体背橄榄绿色，毛基部灰黑色；腹毛灰白色。尾基部约带淡红色，其余部分浅黄色。肛区和后肢内侧呈锈红色，耳后有黄色簇毛。

生境与习性 栖息于针阔混交林中。以树栖为主，也常见它们在地面觅食。白天活动，多采食各种野果和植物嫩枝叶为食，也

4.33a

4.33b

捕捉为食。叫声响亮。

分布 福建、四川、安徽、贵州、湖北、广西、江西、陕西、湖南和云南等省区分布。

种群状况 常见种。

Identification A squirrel with an average weight of 230 g and body length of 180-220 mm. Similar to *Callosciurus erythraeus*, but with a longer snout and slender body. Dorsal surface olive-brown with grey-black hair base; ventral body grey-white; ventral and dorsal tail reddish at the base, remaining buffy. Anal area and inner hindlimb rubiginose; a yellow tuft on posterior surface of the each ear.

Habitat and Behavior Arboreal squirrel occurs in theropencedrymion forests. In the daytime, often seek food on the ground and mainly feed on wild fruit, twigs, leaves, and insect. Sound is quite loud.

Distribution Fujian, Sichuan, Anhui, Guizhou, Hubei, Guangxi,

Jiangxi, Shaanxi, Hunan and Yunnan.

Status of population Common species.

啮齿目 RODENTIA
竹鼠科 Rhizomyidae

4.35 中华竹鼠 *Rhizomys sinensis* Gray, 1831

识别特征 体型较小的竹鼠。上唇为裂唇，门齿发达。眼小，耳隐于体毛中，前后足爪坚硬，适合在土壤中挖洞。体毛细密而柔软，背部呈浅灰褐色或浅灰色，腹部颜色较淡。尾较短，几乎完全裸露，不及体长一半。

生境与习性 山区竹林和芒草丛生处洞栖（故又称：芒鼠），洞道挖在竹林和芒草下。在地洞中挖食未出土的竹笋、竹、芒的根，直接在洞内进食。取食洞道离地面近，安全洞道离地面可达 3 m。

分布 长江以南各省。

种群状况 常见种。

Identification Small bamboo rat. Upper lip has cleft lip, developed incisors. Small eyes, ears hidden in body hair, hard claws, suitable for digging holes in soil. Fine body hair and soft, light grey-brown or grey dorsal hair, darker than abdomen hair. Shorter tail, almost naked, shorter than half the body length.

Habitat and Behavior Lives in areas of bamboo and miscanthus (also called miscanthus rat), burrow hidden under bamboo and miscanthus. Digs not earthed bamboo shoots, and their roots, and eats them in the hole directly. Feeding burrow near the ground, but security holes up to 3 m away from the ground.

Distribution Provinces south of the Yangtze River.

Status of population Common species.

啮齿目 RODENTIA
鼠科 Muridae

4.36 针毛鼠 *Niviventer fulvescens* (Gray, 1847)

识别特征 中型鼠类，体长 137-140 mm，尾长 178-195 mm。口侧和颊部黄褐色，耳棕色；体背自前额至尾基为明亮的黄棕色，背部色较深，杂有白色的刺状针毛（全年可见），毛尖端黑色；体侧、腹面、足背均为乳白色或微黄色；尾上下二色，背面黑褐色，腹面纯白色。

生境与习性 栖于井冈山山间灌丛、田间、丘陵、山谷、溪旁、树根、岩石缝等处。穴居，有 2-3 个口，窝内以杂草、树叶等铺垫。食各种坚果、花生、番茄、秧苗等。善攀缘高处，多在夜晚活动。

分布 陕西、甘肃、河南、西藏、重庆、贵州、安徽、湖南、湖北、海南、江西、云南、四川、广东、广西、福建、浙江等地。

种群状况 常见种。

Identification Medium-size rodent, body length 137-140 mm, tail length 178-195 mm. Sides of mouth and cheek tawny, brown ears; dorsal hair from the forehead to the back of tail bright yellowish-brown, the dorsal pelage color much darker, mixed with white thorn-like aciulum (found all over the year), tip of hair black; body sides,

4.34a

4.34b

4.35

abdomen, back of foot milk-white or light yellow; tail with different color from dorsum and abdomen, the dorsal hair blackish-brown, while the ventral hair is white.

Habitat and Behavior Inhabits shrubs, farms, hills, valleys, streams, tree roots and rock cracks in Mt. Jinggang. Lives in caves, with 2-3 entrances, inside the nest bedded with weeds and leaves. Feeding on many nuts, peanuts, tomatoes and seedlings. Good at climbing high, moves at night.

Distribution Shaanxi, Gansu, Henan, Xizang, Chongqing, Guizhou, Anhui, Hunan, Hubei, Hainan, Jiangxi, Yunnan, Sichuan, Guangdong, Guangxi, Fujian and Zhejiang.

Status of population Common species.

4.37 巢鼠 *Micromys minutus* (Pallas, 1771)

识别特征 体型很小，体长不超过 75 mm，尾略长于体长。耳较短，有耳屏。背部黄褐或暗褐色，毛基深灰色；腹面白色，毛基浅灰色；尾上面略深，下面白色；四足纯白色。

生境与习性 栖息于井冈山农田、灌丛和草地。有筑巢习性，春季筑巢在地面，夏季筑于作物和高草，冬季迁草堆中。巢为圆形，有 2 个出口，巢口关闭，外出时才开启，同时亦筑构造较复杂的地下巢。食各种作物种子、草籽和植物的绿色部分。

分布 辽宁、内蒙古、吉林、黑龙江、四川、云南、贵州、广东、广西、江西、福建、陕西、浙江等地。

种群状况 常见种。

Identification Small body, not longer than 75 mm. Tail slightly longer than body. Short ear, with tragus. Dorsal hair yellowish-brown or dark-brown, and dark-grey hair base, while white ventral hair, with the basal light grey; darker color of the upper part of tail, the lower hair white; limbs white.

Habitat and Behavior Inhabits farmland, shrubs and grassland in Mt. Jinggang. Has nesting habits, in different seasons nests in different places, for example, nests on the ground in spring, crops and high grasses during summer, and moves to weed piles in winter. Nest round, with two exits, keep closed, except for going out, also build complicated underground nest. Feed on crops and grass seeds, also green parts of plants.

Distribution Liaoning, Neimenggu, Jilin, Heilongjiang, Sichuan, Yunnan, Guizhou, Guangdong, Guangxi, Jiangxi, Fujian, Shaanxi and Zhejiang.

Status of population Common species.

参考文献
References

蔡波，王跃招，陈跃英，等. 2015. 中国爬行纲动物分类厘定. 生物多样性，23(3):365-382.（Cai B, Wang Y Z, Chen Y Y, *et al*. 2015. A revised taxonomy for Chinese reptiles. *Biodiversity Science*, 23(3):365-382.）

陈柏承，余文华，吴毅，等. 2015. 毛翼管鼻蝠在广西和江西分布新纪录及其性二型现象. 四川动物，34(2): 211-215.（Chen B C, Yu W H, Wu Y, *et al*. 2015. New record and sexual dimorphism of *Harpiocephalus harpia* in Guangxi and Jiangxi, China. *Sichuan Journal of Zoology*, 32(2): 211-215.）

承勇，宋玉赞，赵健，等. 2010. 江西井冈山国家级自然保护区鸟类资源调查与分析. 四川动物，30(2): 277-282.（Cheng Y, Song Y Z, Zhao J, *et al*. 2010. Investigation and analysis of bird resource in the Jinggangshan National Nature Reserve, Jiangxi Province. *Sichuan Journal of Zoology*, 30(2): 277-282.）

段世华，龙川，龙婉婉，等. 2004. 井冈山自然保护区夏季鸟类多样性分析. 井冈山师范学院学报(自然科学版)，25(5): 12-19.（Duan S H, Long C, Long W W, *et al*. 2004. Diversity analysis of summer birds in Jinggangshan Nature Reserve. *Journal of Jinggangshan Normal College* (*Natural Science*), 25(5): 12-19.）

费梁，胡淑琴，叶昌媛，等. 2006. 中国动物志. 两栖纲. 上卷. 北京: 科学出版社: 1-471.（Fei L, Hu S Q, Ye C Y, *et al*. 2006. Fauna Sinica, Amphibia Vol. 1, Anura Ranidae. Beijing: Science Press: 1-471.）

费梁，胡淑琴，叶昌媛，等. 2009a. 中国动物志. 两栖纲. 中卷. 北京: 科学出版社. 1-958.（Fei L, Hu S Q, Ye C Y, *et al*. 2009a. Fauna Sinica, Amphibia Vol. 2, Anura Ranidae. Beijing: Science Press: 1-958.）

费梁，胡淑琴，叶昌媛，等. 2009b. 中国动物志. 两栖纲. 下卷. 北京: 科学出版社: 959-1848.（Fei L, Hu S Q, Ye C Y, *et al*. 2009b. Fauna Sinica, Amphibia Vol. 3, Anura Ranidae. Beijing: Science Press: 959-1848.）

费梁，叶昌媛，江建平. 2011. 中国两栖动物彩色图鉴. 成都: 四川科学技术出版社: 1-519.（Fei L, Ye C Y, Jiang J P. 2011. Colored atlas of Chinese amphibians. Chengdu: Sichuan Publishing House of Science & Technology: 1-519.）

费梁，叶昌媛，江建平. 2012. 中国两栖动物及其分布彩色图鉴. 成都: 四川科学技术出版社: 1-619.（Fei L, Ye C Y, Jiang J P. 2012. Colored atlas of Chinese amphibians and their distributions. Chengdu: Sichuan Publishing House of Science & Technology: 1-619.）

高正发，侯勉. 2002. 四川攀蜥属一新种——汶川攀蜥. 四川动物，21 (1): 3-5.（Gao Z F, Hou M. 2002. Description of a new *Japalura* species from western Sichuan province, China. *Sichuan Journal of Zoology*, 21 (1): 3-5.）

高正发，秦爱民. 2000. 四川爬行动物新纪录——米仓山攀蜥. 四川动物，19(5):27.（Gao Z F, Qin A M. 2000. *Japalura micangshanensis*—a new reptile record in Sichuan. *Sichuan Journal of Zoology*, 19 (5): 27.）

关贯勋，谭耀匡. 2003. 中国动物志. 鸟纲. 第七卷，夜鹰目　雨燕目　咬鹃目　佛法僧目　鴷形目. 北京: 科学出版社: 241.（Guan G X, Tian Y K. 2003. Fauna sinica. aves. Vol. 7, Caprimulgiformes, Apodiformes, Trogoniformes, Coraciiformes, Piciformes. Beijing: Science Press: 241.）

郭英荣，江波，王英永，等. 2010. 江西阳际峰自然保护区综合科学考察报告. 北京: 科学出版社: 1-245.（Guo Y R, Jiang B, Wang Y Y, *et al*. 2010. Report of scientific survey on mount Yangjifeng Reserve of

Beijing: Science Press: 1-522.）

赵尔宓，赵肯堂，周开亚，等. 1999. 中国动物志. 爬行纲. 第二卷, 有鳞目　蜥蜴亚目. 北京: 科学出版社: 1-394.（Zhao E M, Zhao K T, Zhou K Y, *et al*. 1999. Fauna sinica. Reptilia. Vol. 2, Squamata and Lacertilia. Beijing: Science Press: 1-394.）

赵健，汪志如，杜卿，等. 2012. 江西省鸟类新纪录——云南柳莺、绿背姬鹟. 四川动物，31(3): 447.（Zhao J, Wang Z R, Du Q, *et al*. 2012. New bird records for Jiangxi province: *Phylloscopus yunnanensis* and *Ficedula elisae. Sichuan Journal of Zoology*, 31(3):447.）

赵正阶. 2001. 中国鸟类志(上卷、下卷). 长春: 吉林科学技术出版社.（Zhao Z J. 2001. The avifauna of China (2 volumes). Changchun: Jilin Science and Technology Press.）

郑发辉，陈春泉，邓大吉，等. 2007. 井冈山自然保护区自然资源评价. 福建林业科技，34(3): 159-165.（Zheng F H, Chen C Q, Deng D J, *et al*. 2007. Assessment on natural resources of Jinggangshan National Nature Reserve in Jiangxi province. *Jour of Fujian Forestry Sci and Tech*, 34(3): 159-165.）

郑光美. 2005. 中国鸟类分类与分布名录. 北京: 科学出版社: 426.（Zheng G M. 2005. A checklist on the classification and distribution of the birds of China. Beijing: Science Press: 426.）

郑光美. 2011. 中国鸟类分类与分布名录 (第二版). 北京: 科学出版社.（Zheng G M. 2011. A checklist on the classification and distribution of the birds of China (2nd Edition). Beijing: Science Press.）

郑作新. 2000. 中国鸟类种和亚种分类名录大全. 北京: 科学出版社: 1-322, xxiv.（Zheng Z X. 2000. A complete checklist of species and subspecies of the chinese bird. Bejing: Science Press: 1-322, xxiv.）

郑作新，等. 1978. 中国动物志. 鸟纲. 第四卷, 鸡形目. 北京: 科学出版社: 203.（Zheng Z X, *et al*. 1978. Fauna sinica. Aves. Vol. 4, Galliformes. Beijing: Science Press: 203.）

郑作新，等. 1979. 中国动物志. 鸟纲. 第二卷, 雁形目. 北京: 科学出版社: 143.（Zheng Z X, *et al*. 1979. Fauna sinica. Aves. Vol. 2, Anseriformes. Beijing: Science Press: 143.）

郑作新，冼耀华，关贯勋. 1991. 中国动物志. 鸟纲. 第六卷, 鸽形目　鹦形目　鹃形目　鸮形目. 北京: 科学出版社: 240.（Zheng Z X, Xian Y H, Guan G X. 1991. Fauna sinica. Aves. Vol. 6, Columbiformes, Psittaciformes, Cuculiformes, Strigiformes. Beijing: Science Press: 240.）

钟昌富. 1986. 井冈山自然保护区爬行动物初步调查. 江西大学学报(自然科学版)，10(2): 71-75.（Zhong C F. 1986. Prelemenary survey of reptiles in the Jinggangshan Nature Reserve. *Journal of Jiangxi University (Natural Science)*, 10(2): 71-75.）

钟昌富. 2004. 江西省爬行动物地理区划. 四川动物，23(3): 222-229.（Zhong C F. 2004. Reptilian fauna and zoogeographic division of Jiangxi province. *Sichuan Journal of Zoology*, 23(3): 222-229.）

宗愉，马积藩. 1983. 拟脊蛇属为一有效属称，兼记一新种. 两栖爬行动物学报，2(2): 61-63.（Zong Y, Ma J F. 1983. A new species of the genus *Achalinopsis* from Jiangxi and the restoration of this genus. *Acta Herpetol. Sinica* (*new ser.*), 2(2): 61-63.）

邹多录. 1985. 井冈山自然保护区两栖动物及其区系分布. 江西大学学报(自然科学版)，(9)1: 51-55.（Zou D L. 1985. Amphibians and their faunistic distribution of Jinggangshan. *Journal of Jiangxi University (Natural Science)*, (9)1: 51-55.）

Ananjeva N B, Guo X G, Wang Y Z. 2011. Taxonomic diversity of agamid lizards (Reptilia, Sauria, Acrodonta, Agamidae) from China: a comparative analysis. *Asian Herpetological Research*, 2 (3): 117-128.

Bickford D, Lohman D J, Sodhi N S, *et al*. 2007. Cryptic species as a window on diversity and conservation. *Trends in Ecology & Evolution*, 22: 148-155.

Boulenger G A. 1887. An account of the reptiles and batrachians obtained in Tenasserim by M L Fea, of the Genova Civic Museum. *Ann. Mus. Civ. Stor. Nat. Genova*, 2. Ser. 5: 474-486.

Boulenger G A. 1893. Catalogue of the snakes in the British Museum (Natural History). Voume I. London: Taylor and Francis.

Boulenger G A. 1900. On the reptiles, batrachians and fishes collected by the late M R John Whithead in the interior of Hainan. London: Proceedings of the Zoological Society: 956-962.

Boulenger G A. 1908. A revision of the oriental pelobatid batrachians (genus *Megalophrys*). *Proceedings of the Zoological Society*, London, 78 (2): 407-430.

Bourret R. 1935. Notes herpétologiques sur l'Indochine française. Extrait du bulletin Général de L'instruction publique, Janvier.

Brazil M. 2009. Birds of East Asia (Helm Field Guides). London: Chhristoper Helm Publisher: 528.

Carey G J, Chalmers M L, Diskin D A, *et al*. 2001. The avifauna of Hong Kong. Hong Kong: Hong Kong Bird Watching Society: 563.

Das I. 2010. A field guide to the reptiles of South-East Asia. London: New Holland Publishers: 1-376.

Delorme M, Dubois A, Grosjean S, *et al*. 2006. Une nouvelle ergotaxinomie des Megophryidae (Amphibia, Anura). *Alytes*, 24: 6-21.

Frost D R. 2017. Amphibian species of the world Version 6.0, an online reference: American Museum of Natural History, New York, USA. http://research.amnh.org/vz/herpetology/amphibia/ [2017-11-9].

Funk W C, Caminer M, Ron S R. 2012. High levels of cryptic species diversity uncovered in Amazonian frogs. *Proceedings of the Royal Society B: Biological Sciences*, 279: 1806-1814.

Gill F, Donsker D. 2017. IOC World Bird List (Version 7.3). http:// www. Worldbirdnames.org/ [2017-11-9].

Gosner K L. 1960. A simplified table for staging anuran embryos and larvae with notes on identification. *Herpetologica*, 16: 183-190.

Grismer L L, Wood P L Jr, Anuar S, *et al*. 2013. Integrative taxonomy uncovers high levels of cryptic species diversity in *Hemiphyllodactylus* Bleeker, 1860 (Squamata: Gekkonidae) and the description of a new species from Peninsular Malaysia. *Zoological Journal of the Linnean Society*, 169: 849-880.

Guo Y H, Wu Y K, He H P, *et al*. 2011. Systematics and molecular phylogenetics of Asian snail-eating snakes (Pareatidae). *Zootaxa*, 3001: 57-64.

Hanken J. 1999. Why are there so many new amphibian species when amphibians are declining? *Trends in Ecology & Evolution*, 14: 7-8.

Hoyo J, Elliott A, Sargatal J. 1992. Handbook of the birds of the world (Volume 1). Barcelona: Lynx Edicions.

IUCN. 2017 IUCN red list of threatened species. http://www.redlist.org/ [2017-11-9].

Jiang J, Zhou K. 2005. Phylogenetic relationships among chinese ranids inferred from sequence data set of 12S and 16S rDNA. *The Herpetological Journal*, 15: 1-8.

Kuraishi N, Matsui M, Hamidy A, *et al*. 2012. Phylogenetic and taxonomic relationships of the *Polypedates leucomystax* complex (Amphibia). *Zoologica Scripta*, 42: 54-70

La Touche J D D. 1922. Descriptions of new forms of chinese brids. *Bull Brit Ornithol Club*, 43:20-23.

Li C, Wang Y Z. 2008. Taxonomic review of *Megophrys* and *Xenophrys*, and a proposal for chinese species (Megophryidae, Anura). *Acta Zootaxonomica Sinica*, 33.

Li Y L, Jin M J, Zhao J, *et al*. 2014. Description of two new species of the genus *Megophrys* (Amphibia: Anura: Megophryidae) from Heishiding Nature Reserve, Fengkai, Guangdong, China, based on molecular and morphological data. *Zootaxa*, 3795 (4): 449-471.

Mahony S. 2009. A new species of *Japalura* (Reptilia: Agamidae) from Northeast India with a discussion of the similar species *Japalura sagittifera* Smith, 1940 and *Japalura planidorsata* Jerdon, 1870. *Zootaxa*, 2212, 41-61.

Mahony S. 2011. Two new species of *Megophrys* Kuhl & van Hasselt (Amphibia: Megophryidae), from western

Thailand and Southern Cambodia. *Zootaxa*, 2734: 23-39.

Manthey U, Wolfgang D, Hou M, *et al*. 2012. Discovered in historical collections: two new *Japalura* species (Squamata: Sauria: Agamidae) from Yulong Snow Mountains, Lijiang Prefecture, Yunnan, PR China. *Zootaxa*, 3200: 27-48.

Martens J. 2000. *Phylloscopus yunnanensis* La Touche, 1922, Alstrom laubsänger//Wunderlich K, Martens J, Loskot V M. Atlas der verbreitung palaearktischer vögel, Vol. 19: 1-3.

Mohony S. 2010. Systematic and taxomonic revaluation of four little known Asian agamid species, *Calotes kingdonwardi* Smith, 1935, *Japalura kaulbacki* Smith, 1937, *Salea kakhienensis* Anderson, 1879 and the monotypic genus *Mictopholis* Smith, 1935 (Reptilia: Agamidae). *Zootaxa*, 2514: 1-23.

Mo Y M, Zhang W, Zhou S C, *et al*. 2013. A new species of the genus *Gracixalus* (Amphibia: Anura: Rhacophoridae) from Southern Guangxi, China. *Zootaxa*. 3616 (1): 61-72.

Ngo V T, Grismer L L, Pham H T, *et al*. 2014. A new species of *Hemiphyllodactylus* Bleeker, 1860 (Squamata: Gekkonidae) from Ba Na-Nui Chua Nature Reserve, Central Vietnam. *Zootaxa*, 3760 (4): 539-552.

Nguyen T Q, Le M D, Pham C T, *et al*. 2013. A new species of *Gracixalus* (Amphibia: Anura: Rhacophoridae) from Northern Vietnam. *Organisms Diversity & Evolution*, 13: 203-214.

Nguyen T Q, Schmitz A, Nguyen T T, *et al*. 2011. Review of the Genus *Sphenomorphus* Fitzinger, 1843 (Squamata: Sauria: Scincidae) in Vietnam, with description of a new species from Northern Vietnam and Southern China and the first record of *Sphenomorphus mimicus* Taylor, 1962 from Vietnam. *Journal of Herpetology*, 45(2):145-154.

Ota H. 1989a. A new species of *Japalura* (Agamidae: Lacertilia: Reptilia), from Taiwan. *Copeia*, 1989 (3), 569-576.

Ota H. 1989b. *Japalura brevipes* Gressitt (Agamidae: Reptilia), a valid species from high altitude area of Taiwan. *Herpetologica*, 45 (1): 55-60.

Ota H. 1991. Taxonomic redefinition of *Japalura swinhonis* Günther (Agamidae: Squamata), with a description of a new subspecies of *J. polygonata* from Taiwan. *Herpetologica*, 47 (3): 280-294.

Ota H. 2000. *Japalura szechwanensis*, a junior synonym of *J. fasciata*. Journal of Herpetology, 34 (4): 611-614.

Ota H, Chen S L, Shang G. 1998. *Japalura luei*: a new agamid lizard from Taiwan (Reptilia: Squamata). *Copeia*, 1998 (3), 649-656.

Pan S, Dang N X, Wang J S, *et al*. 2013. Molecular phylogeny supports the validity of *Polypedates impresus* Yang 2008. *Asian Herpetological Research*, 4(2): 124-133

Pope C H. 1928. Seven new reptiles from Fukien province, China. *American Museum Novitat*, 320: 1-6.

Pope C H. 1929. Notes on reptiles from Fukien and other chinese provinces. *Bulletin of the American Museum of Natural History*, 58 (8): 17-20, 335-487.

Pyron R A, Wiens J J. 2011. A large-scale phylogeny of Amphibia including over 2800 species, and a revised classification of extant frogs, salamanders, and caecilians. *Molecular Phylogenetics and Evolution*, 61: 543-583.

Rasmussen P C, Anderton J C. 2012. Birds of South Asia, the ripley guide (2nd Edition). Barcelona: Lynx Edicions: 956.

Robson C. 2011. A field guide to the birds of South-East Asia. London: New Holland Publisher: 544.

Rao D Q, Yang D T. 1997. The variation in karyotypes of *Brachytarsophrys* from China with a discussion of the classification of the genus. *Asiatic Herpetological Research*, 7: 103-107.

Pfenninger M, Schwenk K. 2007. Cryptic animal species are homogeneously distributed among taxa and biogeographical regions. *BMC Evolutionary Biology*, 7: 121.

Smith H M. 1943. The fauna of British India, Ceylon and Burma, including the whole of the Indo-Chinese sub-region. London: Reptilia and Amphibia. Vol. III. Serpentes, Taylor and Francis: 583.

Stuart B L, Inger R F, Voris H K. 2006. High level of cryptic species diversity revealed by sympatric lineages of Southeast Asian forest frogs. *Biology Letters*, 2: 470-474.

Stuart S N, Chanson J S, Cox N A, *et al*. 2004. Status and trends of amphibian declines and extinctions worldwide. *Science*, 306: 1783-1786.

Sung Y K, Yang J H, Wang Y Y. 2014. A new species of *Leptolalax* (Anura: Megophryidae) from Southern China. *Asian Herpetological Research*, 5(2): 80-90.

Taylor E H. 1963. The lizards of Thailand. *Univ. Kansas Sci. Bull.*, 44: 687-1077.

Uetz P, Hošek J. 2017. The reptile database. http://www.reptile-database.org [2017-11-9]

Vogel G, David P. 2006. On the taxonomy of the *Xenochrophis piscator* complex (Serpentes, Natricidae). *Herpetologica Bonnensis*, II: 241-246.

Vogel G, David P, Pauwels O S G, *et al*. 2009. A revision of *Lycodon ruhstrati* (Fischer 1886) auctorum (Squamata: Colubridae), with the description of a new species from Thailand and a new subspecies from the Asian mainland. *Tropical Zoology*, 22: 131-182.

Wang K, Jiang K, Pan G, *et al*. 2015. A new species of *Japalura* (Squamata: Sauria: Agamidae) from Upper Lancang (Mekong) Valley of Eastern Tibet, China. *Asian Herpetological Research*, 6(3): 159-168.

Wang Y Y, Yang J H, Liu Y. 2013. New distribution records for *Sphenomorphus tonkinensis* (Lacertilia: Scincidae) with notes on its variation and diagnostic characters. *Asian Herpetological Research*, 4(2): 147-150.

Wang Y Y, Zhao J, Yang J H, *et al*. 2014. Morphology, molecular genetics, and bioacoustic support two new sympatric *Xenophrys* (Amphibia: Anura: Megophryidae) species in Southeast China. *PLoS ONE*, 9(4): e93075.

Wang Y Y, Zhang T D, Zhao J, *et al*. 2012. Description of a new species of the genus *Xenophrys* Günther, 1864 (Amphibia: Anura: Megophryidae) from Mount Jinggang, China, based on molecular and morphological data. *Zootaxa*, 3546: 53-67.

Wilkins M R, Seddon N, Safran R J. 2013. Evolutionary divergence in acoustic signals: causes and consequences. *Trends in ecology & evolution*, 28: 156-166.

Weigold H. 1922. *Muscicapa elisae* n. sp. *Falco*, 18(1): 1-2.

Yang J H, Wang Y Y. 2010. Range extension of *Takydromus sylvaticus* (Pope, 1928) with notes on morphological variation and sexual dimorphism. *Herpetology Notes*, 3: 279-283.

Zeng Z C, Zhao J, Chen C Q, *et al*. 2017. A new species of the genus *Gracixalus* (Amphibia: Anura: Rhacophoridae) from Mount Jinggang, southeastern China. *Zootaxa*, 4250 (2): 171-185.

Zhang Y Y, Wang N, Zhang J, *et al*. 2006. Acoustic difference of narcissus flycatcher complex. *Acta Zoologica Sinica*, 52(4): 648-654.

Zhao E M, Adler K. 1993. Herpetology of China. Ohio: Society for the Study of Amphibians and Reptiles, Oxford: 1-522.

Zhao J, Yang J H, Chen G L, *et al*. 2014. Description of a new species of the genus *Brachytarsophrys* Tian and Hu, 1983 (Amphibia: Anura: Megophryidae) from Southern China Based on molecular and morphological data. *Asian Herpetological Research*, 5(3): 150-160.

Zheng G M, Song J, Zhang Z, *et al*. 2000. A new species of flycatcher (Ficedula) from China(Aves. Passeriformes: Muscicapidae). *Journal Beijing Normal University* (*Nature Science*), 36(3): 405-409.

Zhou T, Chen B M, Liu G, *et al*. 2015. Biodiversity of Jinggangshan Mountain: the importance of topography and geographical location in supporting higher biodiversity. *PLoS ONE*, 10(3): e0120208.

科拉丁名索引
Index to Families

种拉丁名索引
Index to Species

科中文名索引
Index to Chinese Name of Families

种中文名索引
Index to Chinese Name of Species